World Sustainability Series

Series Editor

Walter Leal Filho, European School of Sustainability Science and Research, Research and Transfer Centre "Sustainable Development and Climate Change Management", Hamburg University of Applied Sciences, Hamburg, Germany

Due to its scope and nature, sustainable development is a matter which is very interdisciplinary, and draws from knowledge and inputs from the social sciences and environmental sciences on the one hand, but also from physical sciences and arts on the other. As such, there is a perceived need to foster integrative approaches, whereby the combination of inputs from various fields may contribute to a better understanding of what sustainability is, and means to people. But despite the need for and the relevance of integrative approaches towards sustainable development, there is a paucity of literature which address matters related to sustainability in an integrated way.

Notes on the quality assurance and peer review of this publication

Prior to publication, the quality of the works published in this series is double blind reviewed by external referees appointed by the editor. The referees are not aware of the author's name when performing the review; the referees' names are not disclosed.

More information about this series at https://link.springer.com/bookseries/13384

Jurgis Kazimieras Staniškis · Eglė Staniškienė ·
Živilė Stankevičiūtė · Asta Daunorienė ·
Joana Ramanauskaitė

Transformation of Business Organization Towards Sustainability

Systems Approach

Springer

Jurgis Kazimieras Staniškis
Kaunas University of Technology
Kaunas, Lithuania

Živilė Stankevičiūtė
School of Economics and Business
Kaunas University of Technology
Kaunas, Lithuania

Joana Ramanauskaitė
School of Economics and Business
Kaunas University of Technology
Kaunas, Lithuania

Eglė Staniškienė
School of Economics and Business
Kaunas University of Technology
Kaunas, Lithuania

Asta Daunorienė
School of Economics and Business
Kaunas University of Technology
Kaunas, Lithuania

ISSN 2199-7373 ISSN 2199-7381 (electronic)
World Sustainability Series
ISBN 978-3-030-93297-8 ISBN 978-3-030-93298-5 (eBook)
https://doi.org/10.1007/978-3-030-93298-5

This Springer imprint is published by the registered company Springer Nature Switzerland AG
The registered company address is: Gewerbestrasse 11, 6330 Cham, Switzerland

Preface

When all the trees have been cut down, when all the animals have been hunted, when all the waters are polluted, when all the air is unsafe to breathe, only then will you discover you cannot eat money.—Cree Indian Prophesy

We have entered a new era where business, technologies, communities, and even pandemic diseases cross borders with unprecedented speed and intensity. The UN Millennium Declaration and its associated Millennium Development Goals have guided the global development goals through the first 15 years of the new century. In pursuit of the Millennium Development Goals, the global community achieved many successes, but also fell short in several ways as it learned important lessons about the opportunity of co-benefits and the inevitability of trade-offs and tough choices. In September, the United Nations Member States decided jointly on a global project—2030 Agenda—in order to shape our common future in a new, better, and more intentional way. Entitled "Transforming our World," this project reflects the global community's high expectations of finally reversing the destruction of our natural and social habitats, and achieving a more balanced and equitable pathway towards the well-being of all. Therefore, not only the Goals and targets, but also interactions among them, are brought into focus in the 2030 Agenda.

However, despite the initial efforts, the world is not on track to achieving most of the 169 targets that comprise the Goals. Limited success in progress towards the Goals raises strong concerns and sounds the alarm for the international community. Worrying is the fact that recent trends along several dimensions with cross-cutting impacts along the entire 2030 Agenda are not even moving into the right direction. Four in particular fall into that category: rising inequalities, climate change, biodiversity loss, and increasing amounts of waste from human activities that are overwhelming the capacities to process them. Thus, advancing the sustainable development must involve an urgent and intentional transformation of socio-environmental–economic systems, differentiated across countries but also adding up to the desired regional and global outcomes to ensure human well-being, social health, and limited environmental impact.

Before leaving office, former Secretary-General of the UN Ban Ki-moon appointed an Independent Group of Scientists (IGS) comprising 15 experts to draft the Global Sustainable Development Report (GSDR). The Report is a key component of the mechanism to follow up and review progress on the recently agreed 2030 Agenda for Sustainable Development and the Sustainable Development Goals (SDGs). It seeks to strengthen the science-policy interface and provide a strong evidence-based instrument to support policymakers in promoting poverty eradication and sustainable development. The document is intended to provide guidance from a scientific perspective that supports the implementation of the Sustainable Development Goals and the 2030 Agenda in ways that integrate economic, environmental, and social dimensions. It is available for a wide range of stakeholders, including businesses, civil society, and the public. In the Report, the experts argue that the value and transformative potential of the 2030 Agenda is more than the sum of its 17 SDGs and 169 indicators. It is not only a unique normative compass; it also represents a vision of how natural resources could be best shared for the well-being of the 9 billion people who will soon populate this Earth. The assessment is based on a total of 65 global assessments comprising the United Nations flagship reports and international scientific assessments, as well as 112 scientific articles published since 2015 with explicit reference to the Sustainable Development Goals. Not all pathways and transformations towards achieving any given goal or target in the 2030 Agenda are equivalent in terms of their implications for the others. The report presents a global system model comprising six major transformations—human well-being and capabilities, sustainable and just economies, sustainable food systems and healthy nutrition, energy decarbonisation with universal access, urban and peri-urban development, and global commons. The selected transformations are critical if the sustainable development goals are to be met by 2030 in ways that will ensure sustainability for both current and future generations. At the same time, means and levers of transformations include governance, consumption and production (including financing), individual and collective action, and society and technology. When we think about the kinds of changes that will be required for the development towards sustainability, it is tempting to focus on the practical issues like financial regulation, taxation, or reduction of carbon intensity. Unfortunately, much more needs to happen, and quite quickly, to bring the required transformative changes about. The thermodynamics makes it clear that humans must find ways to balance the economic subsystem with the Earth's evolutionary and morphological processes, or the planet will use its own mechanisms to restore the balance. Systems analysis applied at the level of organisations, cities, and regional governance buys us time until, among other things, national governments catch up. At any level, it is only a tool to clarify the consequences of our actions, identify our options, and extend our foresight a bit (IGS, 2019).

The world is now closely interconnected by flows of goods, capital, people, and information. These flows, on the one hand, produce some benefits; however, on the other hand, they can also create negative impacts, for example, deepening inequalities, unfair competition, resource depletion, environmental pollution, and destruction.

The subject matter of economics should be the economy—which involves money, work, technology, international trade, taxes, and other things that have to do with the ways in which we produce goods and services, distribute the incomes generated in the process, and consume the things thus produced.

We must make sure our economic analysis is structured in way that allows us to access the risk of such magnitude. All too often, economists are tempted to force everything into a simplistic cost-benefit analysis in which changes are marginal and all relevant effects can be described in terms of a single common denominator such as money. When someone has a hammer, every problem looks like a nail. For all these reasons, a policy analysis must begin with the science of sustainability, including climate change as one of the most important goals. It must examine where we may be going under different assumptions about policy. Worryingly, some plausible assumptions on current intentions suggest we are headed in a very dangerous direction (Stern 2016).

Current development patterns (even those touted as "sustainable") disrupt the social system and ecosystem relations rather than ensuring that the natural resource use by local communities meets their basic needs at a level of comfort that is satisfactory as assessed by those same communities. What is needed is not a common future but the future as commons (Banerjee 2003). Sustainability demands a discontinuous leap from the existing basis of cultural action. Transformation is a very powerful concept, because it denotes a process in which reality in front of us changes its form in an abrupt, discontinuous way.

Current concern about unsustainability has arisen from the observation that both natural and socio-economic systems are losing resilience that is the ability to cope with perturbations created by human activities without the appearance of fundamental, qualitative changes in the functions of these systems. There are great opportunities in the fact that the transformation to the sustainable development based on low-carbon economy coincides with the coming decade of radical structural transformation of the world. If the structural transformation is done well from the point of resource efficiency, responsible consumption and sustainable production, waste and pollution, liveable cities, inequality and poverty, and care of forests and grasslands, it strongly reduces the emissions.

The present book strives to address this issue by adopting a systematic approach to the Sustainable Development Goals, informed by the knowledge of the interactions among them. In this increasingly globalised and hyper-connected world, one goal can lead to unintended consequences for the implementation of other goals, i.e. the chances of progress on achievements in a specific country/region of the world will depend on interventions made in other sectors in distant places. Achieving transformation—a profound and intentional departure from business as usual—will mean carefully taking into account the interactions between the goals and targets.

At the heart of this book is a fundamental belief that the "purpose" of the economic system is to improve the well-being for all within the limits of what the planet can sustain—to produce good lives that do not cost the Earth. The book provides a logical way that links local scale production and service activities to systemic changes in macro-level paradigms. The role of different kinds of associated barriers as well as the complexity and uncertainty of transformations are highlighted. A strong focus is placed on the opportunities and barriers to changing the production-consumption system, driving the environmental degradation. The financial reform should also reorient the investments into incremental and structural innovations towards mitigating or adapting sustainability problems. This book deals with the ways the elaborated system model could be used to search for policy-relevant solutions transforming the society towards sustainability on a global, regional, or country level. While the book as a product focuses on producing knowledge for transformations to sustainable development, IGS views the GSDR also as a process that can advance the collaborations between science, policy, and society.

Sustainability science in this book is understood according to the definition of Sustainability Science Programme at Harvard University: "Sustainability science is problem-driven transdisciplinary scholarship that seeks to facilitate the design, implementation, and evaluation of effective interventions that foster shared prosperity and reduced poverty while protecting environment. It is defined by the problems it addresses rather than the disciplines it employs. It thus draws as needed from multiple disciplines of the natural, social, medical and engineering sciences, from the professions, and from the knowledge of practice" (Harvard Kennedy School 2008).

The book consists of two parts. Part I, which comprises three chapters, mainly deals with sustainability issues at the production level. In Chaps. 1 and 2, the development of methods, mathematical presentation, and systems for unsustainability reduction in industrial organisations are presented. The key element for that is preventive incremental innovations, based on the concept of resource-efficient and cleaner production. An advanced system for the generation, financing, and implementation of innovations is discussed in the context that reduction of unsustainability does not lead to sustainable development as such. Broad discussions on the role and future of high education and sustainability science in transdisciplinary approach and its implementation in practice are presented in Chap. 2. Chapter 3 mainly relates to the sustainable development issue on the regional level. The system for transformations generation and control, comprising feedforward and feedback loops and based on transformation model, is elaborated. The presented transformation model was developed by the UN Independent Group of Scientists and presented at the UN General Assembly as a Global Sustainable Development Report "The Future is Now: Science for Achieving Sustainable Development" in September, 2019. Detailed discussions on different socio-economic systems, approaches and possible transformations, different pathways for their implementation in developed and developing economies are provided.

Part II comprises two chapters and mainly deals with obstacles and drivers to transitions of organisations towards sustainability. In Chap. 4, we provide theoretical insights on organisational transitions towards Corporate Social Responsibility (CSR), reviewing the literature about the historical evolution and concept of CSR, stakeholder theory and its application to CSR, the stakeholders' role in sustainability transitions, and obstacles and drivers of organisations moving towards CSR transitions. Chapter 5 is mainly intended for empirical insights on obstacles and drivers of CSR-committed organisations to sustainability transitions. In this chapter, we present the methodological and empirical part of the conducted studies: the research context, the overview of quantitative research results, and the results of our interviews conducted on a sample of Lithuanian organisations. Finally, we discuss our results from the East European economy in transition.

The key message of the book is that we have to be open stating that future problems cannot be solved within the traditional paradigm of economic growth and reliance on technology and by specific policies intended to attenuate the most unethical behaviour and nudge the consumers, firms, and workers in the "correct" direction, i.e. only calling upon people to behave "right" within "wrong" structures.

Kaunas, Lithuania Jurgis Kazimieras Staniškis

Acknowledgements

This monograph is the product of a group of scientists attempting to deeper understand the type of transformations towards sustainability that should be applied globally and on the level of business organisation to make the world more fair and decent while preserving a beautiful and liveable Earth. Concerning the global problems on economical, environmental and social development, I had a unique possibility to join the United Nations Independent Group of Scientists for Global Sustainable Development Reporting. In the personal letter on 30 December 2016, the UN General Secretary Ban Ki Moon wrote: "I have the pleasure to invite you to serve on the 15-member independent group of scientists selected to draft the quadrennial Global Sustainable Report.

In July 2016, the United Nations Member States decided that a Global Sustainable Report would be produced every four years, to inform on sustainable development when it meets high-level political forum under the auspices of the United Nations Assembly to review the progress made on the 2030 Agenda for Sustainable Development and the Sustainable Development Goals. The Report will strengthen the science-policy interface and provide a strong evidence-based instrument to support policymakers in promoting poverty eradication and sustainable development.

Following an extensive and open selection process, I have selected 15 eminent scientists to be invited to serve on independent group in their personal capacity."

I have to confess that my work in such a transdisciplinary group of highly skilled scientists was the most exiting period in my scientific career. I deeply appreciate the Group's enthusiasm, kindness, dedication, and professional contribution and would like to thank each personally: Peter Messerli (Switzerland), Endah Murniningtyas (Indonesia), Parfait Eloundou-Enyegue (Cameroon), Ernest G. Foli (Ghana), Eeva Furman (Finland), Amanda Glassman (USA), Gonzalo Hernandez Licona (Mexico), Eun Mee Kim (Republic of Korea), Wolfgang Lutz (Austria), Jean Paul Moatti (France), Katherine Richardson (Denmark), Muhammad Saidam (Jordan), David Smith (Jamaica), and Jean-Pascal van Ypersele (Belgium).

The Group produced the Global Sustainable Development Report "The Future is Now—Science for Achieving Sustainable development," the ideas and insights of which helped me a lot, especially, in producing Chap. 3.3 of the book.

My greatest thanks also goes to the APINI team referred to the in book that comprised the Institute of Environmental Engineering, Kaunas University of Technology staff (Irina Kliopova, Žaneta Stasiškienė, Valdas Arbačiauskas, Visvaldas Varžinskas), and to all my former Ph.D. students and other employees. They helped me a lot in developing the systems of preventive innovations generation, financing, and implementation in the country and abroad that significantly reduced the unsustainability of business organisations.

Last but not least, all the authors of the monograph would like to extend their gratitude to colleagues for valuable comments and long discussions on the topic of monograph and to reviewers for their important feedback. The authors would also like to acknowledge the willingness of all interviewees to share their experiences and address the issue of corporate social responsibility, and to thank the Research Council of Lithuania for partly funding this research (grant No. S-MIP-17-122).

On behalf of authors Jurgis Kazimieras Staniškis

Contents

About the Authors

Jurgis Kazimieras Staniškis full member of the Lithuanian Academy of Sciences, professor—emeritus at Kaunas University of Technology, Lithuania, member of the United Nations Independent Group of Scientists for Global Sustainable Development Reporting.

J. K. Staniškis deals with the development of methods and tools based on systems theory for sustainable economy and sustainable industrial development, for instance, resource-efficient and cleaner production, systems for incremental preventive innovations generation, financing and implementation, and systems for transformations of global development towards sustainability. He is a manager and the main expert for more than 80 scientific and implementation international and national projects, establisher of the new transdisciplinary M.Sc. programme "Cleaner production and environmental management" and Ph.D. programme "Environmental engineering."

As an UNIDO/UNEP/OECD expert on unsustainability reduction in companies, J.K. Staniškis participated in the projects/missions in Tanzania, Zimbabwe, Uganda, Vietnam, China, Russia, Ukraine, Macedonia, Albania, Kyrgyzstan, Kazakhstan, Armenia, Azerbaijan, Guatemala, Nicaragua, Peru, etc. As a member of the UN Independent Group of Scientists, he is one of the authors or the Report "The Future is Now—Science for Achieving Sustainable Development" (2019).

Awards: National Science Awards (1979, 1988, 2004), His Excellence the President Dr. Valdas Adamkus Award (2005), the Baltic Sea Award (2010), the Cross of Officer of the Order for Merits to Lithuania (2009).

Eglė Staniškienė a professor at the Sustainable Management Research Group, School of Economics and Business, Kaunas University of Technology (KTU), Lithuania, Ph.D. in Social Sciences. Research interests include sustainable development, corporate social responsibility, responsible consumption and sustainable production, employee well-being, quality system management, and interdisciplinary research. Currently, E. Staniškienė is a member of the expert council of the Lithuanian Responsible Business Association (LAVA). She has published more than 60 scientific publications, participated in more than 20 national and international projects. E. Staniškienė is an expert of the United Nations Economic and Social Council (ECOSOC), a KTU representative in the international COPERNICUS alliance, a member of European Group for Organisational Studies (EGOS), International Sustainable Development Research Society (ISDRS), Academy of Management (AOM), and a European Member of the Academy of Management (EURAM). She uses and develops her scientific competence in the study process and in cooperation with business and public sector organisations. E. Staniškienė teaches the modules of Sustainable Development, Quality Management, Quality and Sustainable Development Management, Circular Economy, and others.

Živilė Stankevičiūtė Ph.D. in Management, is an Associated Professor at the Sustainable Management Research Group, School of Economics and Business, Kaunas University of Technology, Lithuania. Her doctoral dissertation on Sustainable Human Resource Management has been recognised internationally: she received the Baltic University Programme Annual Award for the Best Ph.D. defended in 2015. Her research interests include Sustainable Human Resource Management, Employee Well-being, Digital Transformations, Corporate Social Responsibility, and Sustainable Development. Ž. Stankevičiūtė is a member of the National Expert Council for sustainable development goals at the Lithuanian Responsible Business Association. Lecturing includes courses on Sustainable Development, Employee Performance Management, and Human Analytics.

Asta Daunorienė is an Associated Professor of Quality and Value Chain Management at Kaunas University School of Economics and Business, and a senior researcher at the Sustainable Management Research Group at School of Economics and Business, Kaunas University of Technology, Lithuania. Asta has completed her Ph.D. in the field of Economics at Kaunas University of Technology. Her research interest lies in the area of quality management, sustainable value and supply chain management, and new business models' development. She has actively collaborated with researchers in other disciplines of pedagogy and information technology. Asta currently serves on the ECIU learning and teaching expert group. A. Daunorienė has been involved in the expert activities in Ministry of Education, Science and Sport, contributing her skills to the strategy development for teachers' professional development and study quality improvement. She has comprehensive experience in quality management systems consulting.

Joana Ramanauskaitė is a Ph.D. candidate at the Sustainable Management Research Group at Kaunas University of Technology, Lithuania. She is interested in corporate sustainability, sustainable transitions, and their tensions. She is also a member of the Sustainability Transitions Research Network (STRN). Joana teaches the courses of Sustainable Development and Quality Management. She has been participating in the development of the interdisciplinary course of Sustainable Development for first-year bachelor students at KTU since 2019.

Abbreviations and Acronyms

APINI	Institute of Environmental Engineering at KTU
CSR	Corporate Social Responsibility
ESD	Education for Sustainable Development
GSDR	Global Sustainable Development Report
ICT	Information and Communications Technology
IGES	Institute for Global Environmental Strategies
IGS	Independent Group of Scientists
KTU	Kaunas University of Technology
MDGs	Millennium Development Goals
nef	New Economic Forum
ODA	Official Development Assistance
SCP	Sustainable Consumption and Production
SD	Sustainable Development
SDGs	Sustainable Development Goals
SROI	Social Return on Investment
UN	United Nations
UN-DESA	United Nations Department of Economic and Social Affairs
UNEP	United Nations Environment Programme
UNIDO	United Nations Industrial Development Organisation
WEEE	Waste from Electrical and Electronic Equipment
WHO	World Health Organisation

Part I
Application of the Systems Theory for the Development of Measures Driving Business Organisations Towards Sustainability

Business and entrepreneurship based on inner awareness of the self, and directed to the natural and human environments; on technology design and implementation that meets people's material and non-material needs rather than just the profit motive; on technological change that creates the conditions for human inner growth and development; and on governance and business management instruments that benefit all people.

—Alfredo Sfeir-Younis, Chilean economist

Chapter 1
Sustainability Challenges in an Business Organisation

Jurgis Kazimieras Staniškis

> Almost always the men who achieve these fundamental inventions of a new paradigm have been either very young or very new to the field whose paradigm they change—Thomas S. Kuhn, American philosopher of science

1.1 The Development of Sustainable Development Concept

> Today's problems cannot be solved if we still think the way we thought when we created them—Albert Einstein, German-born physicist

Adam Smith (1776) along with many others at that time saw nature as "no more than a storehouse of raw materials for man 's ingenuity". The eighteenth century was a period that witnessed the birth of the Industrial Revolution, modern growth–based economics and what recently has been termed "consumer revolution". Inventers produced new machines that eventually revolutionised the global economy and changed the course of human history (Caradonna 2014).

The main idea of classical economics is that a free market regulates itself, bringing products and services to consumers and profits to producers and sellers without the active intervention of the state. Adam Smith in "Wealth of Nation" argues that the wealth of nation is essentially the annual product of its labour and must be continually increased. It meant the privatisation of publicly owned land and increased consumption of natural resources. In short, the eighteenth century set the stage for an enduring conflict over social, economic, and environmental costs of economic growth that have played out throughout the nineteenth century and down to present days. Historians have only recently discovered that strong hostility towards greed, consumerism, and growth that later was called the Industrial Revolution was fairly developed in this period.

One of the most famous critics was Jean–Jacques Rouseau who has influenced intellectual life and the history on sustainable development, for instance, by his statement that technological innovation did not make humans any happier or virtuous,

J. K. Staniškis et al., *Transformation of Business Organization Towards Sustainability*, World Sustainability Series, https://doi.org/10.1007/978-3-030-93298-5_1

his critique of social inequality and its link to the natural environment, and admiration for wilderness, rural values, and simple living (Caradonna 2014).

An important moment for sustainability came in 1980 when the UNEP formed the International Union for the Conservation of Nature (IUCN) and produced the report "World Conservation Strategy: Living Resource Conservation for Sustainable Development". Despite the fact that this report emphasised the interrelationship between environmental, economic, and social problems, it said relatively little about social justice, poverty, inequality, faulty economic and financial systems, and other subjects addressed in the UN documents on sustainable development. In 1983, the General Assembly of the United Nations urgently asked the World Commission on Environment and Development (WCED) to create a global framework for sustainable development, "a global agenda for change". The UN Secretary General appointed the members of the independent commission *(Chairwoman—Gro Harlem Brundtland, Vice-Chairman—Mansour Khalid, principal author of the document—Jim MacNeill)* and asked to address the major challenges to the global community:

- to promote long-term environmental strategies for achieving sustainable development;
- to recommend ways concern for the environment might be translated into greater co-operation among the developing countries and between countries at different stages of economic and social development and lead to the achievement of common and mutually supportive objectives that take account of the interrelationships between people, resources, environment, and development;
- to consider ways and means by which the international community could deal more effectively with environmental concerns; and
- to help define shared perceptions of long-term environmental issues and the appropriate efforts needed to deal successfully with the problems of protecting and enhancing the environment, a long-term agenda for action during the coming decades, and aspirational goals for the world community (WCED 1987).

WCED produced an excellent report "Our Common Future" that eventually became known as the Brundtland Report. The report presented sustainable development as the development, which "meets the needs of the present without compromising the ability of future generations to meet their own needs".

This seems like a rather human–centred definition of sustainability; however, it introduced the innovative "intergenerational equity" principle in order to encourage people to think more deeply about the possible future consequences of what they do in the present. The publication of the Brundtland Report reflected the fact that the United Nations had taken the lead in contemplating the global dimensions of the sustainability challenge (Orr 2016).

What is wrong with the sustainable development concept? Robinson argues that sustainable development suffers from three conceptual pathologies: it is vague, attracts hypocrites, and foster delusions, i.e. if development is seen as synonymous with growth, then sustainable development means ameliorating, but not challenging continuing economic growth (Robinson 2004). The problem that many scholars have pointed out is that sometimes "development" is attached to the word "growth", and if that is the case, "sustainable development" is not a concept that favours a steady–state economy or ecological stability, but is rather a convert and greenwashed vehicle for business-as-usual economy. We have to decide whether society would come about through deregulated financial systems and growth-based economics or through regulated economic systems and the cessation of pro-growth policies. What is definitely clear, that "sustainable development should be pursued in the spirit of finding pathways that enable a good life for all, leaving no one behind, while safeguarding the environment for future generations and ensuring planetary justice". Economic activity should be seen not as an end in itself, but rather as a means for sustainability-advancing human capabilities (IGS 2019). In other words, "the ultimate purpose of business is not, or should not be, simply to make money. Nor is it merely a system of making and selling things. The promise of business is to increase the general well-being of humankind through service, a creative invention and ethical philosophy" (Hawken et al. 1999).

The Brundtland Commission's definition embraced two crucial elements of sustainable development such as the meeting of basic needs and recognising environmental limits, where overriding priority should be given to the world's poor and the principles of intergenerational and intragenerational equity. This definition is often criticised as vague, or in the language of some experts, non-operationalisable. Certainly, it is always difficult to put everything in a short definition, but there is no doubt that it presents the main idea of sustainable development. It should also be mentioned that critics, especially when it comes to language, so far have not suggested anything better. It became evident that many countries in the world meet serious difficulties in finding a meaningful translation of "sustainable development" to national languages. Besides that, there is another endless question: are sustainability and sustainable development the same thing? This is a strange question to ask. From the systems theory point of view, "development" is a process, and "sustainability" is the final state of the object. The "development" is NOT a synonym for "growth". Development could be degrowth, stable state or growth, depending on the country/regional economic, environmental and social situation, defined by the system performance index, boundaries and limitations (see chapter on mathematical formulation of the problem). Such formulation of the problem and possible solutions show the understanding of the sustainable development meaning and allow for its operationalisation.

Sachs elaborated his own definition: "Sustainable development is a process, a way of solving our problems peacefully and globally, using our science and technology, our know-how, and our shared global ethics to address our common needs. Our most basic common link is that we all inhabit this small planet, we all breathe the same air, we all cherish our children's futures, and we are all mortal" (Sachs 2006). Brice Lalonde, former Minister for the Environment in France, presented his version: "Sustainable development refers to how the economy should enable us to live better lives while improving our environment and our societies, from now on and within a globalised world."

A new and distinctive definition of sustainability was suggested by Ehrenfeld: "the possibility that human and other life will flourish on the planet forever. Flourishing is the key to the vision of sustainable future, and this way of conceptualising sustainability connects to every kind of audience I have addressed. To me, the most basic symbol of sustainability is that of flourishing. It pertains to all natural systems, both human and other living systems. For humans, flourishing means more than just remaining healthy. It also means the good life, following precepts handed down over the ages by sages and philosophers. We must shift back to the flourishing fullness of "Being" from its impoverished modern form of "Having" (Ehrenfeld 2009).

There is a growing awareness that our global economy is environmentally unsustainable. Economic activity since the industrial revolution has delivered significant improvements in living standards, but has also caused growing environmental pressures. Our prosperity depends on a wide range of resources and services supplied by our planet, from fresh water, metals and minerals to crop pollination performed by bees. In the meantime, the economy is not delivering quality of life for a huge section of the world's population. Most of the resources and environmental services are overexploited and underpriced, or not valued at all in today's economy. Unsustainable forms of development are not evenly distributed across the world. Some lifestyles lead to a greater depletion of the Earth's resources than others, and some people will be more vulnerable to changes in the ecosystems than others. It is necessary to understand which key values engender the feeling of connectivity and foster greater sustainability. Empathy, for instance, is believed to be an important value both in terms of global interconnectedness and long-term thinking. Our deeply ingrained consumerist culture may be challenging to shift, but values-based action is a critical lever of change.

The global campaign to end poverty started ramping up after 2000, in the wake of the United Nations' Millennium Development Goals, but the emergence of global goals to fight poverty was a great spur to accelerated progress and increased action. After debates, the high-income countries were asked to provide just seven-tenths of one per cent of GDP. As of 2015, there were only five countries that met the 0.7 pledge: Denmark, Luxemburg, the UK, and Sweden with the Netherlands following close with 0.65. The arguments of aid foes have been that "aid is not needed (economic growth is available without it to anyone and any place that really tries), and that aid is invariably wasted (governments are corrupt, untrustworthy, incompetent, and therefore unable to channel resources as promised, no matter how nice and worthy the goal)" (Sachs 2006).

While economists are typically emphasising carbon pricing as a policy tool to tackle global warming, natural scientists and transdisciplinary environmental research groups argue for deeper political engagement and proactive economic transition governance (Barnosky et al. 2014)—something akin to the Global Marshall Plan (Gore 1992). This difference in perspective is in part due to the relatively recent advancements in environmental research, measuring faster-than-expected decline in natural ecosystems and taking into account the whole range of human-induced pressures, not just climate emissions (Järvensivu et al. 2018).

As Hall and Klitgaard (2011) have shown, today's dominant economic theories, approaches, and models were built during the era of energetic and material abundance. The theories were only temporarily tested by the oil crises of the 1970s and the 1990s, with no remarkable theoretical or political changes. Thus, the dominant economic theories as well as policy-related economic modeling rely on continued energetic and material growth. The theories and models anticipate only incremental changes in the existing economic order. As such, they have difficulties explaining the current turmoil (Järvensivu et al. 2018).

Standard models take no account of the use of finite resources and environmental constraints, and are blind to social outcomes in terms of equity and, of course, human well-being.

Macroeconomic models are open-ended by nature, with growth being the primary output of interest. Inputs feed in, interact with each other, achieve balance (or equilibrium) and outcomes result. We need to reverse this. That is, to start with the hard outcomes we need: environmental sustainability; equitable social and economic justice; and high levels of human well-being. To link these to relevant economic determinants within the model (aggregate output, income distribution and working hours, respectively, for example) and to reverse-engineer what this would imply for the levels and types of differing inputs (Jackson 2009).

Sustainable development is a new framing concept and radical philosophy to redefine economic paradigm/progress, which is itself the cause of so many environmental and social problems, for instance, inequality. A fifth of the world's population earns just 2% of global income; at the same time, the richest 20% by contrast earn 74% of the world's income. To have any chance of achieving a sustainable economy vision, the financial markets need to allocate capital differently. At the moment, finances do not flow in support of activities that shape a sustainable economy. By sharing our resources more equally, by building better communities and a better society and by safeguarding the natural environment, we can focus on the things that really matter and achieve genuine and lasting progress with higher levels of well-being.

When we think about the kinds of changes that will be required to bring this about, it is tempting to focus just on the practical issues—financial regulation, taxation and welfare policy, or reducing our carbon intensity—and we will be coming to these issues in detail later. Nonetheless, we need to remember that, as important as these are, in a democracy, none of these changes will come about without the will and desire of the people. People are not like the passive automats of economics textbooks. They have goals, beliefs and aspirations and they actively construct the world around them through the ways in which they talk, behave and make meaning. The main process towards sustainability is socio-economic transformations as co-evolutionary processes that include changes in modes of production, work relations and culture. Although technological innovations have played a key role in the marketisation of society, their potential alone to enable transformations to sustainability is questionable. When lowering prices, efficiency gains often lead to increased consumption, undermining environmental benefits and thus undermining environmental benefits and reinforcing the dominant paradigm of consumerism and materialism (Kemp et al. 2007).

The ultimate goal of transforming towards sustainability is a resilient, equitable, low carbon economy and production based on interconnectedness, our shared past, our common future, and environment on which we depend for life.

1.2 Conventional Economy and Sustainable Development. The Dilemma of Growth

> Anyone who believes exponential growth can go on forever in a finite world is either a madman or an economist—Kenneth Boulding, economist, UK

Economic globalisation manifested more clearly in the rising dominance of the multinational corporations by posing the need for global, rather than national or sub-national action.

"The world as an experiment has existed for only 40,000 years. Of those, the Western variant has been with us for only 250 years, and in that speck of time, more has been done to destroy the conditions for life than in the whole of the preceding 39,750. Destroyed conditions of life mean lost opportunities, not only in the present but also in the future. This too is a way of describing globalisation—as an accelerating process of social entropy, which dissolves cultures and finally, if things turn out badly, leaves behind only the bare, undifferentiated will to survive. To be sure, in the actual course of its history—from modern slave labour and ruthless exploitation of the colonies to early industrial destruction of the conditions for human life, which had nothing to do with the project—the free, democratic, enlightened West eventually wrote its counter-history of un freedom, repression and counter-enlightenment. With the future impact of climate change, the Enlightenment will not be able to free itself from this dialect. It will fail because of it" (Welzer 2017).

> The success of our economy has always depended not just on the size of our gross domestic product but on the reach of our prosperity; on the ability to extend opportunity to every willing heart, not out of charity but because it is the surest route to the common good.
>
> US President Barack Obama, Inauguration speech, 2009

Post-Keynesian analysis is historical in nature: markets would not and do not exist without political regulation. Consequently, the Post-Keynesian approach is not a priori wary of the state's role in the market. It does not see markets as always equilibrium-seeking but maintains that capitalist economies have tendencies toward market bubbles and other crises. Markets do not lead to socially and ecologically desirable outcomes on their own but require active political guidance (Järvensivu et al. 2018).

Standard neoclassical economic paradigm is a product of the Newtonian world-view and thus, is exclusively concerned with qualitative outcomes. Only quantitative-based values and measures are utilised to indicate the state of the macro-economy and the success or failure of economic policies. Qualitative terms such as development and human welfare are key elements of the economist's system of thought.

It is now widely accepted that the twentieth century advances in thermodynamics and evolutionary-based theories have greatly exposed the limitations inherent to the Newtonian view. The Newtonian world-view assumes that the nature of the parts of any system and the relations between them remain unchanged and, as a consequence, within the system activity does not alter its underlying parameters, nor those of other systems. Particular aspect, totally ignored by the Newtonian world-view, involves the positive feedback of system dynamic over time. Positive feedback is a form of dynamic disequilibrium and occurs when the creation, modification and introduction of new or modified components alter the system's future dynamics. Neoclassical economic models fail to acknowledge the powerful relationship that exists between human belief systems and the functional operation of economic systems, and the markets of which they are comprised (Lawn 2000).

The conventional economy paradigm suggests that the best way to address the problem of huge disparity is through growth itself. In a world without limits, it would be acceptable to lift the poorest out of poverty by growing the entire economy. However, the existence of ecological or resource limits poses a more pressing moral question. Another possibility would be to achieve substantial technological improvements in the efficiency with which material resources are converted into economic output. In this case, our faith is in the possibility that we can push relative decoupling fast enough that it leads in the end to significant absolute decoupling. Here is the question: how feasible is this? Decoupling of growth and environmental pressures has been the main hope and focus of politics so far and a large part of economics—witness the recent popularity of the notion "green growth". This notwithstanding, decoupling in unlikely to be fast enough in all relevant environmental dimensions, if successful at all, which means that growth may be at stake when we go for a serious sustainability policy (Antal and Van den Bergh 2013). As Jackson (2009) argues and as the climate research of the New Economic Forum has shown, there is absolutely no evidence to support this—quite the opposite in fact: the scale of output continues to outstrip efficiency gains and no economies have dematerialised to any meaningful extent or show any signs of doing so. The reasons for this have long been well understood, though largely ignored. Environmental economist Herman Daly put it like this: "The notion that we can save the "growth forever" paradigm by dematerialising the economy, or "decoupling" it from resources, or substituting information for resources, is fantasy. We can surely eat lower down the food chain, but we cannot eat recipes" (Daly 1997).

The focus on life-improving and emissions-reducing goals rather than abstract economic goals also characterises the relations between the developing and developed countries: economic activity between them consists of bidirectional learning in order to build new locally suitable infrastructure and practices at both ends. This kind of proactive state-led economic governance toward self-sustained, low-emission production and consumption runs contrary to the currently dominant world political order, which has been organised around international free trade. In the modern global economy, states are the only actors that have the legitimacy and the capacity to fund and organise large-scale transitions. The most emblematic event of a polarised view of industrial capitalism period came in 1981, when Reagan removed the solar panel from the roof of the White House that President Carter had so gladly and symbolically installed.

The main features of conventional economics are:

- Strong emphasis on the efficiency with which the main inputs to production, i.e. capital, resources, and labour, are utilised. Efficiency stimulates demand by driving down costs and contributes to a positive feedback and production expansion at the same time.
- When economic growth is less than increase in labour productivity, someone somewhere loses their job. From an environmental point of view, this could be even desirable because it leads to lower resource use and fewer polluting emissions. However, from the existing macroeconomic system point of view, such situation leads to recession, because growth equals jobs.
- Conventional macroeconomic system model based on growth and competition does not have a steady state regime (weak resilience) and is continuously pushed towards one of the two dynamic states: expansion or recession.
- The key issue of an economic system is profit, which stimulates a permanent search for newer, cheaper products and services. This process of "creative destruction" (Joseph Schumpeter 1942) is a fundamental feature of capitalism, driving the economic growth forward. The restless desire of the consumers is perfect complement for the restless innovation of the entrepreneur (Jackson, 2017). This means that an economic system remains viable as long as consumption rises.
- According to the research of Mazzucato (2018), achieving system-level transition has required and will require proactive mission-oriented innovation—it is not enough for the state to reactively fix the "market failures". Many economists have settled for carbon pricing as the least interventionist, economically most efficient "first-best" policy to cut the greenhouse gas emissions.
- Many economists and politicians hope that carbon pricing can be accomplished via carbon taxes or emissions caps and permit trading ("cap-and-trade"). As a policy tool, carbon pricing lacks the crucial element of coordinating a diverse set of economic actors toward a common goal. Individual actors would have an incentive to decrease carbon emissions, but they would still compete through their own business logics with nothing to ensure that a particular business logic supports the sustainability transition on a systemic level.
- Incomplete and wrong indicators, for instance GDP, are used to determine the economic progress.
- The assumption that if the developing countries were to implement conservative macroeconomic policies while expanding the role of the private market at the expense of the state, they would then achieve sustained high growth rates on their own as well as the statement that if such country is failing to grow, the problem must be either macroeconomic mismanagement or a hindering of private market expansion in the country, usually attributed to corruption or more broadly "bad governance" are unsatisfactory (McCord et al. 2005).

The main critique of conventional economics—the ideas that underpin the rules by which the world is run—lies in the fact that it is primarily critical of the way that money measures the world:

- It ignores the planet (and people);
- It measures the wrong thing (GDP);
- It misunderstands the real life (rationality, invisible hand, dynamics);
- It encourages vulnerability (poorly defended due to subsidies);
- It colludes with short-termism (short electoral cycle);
- It overvalues owners;
- It remains blind to values (ethics behind a product);
- It encourages consumption for its own sake (fuel for growth);
- It encourages and relies on debt and indentured legal agreements.

Taken together, these criticisms reveal not just an economic system that is partially blind, but one that has no moral compass and is destructive of the environmental conditions on which the civilisation depends. The numbers from numerous studies revealed that once societies move past approximately 15,000 USD per capita income, neither objective measures of quality of life, nor subjective measures like life satisfaction show any material improvement. Our current system is designed for growth; that is what keeps us employed, services flowing from government via taxes, and the poor believing that they can escape from poverty. Without growth, there is a danger that the whole house of cards will come crashing down.

The global economy output is now almost ten times bigger than it was in 1950. If it continues to expand at the same average rate, the world economy in 2100 would be more than 20 times bigger than it is today: a staggering 200-fold increase in economic scale in the space of just few generations. It is totally at odds with our scientific knowledge of the finite resource base and the fragile ecology on which we depend for survival and it has already been accompanied by the degradation of an estimated 60% of the world's ecosystem (Jackson 2017).

Current rates of carbon emission are thought to be higher than at any time in the last 65 million years. The concentration of carbon dioxide in the atmosphere has risen sharply since 1850 and is now around 410 parts per million. One of the biggest signs of our time will be the presence of three things we use every day: concrete, plastics and aluminium. We have now produced around 500 million tonnes of aluminium, about 50 billion tonnes of building materials and we now produce more than 300 million tonnes of plastics a year (New Scientist 2018).

The critical factor of economic and financial crisis was a massively "over-leveraged" private sector. Households and firms were simply carrying on an unsustainable amount on debt. However, the most striking aspect of this over-indebtedness is just how long it had been going on. Indeed, it is a feature of the system of debt that for one part of the global economy to be indebted, another part must be saving hard (Hall and Soskice 2001). For instance, the so-called liberal market countries led the march towards liberalisation, competition, and deregulation in the period of 1980–1990. The coordinated market economies—the countries of "old" Europe and Scandinavia—were much slower to deregulate and tended to depend heavily on strategic interactions between firms—rather than competition—to coordinate the economic behaviour (Jackson 2016).

One of the most basic and important ideas of economics is that business and government have complementary roles as part of a "mixed economy". Paul Samuelson provided the intellectual underpinnings of the modern mixed economy. The five core ideas of modern mixed capitalism are as follows:

- Markets are reasonably efficient institutions for allocating society's scarce economic resources and lead to high productivity and average living standards.
- Efficiency, however, does not guarantee fairness (or "justice") in the allocation of incomes.
- Fairness requires the government to redistribute income among the citizenry, especially from the richest members of the society to the poorest and most vulnerable members.
- Markets systematically underprovide certain "public goods", such as infrastructure, environmental regulation, education, and scientific research, whose adequate supply depends on the government.

The market economy is prone to financial instability, which can be alleviated through active government policies, including financial regulation and well-directed monetary and fiscal policies. David Hume (1711–1776), a Scottish contemporary of Adam Smith, argued that trade between nations is one way to foster peace. A problem with free trade areas is that economic objectives, i.e. reduced prices and access to markets, tend to supersede environmental issues, and capital mobility is given precedence over labour mobility. A problem in terms of sustainability is that large corporations, especially in agriculture, can produce at a much lower average total cost than smaller firms, resulting in further consolidation, internal and external migration as displaced agricultural workers see work elsewhere; and further intensification of already marginal land like hill slopes, forest margins, and vulnerable arid lands. Many far-right parties in Europe share with Trump anti- globalisation policy and the Brexit, a dislike of elites, a preference for increased political and economic barriers, anti-immigration and anti-globalisation stance (Reardon et al. 2018).

Advocates of sustainability are not opposed to industrialisation per se and do not seek a return to the Stone Age. Nonetheless, they question whether the Industrial Revolution has jeopardised the humankind's ability to live happily and sustainably upon the Earth, because it was based on such statements as: economic growth at the expense of everything, including the integrity of the environment, social inequality for the sake of private wealth and assumption that technological innovations are always a positive thing (Caradonna 2014).

The Great American Dream, built upon the foundation of economic growth, suggests that anyone who works hard can improve himself or herself and increase their wealth. In this context, many believe the poor are at least to some degree lazy or incompetent. They are poor by their own actions or lack thereof. Accepting the end of economic growth means that this idea, at best highly debatable, can no longer be argued. If the amount of wealth cannot increase, you can improve your wealth only by taking it away from someone else. The American Dream is dead. The only way to lift the bottom is to drop the top. Not only do we have to face the end of economic growth, but now we have to consider the most heretical idea of all: redistribution. Ronald Reagan argued that "there are no great limits to growth because there are no limits on human capacity for intelligence, imagination and wonder" (Gilding 2011).

The dilemma: to resist growth is to risk economic and social collapse. To pursue it continuously is to endanger the ecosystem on which we depend for a long-term survival. It is clear that strong political governance is required to accomplish the key transitions. Market-based action will not suffice—even with a high carbon price. A comprehensive vision and plans with tight coordination are needed.

The transformation pathways:

- The first step in achieving sustainable economy is urgent reform of national accounting system so that we measure what really matters, i.e. integration of ecological and social variables into national accounts and end with GDP as the main and unambiguous economic indicator.
- It is a mistake to assume that human motivations are all selfish. Each society strikes the balance between altruism and selfishness and also between novelty and tradition. Government, schools, the media, and religious and community institutions should promote cooperation, instead of competition, reward people who sacrifice their own gain to serve others, i.e. support more altruistic and more conservative values.
- The perverse incentives for unsustainable (and unproductive) status competition (advertising, corruption) should be dismantled or corrected. New structures that provide capabilities for people to flourish and particularly to participate fully in the society in a less materialistic way should be established.

Thus, the vision for sustainable development economy is to create a strong, healthy and just post-carbon society operating/living within ecological limits of a finite planet. In sustainability, the associated values of sufficiency, equity and efficiency become the central organising principles of the economy. The technical and economic problems involved in achieving sustainability are not that difficult. The hard problem is overcoming our addiction to growth as a favoured way to assert our creative power, and "idolatrous belief that our derived creative power is autonomous and unlimited" (Soddy 1926). Shortly: living well on a finite planet cannot simply be about consuming increasingly more stuff. Nor can it be about accumulating increasingly more debt.

References

Antal M, Van den Bergh JCJM (2013) Macroeconomics, financial crisis and the environment: strategies for a sustainability transition. Environ Innov Soc Transit 6:47–66

Barnosky AD, Brown JH, Daily GC et al (2014) Introducing the scientific consensus on maintaining humanity's life support systems in the 21st century: information for policy makers. Anthr Rev 1:78–109

Caradonna JL (2014) Sustainability: a history. Oxford University Press, New York, NY
Daly HE (1997) Beyond growth: the economics of sustainable development. Beacon Press, Boston
Ehrenfeld JR (2009) Sustainability by design: a subversive strategy for transforming our consumer culture. Yale University Press
Gilding P (2011) The great disruption: How the climate crisis will transform the global economy. A&C Black
Gore A (1992) Earth in the balance-ecology and the human spirit
Hall CAS, Klitgaard KA (2011) Energy and the wealth of nations. Springer
Hall PA, Soskice D (2001) Varieties of capitalism: the institutional foundations of comparative advantage. OUP Oxford, Oxford
Hawken P, Lovins AB, Lovins LH (1999) Natural capitalism: the next industrial revolution. Routledge, Boston
Independent Group of Scientists (IGS) appointed by the Secretary-General:, Messerli P, Murniningtyas E, et al (2019) Global Sustainable Development Report 2019: The Future is Now–Science for Achieving Sustainable Development. United Nations, New York
Jackson T (2009) Prosperity without growth: economics for a finite planet. Routledge
Jackson T (2016) Prosperity without growth: foundations for the economy of tomorrow. Routlege, London and New York
Jackson T (2017) Beyond consumer capitalism: foundations for a sustainable prosperity. In: Handbook on Growth and Sustainability. Edward Elgar Publishing
Järvensivu P, Toivanen T, Vadén T et al (2018) Invited background document on economic transformation
Kemp R, Schot J, Hoogma R (2007) Regime shifts to sustainability through processes of niche formation: the approach of strategic niche management. Technol Anal Strateg Manag 10:175–198. https://doi.org/10.1080/09537329808524310
Lawn PA (2000) Toward sustainable development: an ecological economics approach. CRC Press
Mazzucato M (2018) The value of everything: making and taking in the global economy. Hachette UK, New York
McCord G, Sachs JD, Woo WT (2005) Understanding African poverty: beyond the Washington consensus to the millennium development goals approach. Africa World Econ 23
New Scientist (2018) This is planet earth: your ultimate guide to the world we call home. Nicholas Brealey Publishing
Orr DW (2016) Dangerous years: climate change, the long emergency, and the way forward. Yale University Press
Reardon J, Madi MAC, Cato MS (2018) Introducing a new economics. Pluto Press
Robinson J (2004) Squaring the circle? Some thoughts on the idea of sustainable development. Ecol Econ 48:369–384. https://doi.org/10.1016/j.ecolecon.2003.10.017
Sachs JD (2006) The end of poverty: economic possibilities for our time. Penguin books, New York
Schumpeter JA (1942) Creative destruction. Capital Social Democr 825:82–85
Soddy F (1926) Virtual wealth and debt: the solution of the economic paradox
WCED (1987) Our common future. World Commission on Environment and Development, Oxford
Welzer H (2017) Climate wars. Why people will be killed in the 21st century. Polity Press

Chapter 2
Reduction of Unsustainability of Enterprises

Jurgis Kazimieras Staniškis

> There is no simple solution to the problem of planetary limits. It is clear that better technology, intelligent policymaking and regulation are part of the answer. But unless we grapple with the necessity of radically reforming the underlying economic system, these are unlikely to be enough.—Liz Murray, Scottish economist

2.1 The Concept of Resource-Efficient and Cleaner Production

> Adam Smith (1776) along with many others at that time saw nature as "no more than a storehouse of raw materials for man 's ingenuity".

The concept and the word "sustainability" in relation to industrial processes and products have been heavily overused, with too many different meanings applied to it (Aras and Crowther 2009). Sustainability takes much more than simple problem-solving or some improvements in the existing consumption and production system. It needs a discontinuous leap from the existing level of actions.

Even though reducing unsustainability is not the same as creating sustainability, it still makes great sense to prevent and remove the causes and sources of unsustainability; however, at the same time it should be recognised that this is just attacking the symptoms. In this context, reduction of unsustainability of industrial development may be considered as a process of continuous improvement of environmental, economic and social performance in industry. Such approach allows the specialists to identify particular process performance parameters that could be controlled and managed. Integration of sustainability management into the overall business planning is a very important aspect to be tackled, because the efficiency of management systems largely depends on connections between the management systems and strategic/financial decision-making. Enterprises often lack explicit information about their activities, particularly reliable quantitative information on technological processes and various sustainability aspects. Moreover, the existing information is seldom systemised in a form suitable for effective decision-making.

J. K. Staniškis et al., *Transformation of Business Organization Towards Sustainability*, World Sustainability Series, https://doi.org/10.1007/978-3-030-93298-5_2

The Institute of Environmental Engineering (APINI) of Kaunas University of Technology in 1995 initiated wide research on preventive environmental innovations generation and implementation in production companies aiming at unsustainability reduction.

The main goal of this research was the evaluation and analysis of the performance of incremental innovations and their introduction in two types of management systems (single-loop and cascade), providing continuous reduction of industrial enterprise unsustainability. The innovations were developed and implemented by the APINI scientists together with companies' experts using a modified resource-efficient and cleaner production (RECP) approach (Staniškis et al. 2010).

In practical terms, RECP means the continuous application of preventive environmental strategies to processes, products and services in order to increase the efficiency and reduce risks to people and environment. RECP comprises three sustainability dimensions: a) increased economic performance through the efficient use of energy and material resources, b) minimised industry's impact on the natural environment, and c) enhanced social dimension by providing jobs and protecting the well-being of workers and local communities. The research on the performance analysis comprised 177 resource-efficient and cleaner production incremental innovations implemented by the APINI team in industrial companies in the period of 1995–2015.

To ensure effective decision-making aimed at reduction of performance unsustainability, a cascade management system based on three hierarchical levels (process, activity and strategic decision-making) is recommended for use in the enterprises because it introduces the social component of sustainability, ensures involvement of decision-makers at all management levels and enables effective decision-making at different managerial and operational levels.

The procedure for unsustainability management involves the compliance of enterprise's processes, activities, and strategic decision-making with legal and other requirements (life quality, etc.), new scientific knowledge, and stakeholder expectations, which all together constitute the system's input information. Decision-makers at different management levels were provided with feedback information based on the indicators of enterprise's sustainability performance. The results are based on the theoretical and practical research conducted by the APINI team in the course of different national and international projects, which were focused on unsustainability performance improvement and were implemented jointly by the APINI researchers and industrial enterprises.

Although each company can make the choice for itself to reduce unsustainability or to adhere to the more traditional model, changing social and investor expectations will only increase the pressure on them to adopt the sustainable model (Eccles et al. 2012). Sustainable development at an organisational level is usually described as using a triple bottom line that divides the performance into economic, environmental and social dimensions (Elkington 1998; Topfer 2000). At the same time, "sustainability is a normative concept referring to an ideal state of being in which humans are able to flourish within the ecological thresholds of the planet alongside other living entities for perpetuity" (Williams et al. 2017). Thus, sustainability management can be defined as "a profit-driven corporate response to environmental and social issues that are caused through the organisation's activities" (Salzmann et al. 2005). In this context, "innovating for sustainability is a systemic, dynamic and nonlinear process that faces many uncertainties" (Foxon and Pearson 2008). In fact, business and societies can find approaches that will move towards reduction of unsustainability and simultaneously achieve three goals—environmental protection, social well-being and economic benefit. Reduction of unsustainability is a good business in itself. It creates opportunities, for suppliers of "green products", developers of environmentally safer materials and processes, companies that invest in eco-efficiency, and those that engage in social well-being (Staniškis et al. 2005).

One of the key approaches to reduce the unsustainable performance of companies is resource-efficient and cleaner production management system. However, the application of RECP is (in general) comparatively slow despite the good results achieved (Bonilla et al. 2010). It could be also stressed that management systems are often implemented with a "certificate-oriented" approach whose efficiency in terms of sustainability performance improvement is low and there is often a lack of motivation to maintain the system after the certification (Pedersen and Nielsen 2000; Iraldo et al. 2009). There are studies urging the necessity of the human resource dimensions for a proactive environmental management. A model that establishes the relationship between these dimensions and the typical phases of an environmental management system was proposed (Jabbour and Santos 2008). Sustainability analysis of a system should include specific indicators used to quantitatively represent the system from the viewpoint of environmental, economic, and societal impacts, in accordance with the basic principles of sustainability (Krajnc and Glavič 2003; Sikdar et al. 2012).

One of the key issues for successful management is the availability of accounting tools capable of monitoring and tracking the overall performance and sustainability from a qualitative and quantitative viewpoint (Perrini and Tencati 2006; Staniškis and Arbačiauskas 2009). Sustainability management will succeed only if managers and personnel recognise that the actions of the system create value for them. Cultural changes must be accomplished within the business in order to provide sustainability management-based benefits, seizing the opportunities (Yilmaz and Flouris 2010). It is important to stress that to overcome the shortcomings of existing management accounting methods and techniques, new accounting systems should be developed or the traditional ones redesigned.

There are several internationally acknowledged sustainability/environmental performance evaluation/reporting initiatives and methodologies, for instance, international standard ISO 14031 for environmental performance evaluation; a sustainability performance evaluation initiative of the Britain's Institution of Chemical Engineers (Sikdar 2003); Eco-Efficiency Assessment; and the Global Reporting Initiative (GRI, 2015), which was intended to assist the enterprises in assessing the performance and improving the communication with stakeholders (Bass and Dalal-Clayton 2012). For the selection of initial environmental indicators, indicator systems specific for particular industrial branches could also be used (Pohjola 2005; Enroth 2006; Viluksela 2009).

The possibility to use benchmarking is one of the main strengths of the abovementioned methodologies in the context of sustainability performance management;

however, at the same time, its focus on external reporting and underestimation of the internal information needed for decision-making represents a significant drawback. For example, a review of the frameworks of business sustainability indicators has shown that they present simple lists of indicators with little or no guidance as to how to apply them over time to become more sustainable (Veleva and Ellenbecker 2001).

These shortcomings can be partly explained by the fact that pressure on an industrial enterprise from external stakeholders to publish the sustainability performance information often is one of the main driving forces for sustainability performance evaluation. It could also be related to the establishment of "socially responsible" investment funds and investment rating systems, e.g. Dow Jones Sustainability Index (Ballou et al. 2006). It should be stressed that the efficiency and value added of the performance evaluation system for an enterprise depends mainly on the strength of internal motivating factors and ability of enterprises to apply the sustainability performance indicators properly. They are to be focused more on the information needs for decision-making at an enterprise level (Staniškis and Arbačiauskas 2009). Generally, the main drivers for enterprises to act in sustainable ways include market demands, changes in corporate procurement, government legislation and regulation, rise of socially responsible investments, competitors' actions and changing expectations of employees (Pryce 2002; Epstein and Buhovac 2014).

The social sustainability challenge is still receiving relatively little attention within the sustainability management. An enterprise can be described as socially effective when it has effectively reduced the absolute level of negative social impacts and has succeeded in keeping it low, and provided that these also give rise to important positive social impacts and benefits (Schaltegger et al. 2002). In order to incorporate unsustainability reduction management, companies need to work in partnership with other stakeholders and organisations, which have an interest in the companies' activities and their social and environmental impacts.

Findings from the review of the innovation research topic clearly show that the major research efforts are still mainly focused on the company production level, on the development of new processes and products in isolation of other developments in environmental and social areas. Therefore, future research on sustainability management and innovations should take broader scope on the systems with feedback/feedforward loops and on other dimensions of sustainable development (Williams et al. 2017). Experience from leading companies suggests that embedding sustainable innovations into the existing processes and systems is essential, but the different levels of innovations (incremental, re-design, functional, systems) need to be recognised and different approaches are needed to stimulate each level and type (Charter 2006).

A production organisation's system consists of operations and processes, management and strategy, organisational systems, procurement and marketing, and assessment and communication (Lozano 2012). Taking into account that there is no particular standard for unsustainability reduction management, researchers and practitioners try to fill this gap. The number of papers on corporate social responsibility clearly shows a necessity of its integration into the business process. One of them (Asif et al. 2013) presented the concept of simultaneous "top-down integration" and "bottom-up community-related indicators development" approaches to corporate social responsibility; however, the studies on how these approaches unfold in practice are still missing.

The encouragement to develop certified management systems (in particular, environmental systems) is the first step towards sustainability. After that, enterprises should be encouraged to broaden their sustainability focus by widening environmental issues, internal social issues such as staff training and development, local community relations and the performance of suppliers (Holton et al. 2010). It is necessary to analyse the advantages and disadvantages related to each environmental tool; however, their potential can be increased by integrated utilisation (Seiffert and Loch 2005).

Many companies have already recognised the importance of environmental sustainability issues including their products' environmental impact and, therefore, began to incorporate significant environmental aspects into their product development and production processes. However, the understanding that systematic avoidance of waste and pollutants increases process efficiency, enhances product quality and minimises the costs is missing. The main result of the APINI research mentioned above is the methodology and systems for the generation, control and implementation of preventive incremental innovations that reduce the unsustainability in companies (Staniškis and Katiliūtė 2019).

2.2 Mathematical Formulation of Unsustainability Reduction Problem

> No matter how complex global problems may seem, it is we ourselves who have given rise to them. They cannot be beyond our power to resolve—Daisaku Ikeda, Japanese philosopher

The structural innovations management models present the key elements of environmental management system and other sustainable industrial development tools in a sequence of integration. Distinctive features of the proposed models include the integration of sustainability aspects and criteria at operational level, and the shift of conventional management system to sustainability management system. Hierarchical procedure for sustainability management covers process, activity and strategic decision-making with legal and other requirements as well as new scientific knowledge and stakeholder expectations constituting the system input information. Decision-makers at different hierarchical levels are provided with the feedback information based on the indicators of enterprise's sustainability performance.

Modern industrial development tools generally address different elements of an enterprise system: operations and processes, management, communication, etc. At the same time, these tools contribute to different dimensions of sustainability (environmental, economic and social). Lozano (2012) proposed a framework for the selection of tools (Corporate Integration of Voluntary Initiatives for Sustainability (CIVIS) framework). The main criteria include full coverage of the enterprise's elements and all dimensions of sustainability (including the time dimension).

Analysis of different approaches/ tools suggests that application of the different approaches/ tools covering all key elements of a production company system (production, products, waste management, and communication with internal and external stakeholders) could be used to ensure the unsustainability reduction performance improvement in enterprises.

Any management system aimed at reducing unsustainability in the performance criterion must include all three sustainable development dimensions: economical, environmental and social, i.e.:

$$I = f(E, R, S),$$

where:

I—performance criterion;
E—economic savings;
R—reduced amount of generated wastes and/or resource/energy savings;
S—social benefits.

The object of unsustainability reduction management are production processes of a company or its part. In this case, the management or control task formulated as follows:

To find the control input vector ***U*** by which the state of processes ***X1*** should be transformed to state ***X2*** within the system boundaries ***B*** and constraints ***C*** of input vector ***U*** and maximising the performance criteria ***I--max***. The state of the object is the set of data on process variables giving complete and unambiguous information about the object history necessary for the determination of its behaviour in the future under certain management or control actions (Staniškis and Katiliūtė 2019).

One of the key approaches to increase the unsustainability reduction performance of companies is the resource-efficient and cleaner production approach. RECP means the continuous application of preventive environmental strategies to processes, products and services in order to increase the efficiency and reduce the risks to people and environment. It comprises three sustainability dimensions: (a) increased economic performance through the efficient use of energy and material resources, (b) minimised industry's impact on the natural environment, and (c) enhanced social dimension by providing jobs and protecting the wellbeing of workers and local communities. This could be achieved by different levers of ***U*** *{u1,u2,u3,…,un}*:

- *Process input* substitution: renewable, less toxic or less critical materials (main or/and ancillary), materials having longer service life-time, also including good housekeeping;
- *Improved technological process control*: for instance, introduction of new measurement systems, advanced control algorithms, improved parameters' control structure by implementing new feedforward/feedback loops;

- *Equipment modification and/or technology change*: modernisation/change of equipment and process sequence, replacement of technology and product redesign in order to minimise the environmental impact during the entire product life-cycle;
- *On-site reuse of waste materials*, which involves the return of waste materials to the originating process or to another process at the same enterprise as an input;
- *Waste recycling/recovery* outside the company (Staniškis and Katiliūtė 2019).

"For the implementation of the methods listed above, a single loop feedback management system—the Deming's cycle or PDCA (Plan–Do-Check-Act) cycle is used. However, prevention, resource efficiency, cleaner production, life cycle assessment, eco-design, circular production and sustainability reporting are seldom sufficiently integrated in the management systems, despite the fact that the efficiency of these tools largely depends on their integration level in the overall strategy of enterprises and everyday activities" (Staniškis and Katiliūtė 2019).

Extensive research of the APINI team suggests that one of the ways to increase the motivation for maintaining the unsustainability reduction management system and ensuring its efficiency is the systematic/integrated use of modern industrial development tools listed above, which enable an increase in their efficiency and lead to cost savings associated with the more efficient use of human and natural resources, improved product characteristics, more effective operational procedures, reduced waste generation, and harmful emissions to the environment (Fig. 2.1.) "The system comprises two closed loops: feedback and feedforward, where feedback presents the reactive approach and feedforward—preventive. In the figure:

X—vector of the process state,
Xr—vector of reference state of the process,
X*m*—vector of measured/assessed state,
U—integrated control actions/tools,
D—vector of disturbancies,
D*m*—measured/assessed disturbances,
ΔX—vector of deviations.

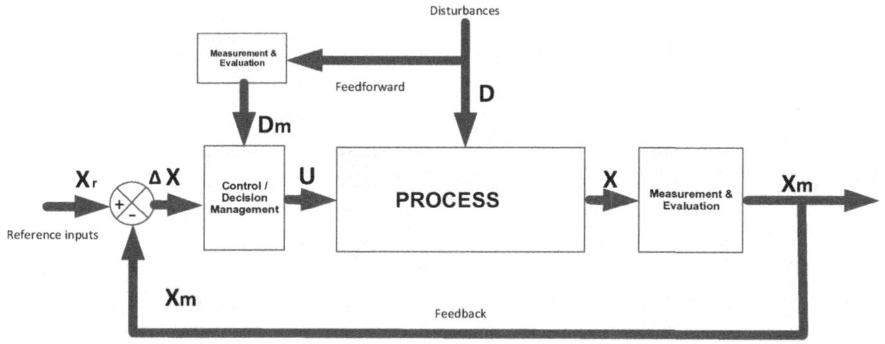

Fig. 2.1 The general feedback-feedforward control system for unsustainability reduction (Staniškis 2020)

Integration of different environmental and other management tools is not particularly novel; however, practical models for enterprises to meet the sustainable development requirements are lacking. We claim that knowledge about the sustainability management tools is a crucial factor between the strategy and implementation (Staniškis and Katiliūtė 2019).

Knowledge is of utmost importance for the extent to which a company applies the sustainability management tools, especially having in mind a difference between SMEs and large companies (Hörisch et al. 2015). "In the given case, environmental management system is taken as a basis for the proposed integration of sustainable industrial development measures (see Fig. 2.2.). The structural model presents the key elements of environmental management system and other sustainable industrial development tools in the sequence of integration. The level of environmental performance improvement depends largely on the planning phase, when the potential for performance improvement is systematically analysed and preventive measures are developed. To identify the preventive performance improvement options the proposed

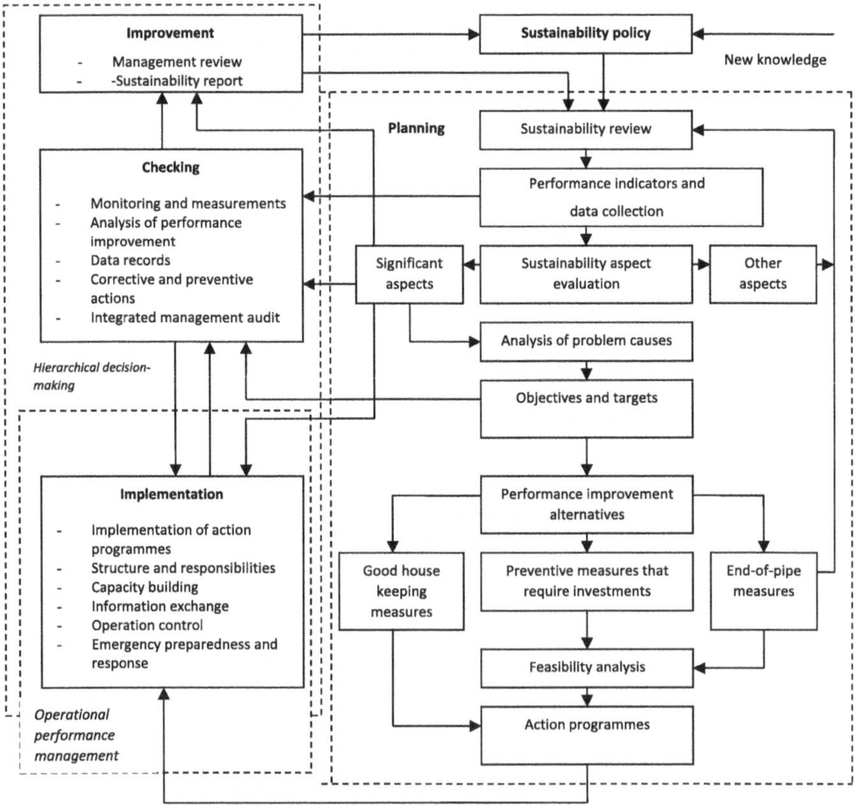

Fig. 2.2 Structure of a model integrating the key tools of the industrial development into the single-loop management system

methodology was used, when a set of alternatives were developed for each significant aspect. Environmental, technical and economic feasibility analysis leads to the development of action programmes for implementation. In the given case, both process and product improvement options are considered. A set of sustainability performance indicators has been developed in the planning stage also seeking to ensure effective decision-making (Staniškis and Katiliūtė 2019).

2.3 Methodology and Systems for Unsustainability Reduction in Industrial Organisations

> You never change things by fighting the existing reality. To change something, build a new model that makes the existing model obsolete—Richard Buckminster Fuller, system theorist, US.

The methodology for synthesis and analysis of incremental innovations was divided into five main steps: (i) organisation, (ii) pre-assessment, (iii) assessment, (iv) feasibility analysis, and (v) financing and implementation.

The *organisation* step is aimed at ensuring the top management commitment towards sustainability, generation of policy/declaration, and presenting the unsustainability reduction team together with the main goals and activities. The team consisted of persons responsible for the major facilities and environmental issues in the enterprise, research or development staff and external expert(s) from outside the company. Profound knowledge generally comes from outside; therefore, participation of external business development experts is crucial. A system cannot understand itself without professional consultancy from outside, because prior experiences will bias objectivity, preventing critical analysis. Unsustainability reduction process is the optimisation of performance of the production system where individual components should reinforce, not compete which each other to accomplish the goals of the entire system. This process can only occur when all interconnected components are directed to achieve the company's goal. The people, free from fear and competition within the system, will be empowered to have ownership in the change towards sustainability. In most cases, training in creativity, communication and interpersonal skills was necessary (see Sect. 2.6).

The main aim of the *pre-assessment* is to review the production and environment aspects having in mind the unsustainability reduction of the entire enterprise. A flow chart method showing inputs, outputs and disturbances is the best representation of the production processes and environmental problems. Producing a flow chart is a key issue in the assessment. Moreover, it was the basis for material and energy balances, which occur later in the assessment. This step is implemented by the unsustainability reduction team visiting all organisation's areas for a perspective of the impacts, to the understand layout of the processes, especially the main flows and units. The most important decision of this step is on where the system boundaries

B and constraints C should be set, because these strongly influence the subsequent phases of the methodology.

The *assessment* step comprises the collection of data within the system boundaries B, building of material and energy balance, generation of resource-efficient and cleaner production innovations and feasibility analysis of selected innovations for unsustainability reduction. Once all control inputs, outputs and disturbances are identified and validated, their quantification begins. Most data are available within the enterprise recording system, for instance, stock records, accounts, purchase receipts, waste disposal documents and production data. Where information is not available, estimates or direct measurements are performed. Based on the collected data material/energy balance is developed which is simply expressed as:

Total materials/energy in = materials/energy in product + materials/energy wasted (wastes and emissions) + accumulated materials/energy

The *assessment* step was used as a systematic assessment of the flows and stocks of materials within the defined production and time period. The result of assessment is a process flow diagram including the identification of all sources of waste and emission generation, the volume and composition of which was examined by compiling a detailed material and energy balance. Besides that, environmental performance indicators were developed from the balance data, for instance, by dividing the quantity of a material input or waste generated by production over the same period. The most creative part of the assessment step is the unsustainability reduction innovations generation during brainstorming sessions when managers, engineers, process operators as well as outside experts work together. The team focuses on prevention practices by carefully analysing the cause of the problem and having in mind all five different methods mentioned above for possible solutions/incremental innovations generation. Innovations could be associated with good housekeeping and prevention (these are the most desirable, because "prevention always pays") or end-of-pipe—less desirable, because in most cases they are economically nonviable. It is crucial at this moment to integrate into decision the advanced industrial development tools, for instance, life cycle assessment, eco-design, circular production, sustainability reporting and others. Finally, the generated incremental innovations are sorted into those than can be implemented directly and go through to feasibility analysis, and those that require further investigation and go back to the Planning phase. Important in this step is understanding that the main barrier to generation of unsustainability reduction innovations is mostly related to human rather than technical factors. The reactive end-of-pipe approach is well known and accepted by industry and engineers. Existing government policies and regulations often favour end-of-pipe solutions. As usual, there is a lack of communication between those in charge of production processes and those who manage the wastes that are generated. Managers and workers who know that the company is inefficient and wasteful are not rewarded for suggesting improvements (Staniškis et al. 2005).

Feasibility analysis of the selected innovations for unsustainability reduction consisted of technical, economic, and environmental evaluation. The technical evaluation comprised two parts: (a) availability and reliability of new equipment and its effect on process/product quality and productivity; (b) changes in material and energy balance, reflecting the input and output flows after the innovation implementation. Data for economic evaluation (payback period, net present value and internal rate of return) were based on the technical evaluation results. Comprehensive environmental evaluation was traced to the responsible product and included all costs of materials used and other related product or operational costs such as energy, water, etc. The same procedure was performed for waste flows. To avoid the situation where the proposed innovation reduces one environmental impact and creates another, the whole life-cycle of a product or service was taken into account. Furthermore, it is important to point out that for economic evaluation the principles of environmental management accounting (EMA) were applied. EMA is gaining importance not only in environmental management decisions, but also in all types of routine management activities, such as product and process design, cost allocation and control, capital budgeting, product pricing and performance evaluation of unsustainability reduction process. Environmental costs' assessment and environmental management accounting are a company's internal tools and especially useful for capital budgeting of environment–related investments. At the end of feasibility analysis, action plan for the implementation of selected preventive innovations was drawn up (Staniškis et al. 2008; Staniškis and Katiliūtė 2019).

A stable and secure source of raw materials is the key to any successful industrial activity and very important for success of unsustainability reduction process. Resource criticality is often discussed in the context of the impact on the economies of certain geographic regions. However, the availability of required resources first of all concerns the competitiveness of industrial companies, especially in those countries, which do not possess abundant natural resources. The Lithuanian economy relies heavily on imports since the country is not rich in natural resources. Malinauskienė et al. introduced resource criticality as an additional dimension for evaluating and prioritising resource-efficiency improvement options. Evaluation of resource criticality was integrated into the methodology for evaluation of RECP. Simple additive weighting (SAW) was used to solve the multi-criteria decision-making problem. A typical metal processing company in Lithuania was selected for the detailed investigation of technological processes and cleaner production possibilities. The selected company processes about 3,000 tons of various metals per year. The results of process material flow analysis showed that most of the metal waste is generated during the metal plate cutting process (about 30.3% of total metal consumption). Three resource-efficiency improvement alternatives were evaluated and compared. The suggested decision support system was tested in order to decide on a definitive solution. The results revealed that evaluation of resource criticality in terms of geostrategic supply risk and economic importance could be used as an advantageous criterion to support the prioritisation of resource-efficient and cleaner production alternatives (Malinauskienė et al. 2018).

2.4 Systems for Unsustainability Reduction Innovations Generation and Implementation

> People change as a result of what they notice, not just what they are told—Michael Mandelbaum, lawyer, US

Organisation, pre-assessment, assessment, innovations generation and feasibility analysis steps comprise the planning phase, which is the key phase of single-loop management system (see Fig. 2.2).

The *financing and implementation* of innovations was performed by the APINI scientists who developed a special innovations financing and implementation system, which consisted of: (a) the pool of business development experts on preventive innovation generation and financial engineering, (b) financial institution—soft credit loan provider, and (c) industrial company (Staniškis et al. 2008). The system itself is very flexible and can be applied to any production enterprise. The key issues that can be encountered in this include the creditworthiness of companies and proper identification of costs and benefits of an innovation. The most important step in the methodology of unsustainability reduction in companies is financing of the innovations' implementation. Therefore, the sources and systems for incremental preventive innovations' financing are presented in a separate Sect. 2.5.

In the *checking* phase, it is important to ensure proper monitoring of the effectiveness, i.e. improved performance of the implemented innovations and comparison with the calculated values on:

- Reduction of wastes and emissions per production unit;
- Reduction of resources (materials & energy) per production unit;
- Economical savings.

Achievement of continuity of the enterprise performance/unsustainability reduction improvement process is the main task of system management in an *improvement* phase (see Fig. 2.2). It should be stressed that the main focus in our methodology was on the Planning phase, because all major decisions have a direct influence on the continuation of the remaining phases of the management system cycle (da Silva et al. 2013).

In the period from 1995 to 2015, the Institute of Environmental Engineering, Kaunas University of Technology supported 85 Lithuanian companies in generating and implementing the incremental preventive innovations. Research programmes and projects were financed by different funds and sponsors and in cooperation with foreign and local partners. The most important role has been played by the Revolving Facility established at the Nordic Environment Financing Corporation (NEFCO). The main task of the Revolving Facility was to enhance the financing on favourable terms for the implementation of preventive incremental innovations in industrial enterprises aiming at reduction of unsustainability (see Sect. 2.5).

APINI has fulfilled the role of the business development service provider according to the management method and system presented in Fig. 2.2, covering:

- Technical development of incremental innovations (environmental review, development of process flow diagrams, material and energy balance, and feasibility analysis of innovation);
- Preparation of loan application, including cost savings and environmental benefits;
- Assisting the Revolving Facility in communication with the industrial company;
- Preparation of innovation project implementation plan, progress and completion report;
- Assisting the company on incremental innovations monitoring including the environmental, economic and social indicators during the pay-back period after the completion.

Incremental innovations generation, financing and implementation system was successful from the companies' point of view and that of the Revolving Facility as well. The main economic and environmental results of implemented innovations are presented in Tables 2.1 and 2.2. Without going into details, data clearly shows that the use of integrated management of unsustainability reduction by application of preventive incremental innovations to processes, products and services significantly increases the production efficiency through optimisation of natural resource consumption (materials, energy, water) at different stages of production, minimises the generation of waste and emissions, and fosters safe and responsible production which leads to economic, reputational, and regulatory benefits. Besides the promotion of safer, responsible work environment, the proposed management system ensures the engagement and training of employees by two-way communication to encourage participation and raise awareness (see Sect. 2.4).

The accounting system plays a very important role when analysing the economic and environmental efficiency of innovations. There is a conflict between accounting and sustainability, because accounting assumes that the enterprise is run on behalf of its stakeholders and that outputs consist only of goods or services for sale and profit for distribution. Actually, there is always a variety of other stakeholders who are concerned with the company's activities and are affected by them (Aras and Crowther 2009). The APINI team was using environmental management accounting, which is a more comprehensive approach to management accounting, with a particular focus on costs related to raw materials wasted and other environmental issues. Environmental management accounting is an emerging discipline, which will impact production companies sooner or later. It may be driven by companies' own proactive initiative, the growing requirements of financial auditors or the demands for greater transparency by the public. In the final analysis, environmental management accounting is critical, because an organisation can only manage effectively what it measures and is measured on what it reports (Staniškis et al. 2008).

To assess and evaluate the progress towards unsustainability reduction process, it is essential that proper monitoring and management of the economic, environmental and social impacts are carried out on a regular basis and that the results are used to help focusing the attention of the society on ways to make further improvements towards sustainable societal lifestyles (Klemeš et al. 2012).

Table 2.1 Environmental benefits of incremental preventive innovations implemented in enterprises in the period of 1995–2015

No	Production/service sector	Number of innovations/companies	Water savings, thousand m^3/year	Decrease of solid waste t/year	Energy (electricity + heat) savings, MWh/year	Decrease of air pollution, t/year
1	Textile products	42/15	331.3	1.9	53,000	5,400
2	Food and beverages	30/14	201,7	38,255	12,500	4,250
3	Electricity and heat supply	21/8	28	765.5	77,800	24,850
4	Chemicals	15/6	15.6		16,200	3,750
5	Machinery and equipment	7/6	36	6.0	20,700	4,350
6	Furniture	12/6	23.5	334.6	47,600	8,960
7	Non-metal and mineral production	8/4		27,000	26,900	8,050
8	Wood products	8/4	77.8	38,546	−990	22,350
9	Transport and communication	6/4	12.6	7.5	550	410
10	Other metal products	8/3	2,6		5,400	565
11	Oil products	2/3	4		–	230
12	Leather and leather products	2/3	87	807	340	85
13	Radio, TV and communication equipment	5/2	1.9	176	9,800	1.0
14	Medical, precise and optical equipment	5/2		0.5	2,800	0.3
15	Refined oil products	2/1		1,898	6,300	505
16	Glass and glass products	2/1			10,600	
17	Agriculture and forestry	1/1		8,902	280	360
18	Municipal services	4/1			–	5.0

(continued)

Table 2.1 (continued)

No	Production/service sector	Number of innovations/companies	Water savings, thousand m^3/year	Decrease of solid waste t/year	Energy (electricity + heat) savings, MWh/year	Decrease of air pollution, t/year
19	Gas and water supply	1/1			–	
20	Wet and dry cleaning	1/1	4.7		1,100	240
	Total	177/85	826.7	116,700	290,880	84,361.3

Distinctive features of the proposed model include the integration of unsustainability aspects and criteria at operational level, as well as the shift of the conventional management system to the sustainability reduction management system (see Fig. 2.2); however, the social dimension is still reflected only partially.

One possible solution towards deeper integration of the social and outside stakeholder dimension into the management system could be introduction of new feedback into the so-called cascade closed-loop management system (Fig. 2.3). The system theory views this control system as a hierarchy of successively higher-order systems in which the control of the smaller, more immediate events employing less energy in the lower-order systems brings about larger, more distant, events involving more energy in the outer higher-order system loops. Although the processes of the inner loops must be of lower magnitude and of higher frequency than those in the outer loops, management of the inner loops is more possible and, by influencing the rates of change of the higher-order loops, provides a wider range of choice in influencing the behaviour of the outer loops. Thus, the output from the inner loops proceeds to the outer loops as a hierarchy of management ascended. The inner loops are susceptible to short-range planning in order to bring about longer-term changes in the outer loops (VanDoren 2014; Staniškis and Katiliūtė 2019).

The APINI team has modified and developed a classical cascade system for sustainability management in enterprises (Fig. 2.3). This system helps to increase the effectiveness of the decision-making process and to facilitate the reorientation of a problem-solving approach from reactive to proactive/ preventive including the social dimension.

The proposed system of sustainability performance management based on a hierarchical approach to decision-making. The system covers process, activity and strategic decision-making with legal and other requirements as well as new scientific knowledge (stakeholder expectations and scientific knowledge) constituting the system input information. The control system also relies on feedback information from several process stages. Decision-makers at different hierarchical levels are also provided with the feedback information based on the sustainability performance indicators of the enterprise. Such approach enables an increased participation of employees in problem-solving at different levels of enterprises. Process indicators

Table 2.2 Investments/economic benefits of incremental innovations implemented in the period of 1995–2015 (EUR, thousands)

Production/service	Input substitution		Technology change		Process optimisation		On-site waste reuse		Waste recycling	
	Investment	Saving	Investment	Saving	Investment	Saving	Investment	Saving	Investment	Saving
Textile products		261	146	99.5	2,564	1,998	372	226	1.45	2.6
Food and beverages	82.3	23.2	1,151	559	620	536	169	149	810	627
Electricity and heat supply	2,358	560	1,046	307	1,987	1,599	103.5	55.4		
Chemicals			20.2	47	135	347	280	98.9		
Machinery and equipment			938	359	93	30.5	120	145		
Furniture	394	108	452	198	341	114	238	101		
Non-metal and mineral production			1.5	2.5	99.7	207			642	221
Wood products		0.3	23.2	72.5	242	240	1,304	868		
Transport and communication		0.6	244	240	67	33				
Other metal products		127	152	63.2	202	137	58	38.3		
Oil products					513	105.2				
Leather and leather products		0.3		53	35	87.2	2.6	16.8	141	68.7
Radio, TV and communication equipment			307	177	1 018	335	200	101		
Medical, and optical equipment			16.2	5.8	335	146				
Refined oil products			70	23.2	1,230	363				
Glass and glass products							215	100		
Agriculture and forestry									665	205
Municipal services		0.6	120.6	49						
Gas and water supply					115	46.4				
Wet and dry cleaning			410	109.3						
Total	2,834.3	1,081	5,097.7	2,255.7	9,596.7	6,324.3	3,062.1	1,899.4	2,259.45	1,124.3

Total investments: 22,850,250 EUR, total savings per year: 12,684,700 EUR

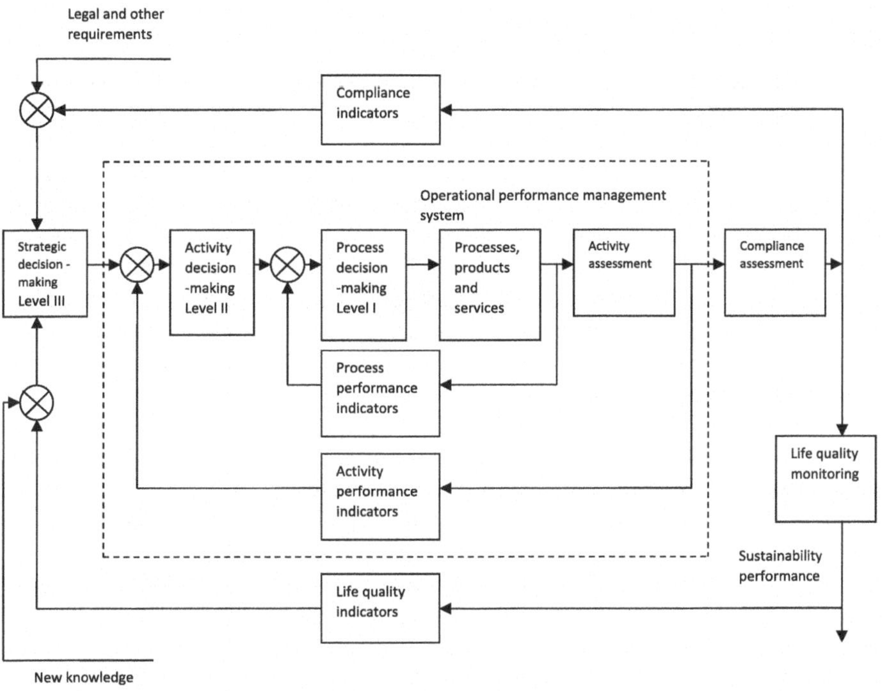

Fig. 2.3 Cascade system for unsustainability reduction management

(level I) provide information to the enterprise personnel on the process efficiency and help identifying both the deviations from technological specifications and the measures for improving the process efficiency. In the context of management systems, process indicators facilitate operational control. Activity performance indicators (level II) used at process, product, department and enterprise levels and present the "digested" information obtained from a detailed analysis of processes, products and services. These indicators are particularly useful for decision-making in terms of identifying the priority areas for sustainability performance improvement and generating particular improvement measures. It is very important to monitor the performance indicators to make the right decisions on time. This level is identical to single-loop management elaborated above On level III, compliance and life quality indicators are used. Compliance indicators cover the following areas: legal compliance, compliance with the other requirements as defined in the management system standards (e.g. requirements specified in the contracts with other organisations), and achievement of objectives and targets specified in the management system documentation or other documents (Staniškis and Katiliūtė 2019).

The so-called bottom-up approach is used for community-related indicators. Identification of such indicators begins with the active involvement of community members who help visualising the picture of quality of life and possible approaches of realisation through attaining the defined indicators. These indicators cover the natural environment, working environment, and social and economic spheres. They may reflect the air quality in the vicinity, biodiversity in nearby water bodies, quality of the working environment and other health and safety issues, employment rate in the local community, etc. Life quality indicators may be used to identify the strategic directions of the enterprise's efforts to reduce the environmental impact and, more generally, to improve the sustainability performance. A comprehensive list of indicators, which can be selected by the companies for their community-related sustainability initiatives, could be found in literature (Asif et al. 2013). The complete list of social indicators for any company goes beyond these life quality indicators and could be defined as an outcome of monitoring or statistical activities performed by the enterprise or other organisations, for instance, how the company's products and services contribute to the well-being of individuals or human development in general (Staniškis and Katiliūtė 2019).

It is important to note that legal and other strategic requirements or tasks for the cascade unsustainability reduction system are coming from transformations towards the sustainability management/control system (see Fig. 3.1, Sect. 3.3).

Taking into account the nature of decision-making (e.g. strategy development, innovation generation), performance indicators can be defined at enterprise, department or process levels. Moreover, to ensure legal compliance and adequate response to negative changes in the environment in relation to the enterprise's activities, a set of compliance and environmental condition indicators has to be selected. However, it should be stressed that although the number and nature of the indicators chosen may differ from one system type to another, the alternatives to be compared must use the same set of indicators (Diakaki et al. 2006; Sikdar et al. 2012).

As it was stated above, all incremental preventive innovations presented in Tables 2.1 and 2.2 demonstrate economic and environmental results; however, the social dimension of sustainability is reflected only partially. Process efficiency allows saving materials/resources in a way respecting the limitations of the planet and the rights of future generations to the ecosystem services. At the same time, the level of human well-being is the most challenging intervention, especially when the business activities have a negative impact on the outside environment quality and communities' well-being. The potential of such innovations is still insufficiently explored and the efforts of scaling up are extremely limited (Lorek and Spangenberg 2014). That is why the cascade management system was introduced in several heat energy production units based on alternative fuel (see Fig. 2.4). This illustrates the cascade of emissions into atmosphere producing an air quality, which is monitored not only as emissions at the stack, as it was in innovation case, but also the air quality in the atmosphere and in terms of public health, each of which produces a negative feedback loop designed to improve the life quality. The system consists of three loops, where time constant/inertia of the first loop T1 is lower than that of the second loop T2 and time constant/inertia of the second loop T2 is lower than that of the third loop T3 ($T1 < T2 < T3$).

In the figure:

Δe—deviation between standard and monitored emissions;

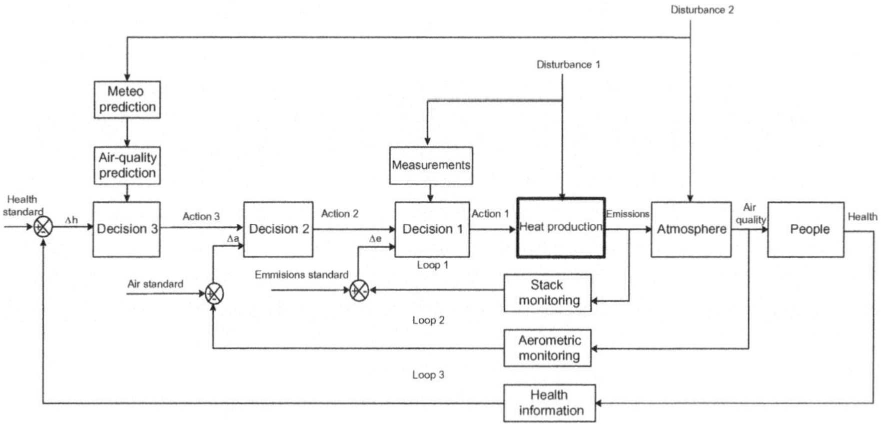

Fig. 2.4 Cascade unsustainability reduction management system for a heat production unit

Δa—deviation between standard and monitored air quality;

Δh—deviation between standard and monitored health quality.

Each of the three feedback loops suffers from the same disadvantage that they correct for system deviations from the target/standard values after a deviation has occurred. Whilst such management systems are effective in industrial applications, they are less appropriate to the environment. Therefore, it is preferable in most cases to detect and measure the disturbances before they occur and before a deviation from the target/standard value can begin. The measuring (disturbance 1) and forecasting (disturbance 2) shown in Fig. 2.4. detect disturbances, which will enter the system, before they have had time to affect the system response. In the case of heat production management system, disturbance 2 refers to changes in meteorological conditions, whilst disturbance 1—to uncontrollable changes in fuel characteristics, for example. This is the so-called feedforward control or preventive management. The main advantage of preventive management is that it uses the human ability to anticipate the course of events before they occur and, theoretically, it is possible to achieve ideal management; however, the method requires the measurement/prediction of all disturbances. This implies deep knowledge of production and environmental and social processes. The disadvantage is that there is no possibility to compare the system output with the target/standard in order to determine the management system effectiveness degree. At the same time, our investigation has shown that the feedforward control combined with the cascade management complement each other since serious disturbances can be detected prior to their effect on the system and the cascade can counteract any imperfections in the feedforward control due to unmeasured or unmeasurable disturbances (Staniškis et al. 2012).

The proposed structure for unsustainability reduction management in enterprises allows using all possible methods of systems theory to analyse its dynamics, stability and resilience.

The cascade management system was used in several heat production companies, which have switched from fossil to alternative fuel and where resource-efficient and cleaner production management system was introduced (see Fig. 2.2, Table 2.1). The economic and environmental benefits in most cases were the same (sometimes better); however, the social effect, i.e. better quality of life, was evident (less sick leaves, higher trust and satisfaction of employees, better image in the region, better life quality for the neighbouring community, etc.) (Staniškis and Katiliūtė 2019). Different studies have shown that both the improvement in the services that are provided as well as the way they are provided from the environmental and social perspectives influence one 's perception of the quality of life (Bayulken and Huisingh 2015).

The proposed cascade management system requires relative simplicity and takes life quality including the social dimension perspective into account. The question for cascade management system usually concerns the types of indicators that should preferably be used for a particular purpose (level of decision-making). For example, absolute process indicators (technological parameters) are best suited for process decision-making as they enable efficient process control. Relative compliance and life quality indicators are effective in strategic decision-making; however, these are not useful for decision-making at process level. Generally, indicators at higher hierarchical levels (from process to quality of life indicators) are more useful in strategic decision-making (see Fig. 3.2 in Sect. 3.11). When developing the set of sustainability performance indicators, application of a hierarchical approach that corresponds to the level of the enterprise's ambitions in its performance evaluation may be useful as it helps the managers to keep a clear and relevant structure/ composition of the performance evaluation system. The enterprise can start from the evaluation of compliance/ resource use efficiency and proceed, while gradually gaining experience, to a more sophisticated performance evaluation.

A particularly important aspect in selecting the unsustainability reduction environmental design is a proactive approach for integrating resource efficient and cleaner production performance indicators in a life cycle approach.

In the past, products were designed and developed without considering their adverse impacts on environment. Typical factors considered in the product design included function, quality, cost, ergonomics, and safety. However, no specific consideration was given to the environmental aspects of a product throughout its entire life-cycle. Conventional end-of-pipe regulation focused only on emissions from the production processes of a product. Often adverse impacts on the environment occurred from the other life-cycle stages such as use, end-of-life, distribution, and raw material acquisition. Without addressing the environmental impacts from the entire life-cycle of a product, one cannot resolve the unsustainability reduction tasks rising due to both production and consumption of a product. The product environmental design is a systemic approach that requires close cooperation between industry and research involved in the product development. APINI has developed and implemented product innovations based on life-cycle assessment in 15 industrial companies for products like electricity meter, refrigerator, firewood stove, buckle switch, corner sofa, etc. (Staniškis and Varžinskas 2010; Varžinskas et al. 2012).

Frequently, enterprises limit their performance analyses to the production and other internal processes, sales and general economic indicators. The requirements for a good indicator are the following: target orientation, comparability, measurability (access to data), meaningfulness (scientific reliability or analytical soundness), integrity (capability to relate to other indicators), continuity, clarity, and efficiency. The key attributes like unit of measurement, type of measurement (absolute or adjusted), period of measurement and boundaries should be defined for each indicator (Veleva and Ellenbecker 2001). In practice, enterprises seek to have a manageable number of indicators that are clear and easy to measure/ monitor, and that provide the possibility to compare with the best practices.

International sustainability performance evaluation systems, for instance, Global Reporting Initiative, should be preferably used as reference materials. This will help any enterprise develop a functional and effective performance evaluation system that fully reflects its values and needs. This recommendation may be supported by the findings of other researchers (Keeble et al. 2003; Searcy et al. 2005).

Discussion and Conclusions

> Humans are clearly a resilient species, but there is no guarantee that we will ever learn lessons from our mistakes.—Peter Townsend, British physicist

Investigations and experience clearly show that unsustainability reduction management systems are often implemented in a formal way and their efficiency in terms of sustainability performance improvement is rather vague. Moreover, many companies still rely on the end-of-pipe approach when dealing with environmental issues. Integration of unsustainability reduction management into the overall business planning and integrated application of sustainable industrial development tools lead to environmental, economic and social benefits. One of the side-effects observed is a positive change of employee thinking and improvement of the work culture. Moreover, integrated application of management tools ensures continuous improvement of the sustainability performance. The combination of socio-economic and biophysical indicators is essential to provide a better understanding of the limits of economic growth while ensuring sustainable societal well-being (Frugoli et al. 2015).

The main results of the research on unsustainability reduction in enterprises by incremental innovations management could be summarised as follows:

1. Analysis of the implemented cases clearly shows that the generation and implementation process of innovations based on prevention will be incomplete unless it also involves continuous improvement. The proposed systems for unsustainability reduction in production companies are based on the Plan-Do-Check-Act closed loop, which provides for a continuous process of improvement along the social, environmental and economic dimensions when generating and implementing the incremental innovations.
2. A number of tools are available to industrial enterprises to be applied; however, the best results are achieved by applying the key tools in an integrated way, seeing that particular unsustainable reduction development tools are mutually

supportive. The best place for the integration of different tools is the Planning phase in the management system where all major decisions take place.

3. To ensure effective decision-making aimed at improving the production process performance, an unsustainability reduction management system based on three hierarchical levels (process, activity and strategic decision-making) is recommended for use in enterprises, seeing as it ensures the involvement of decision-makers at different levels and enables the collection of information needed for effective decision-making at different managerial and operational levels.
4. Since there is no standard set of unsustainability reduction performance indicators, enterprises have to develop their own sets of indicators that reflect their profile and needs. To satisfy the needs of decision-making in enterprises aimed at continuous improvement of unsustainability reduction performance, four categories of performance indicators should be used: (i) process performance indicators, (ii) operational performance indicators, (iii) compliance indicators, and (iv) life quality indicators. Relative indicators are particularly useful in decision-making as they enable the specialists to observe the changes of particular values (e.g. pollution) in relation to a common denominator (e.g. raw material or production unit). Aggregated indicators may also prove to be valuable in assessing the unsustainability reduction of an industrial company.
5. It is recommended to build up a complex set of feedback-feedforward loops of unsustainability reduction management capable of compensating the whole range of system disturbances and changes, which may occur (see Fig. 2.3). Human behaviour and resource management of the man-environment interface is the example of complex management system comprising the economic, environment and social dimensions (Staniškis and Katiliūtė 2019).

It is to be hoped that the application of systems theory to unsustainability reduction management will encourage planners to give more attention to the structural character of the management system, which enables including all dimensions of sustainable development in the performance index. The unsustainability reduction based on incremental preventive innovations policies could be emphasised more explicitly as this is one of the opportunities to create more competitive products and services, and eco-design can be used as one of the tools for product improvement and development. As this system is implemented, a new identity can emerge: that of a more sustainable business model and enterprise. To achieve this state, it is not enough to maximise the material and energy efficiency and reduce the relative environmental impact. For this purpose, there is a need to switch from efficiency to sufficiency, from ownership to functionality, from fossil to renewables and natural resources, and to repurpose the business for society/environment rather than prioritising economic growth (Bocken et al. 2014).

2.5 Raising Awareness and Building Competency for Sustainability in Organisations

> Never doubt that a small group of thoughtful, committed citizens can change the world. Indeed, it is the only thing that ever has—Margaret Mead, anthropologist, US

The social sustainability challenge is still receiving relatively little attention, but in near future, the social aspects will become more important as a result of the growing expectations and demands addressed to business enterprises by employees and external stakeholders. Profit-oriented business companies run primarily for economic purposes; therefore, environmental and social issues in most cases conflict with the goals of making profit or at least minimising operating costs. As usual, environmental and social aspects from management and organisational point of view are dealt with separately. Therefore, APINI initiated a new research on possibilities how to introduce social aspects into the management performance index of production processes. The aim of the mentioned research carried out at APINI was to investigate the integration challenge of ecological and social dimensions in the context of economic activity by implementation of different preventive environmental strategies and programmes in the companies. The research method is based on development and implementation of special training on preventive innovations generation methodology for company's management and employees. There are many preventive environmental strategies, for instance, pollution prevention, waste minimisation, cleaner production, etc. For this particular investigation, a resource-efficient and cleaner production (RECP) strategy, which applies to industrial processes, products and services, was mainly introduced due to its potential to achieve production efficiency through optimisation of productive use of natural resources, to minimise the adverse impacts on environment and to reduce risks to employees and communities. Sustainable development strategy requires behavioural change by individuals as well; therefore, employee engagement is crucial. At the same time, resource-efficient and cleaner production concept, besides the potential mentioned above, provides the possibilities to support their personal development. The methodology for preventive environmental innovations generation and implementation was introduced in more than 150 industrial companies and economic, environmental and social benefits were identified and assessed. Companies, which implemented successful preventive environmental programmes, benefitted by significantly lowering operating costs, enhancing the safety of their employees, improving product quality and productivity, reducing environmental liability, improving public image, and chances for broader cooperation. Furthermore, companies managed to implement the employee engagement strategy by (1) communicating the professional impact the employee contribution had on the company, (2) enabling cross-functional communication and idea exchange. From another point of view, these programmes helped financiers secure better clients and make more profitable investments and assisted government agencies in their pursuit of a safer and healthier environment.

Society is facing many environmental challenges, including the depletion of natural resources, pollution, climate change, and collapse of the whole ecosystem. These changes dramatically increase the threats to human security, and health and well-being of society. Environmental and social responsibility is a precondition for the prosperity of human beings and an is the most important aspect of sustainability. According to the ISO 26000 standard, social responsibility refers to the responsibility of an organisation for the impacts of its decisions and activities on society and environment, through transparent and ethical behaviour that:

- Contributes to sustainable development, including quality of life;
- Incorporates social and environmental considerations in its decision making process;
- Is accountable for the impacts on society and environment;
- Is in compliance with the applicable law and consistent with international norms of behaviour;
- Social responsibility is practised in its relationships with stakeholders and takes into account their interests (ISO 26000 2010).

In the past 20–25 years, new ideas have emerged to reduce emissions to the environment at the source. These waste minimisation, pollution prevention and resource-efficient and cleaner production strategies appeared to be necessary to reduce the enormous costs of clean-up actions, certainly from the moment that the 'polluter pays' principle was brought into legislation. A preventive approach means that environmental problems are addressed before they arise when choices are made concerning processes, raw materials, design, transportation, services etc. Such an approach effectively addresses the wasting of natural resources since pollution not only leads to environmental degradation but is also a sign of inefficient production processes or management.

As usual, there is a need for extra investments to replace waste generating industries with new clean technology facilities. However, large gains can often be achieved by simple "good housekeeping" measures, improved inventory control, better maintenance, repairing leaks, simple equipment or process modification, installing measurement and control systems, and improving enterprise and its process control.

UNEP and UNIDO define resource-efficient and cleaner production (RECP) as continuous application of preventive environmental strategies to processes, products, and services to increase efficiency and reduce risks to humans and the environment. RECP works specifically:

- production efficiency—through optimisation of productive use of natural resources (materials, energy, water) at all stages of the production cycle;
- environmental management—through minimisation of the adverse impacts of industrial production systems on nature and environment;
- human development—through minimisation of risks to people in the communities and support to their development.

Of key importance for resource-efficient and cleaner production promotion and implementation is capacity building that should enable:

- Further human development by using workplace policy and addressing important social concerns;
- A common understanding of the RECP concept;
- Incorporation of RECP concepts in policy frameworks;
- Integration into enterprise policies and operations;
- Indicators to measure progress in RECP implementation;
- Providing information about both the technology involved and the environmental management and social tools needed (Staniškis and Staniškienė 2019a).

The second most important condition for sustained RECP is the moment when it becomes part of the management through company's formal environmental, quality, and social responsibility management approach. An environmental management system provides a decision-making structure and action plan to support continuous environmental improvements, such as the implementation of RECP. If a company has already established an environmental management system, the RECP can be an effective tool for focusing attention on specific environmental and social problems. If, on the other hand, the company establishes RECP procedures first, this can provide the foundations of an environmental management system (Staniskis and Katiliute 2017).

Experience of RECP programmes in Lithuania demonstrated that long-term training programmes emphasising on-the-job training are the most effective as a way to create domestic professional capacities. Furthermore, the only cost- and time-efficient way of fulfilling this need is to adopt a train-the-trainer approach. Presented below is the methodology of the human development and training in a workplace that includes social responsibility, and is based on the detailed structure of RECP assessment system.

Despite the potential benefits of RECP programmes, the dissemination process is quite slow, which could be caused by many difficulties and problems. Such barriers could be divided into four groups: (I) policy and market, (II) financial and economic, (III) technical and informational, and (IV) managerial and organisational. Besides the barriers mentioned above, such programmes are seldom implemented systematically resulting in the lack of RECP continuity, i.e. options are implemented but not monitored and extended. Moreover, the analysis of existing RECP assessment systems revealed the lack of a more comprehensive RECP method, as well as the lack of information and clear description of the activities that should take place in each step of RECP programme (Silva et al. 2013).

The comparative analysis of the selected nine RECP programmes (Silva et al. 2013) has shown that all of them had the same basic steps: development of environmental and social policy, organisation and planning, assessment/audit, identification of options for improvement, evaluation of options, implementation and monitoring, and review that also resembled the main phases of the Deming's cycle or the so-called PDCA (Plan-Do-Correct-Act). Based on the studied programmes, the authors developed a standard approach where the most of improvements suggested and explained were related to the planning phase of the PDCA cycle. The explanation is that "all major decisions taken in this stage have a direct influence on the continuation of the remaining cycle phases, and when a good plan is established it is more likely that the other phases will also have good implementation success. Therefore, it can be expected that some benefits of a RECP programme will be maximised and at the same time consumed by each activity will be reduced by applying this proposed method" (Silva et al. 2013). This is true, but not always; in other words—there are more conditions for success.

To make a sound business plan for RECP innovation, the information on the environmental and social aspects of the relevant product, raw material or constituent part of the process should be gathered. This information should consider the environmental effects not only of the production phase and product life cycle, but also of extracting and transporting the alternative raw materials and of treating any avoidable waste. Innovations with significant capital costs will require a more detailed analysis. There is a number of factors, which make the RECP costs and benefits difficult to calculate for many business plans. Many of those costs are hidden or probabilistic, although the risks are real and it is difficult to predict even the occurrence date from past experience. Total Cost Analysis suggests some approaches to calculating indirect and probabilistic costs so that their full impact could be included in economic feasibility assessments.

Business plan for RECP innovation should provide the basis for obtaining funding for the RECP project. Projects of this type should not be sold on their technical merits alone; a clear description of both tangible and intangible benefits could help the proposed project obtain funding. There will probably be other projects, such as expanding production capacity or moving into new product lines that will compete with the RECP projects for funding. Some companies may have difficulty raising funds internally for capital investment. The company then has to look for outside financing, for instance, private sector financing that includes bank loans and other conventional sources of financing. Revolving financial funds that provide soft loans for short pay-back projects are extremely attractive for financing of the RECP innovations.

A RECP project does not end with implementation. Tracking of its effectiveness versus the claims made—technical, economical, environmental, and social—should follow. Short description of the RECP assessment, including building capacity for social responsibility:

The recognised need for RECP: top management decision, policy statement, consensus building. Understanding the benefits of social and environmental responsibility should start at the top of the organisation (Staniškis and Staniškienė 2019b).

First step–planning and organisation: name the assessment leader and group, define goals. It would be useful to focus initial efforts on those parts of the organisation that are more interested and receptive in taking actions on social responsibility.

Second step—pre–assessment: collect data on all air, water, and solid waste emissions and releases, identifying their sources and quantifying the true costs of pollution control, treatment, and waste disposal; site visits, signing priorities to process, operations, and materials. Identify the need for changes in decision-making process and governance that would promote more freedom and motivation to suggest new approaches and ideas.

Third step—write the programme plan: contact external groups from outside the company including the surrounding community; define the objectives in quantitative terms and target dates; identify and divide the potential obstacles into four broad categories: economic, technical, social, regulatory, and institutional/human; define the cost–benefit analysis procedures; develop a schedule addressing some social responsibility issues in the short term and some over a longer period of time.

Fourth step—assessment: define additional staff to comprise the detailed assessment team(s) with relatively specific focus; review pre–assessment data, organise a thorough site review and interview workers; prepare material and energy balances as a means of analysing pollution sources, their environmental and social impact and opportunities for eliminating them.

Fifth step—define RECP innovation options: generate options encouraging creativity and independent thinking of each assessment team member, use brainstorming sessions; screen options by procedure (for example, weighted sum method) when the assessment team selects the options that appear the best after examining each of them, taking account of social responsibility when the organisation is conducting its operations.

Sixth step—feasibility analysis: prioritised list of RECP options/innovations should be examined to determine which are technically, environmentally, economically and socially feasible and selected for implementation.

Seventh step—assessment report: write and review the report, which should be the basis for innovation(s) business plan preparation; report has to be reviewed by people affected by the proposed project(s) in order to eliminate inaccuracies or misunderstandings.

Eighth step—business plan of RECP innovation: content: company description, feasibility data, procurement of equipment, investment analysis (payback period, net present value, internal rate of return, risk analysis, sensitivity analysis), company's economic "health" analysis (profit-loss statement, balance sheet, economic ratio analysis), project implementation schedule, supervision and progress monitoring. The plan should confirm that the principles of social responsibility are adequately applied in its governance and reflected in the plan structure.

Ninth step—financing: select the source of external capital: commercial bank, shareholders, leasing companies, revolving facility, environmental fund, etc.; adapt the RECP project business plan to the banking institution requirements.

Tenth step—implementation: procure and install all the equipment listed in the RECP project business plan; at the same time install extra measuring devices for materials and energy savings and prepare social responsibility documentation. Indicators should be quantitative or qualitative, and allowing for comparison and demonstrating changes over time. Our experience has confirmed that if any financing of RECP is to be provided, a revolving facility is a more appropriate mechanism. Such facility may offer low interest rates, grace periods or other incentives to borrowers. However, it is not necessarily easier to get financing from a revolving facility than from a commercial bank. A revolving facility might even ask for additional information or participation of borrower in specific, for instance RECP programmes or other support systems. It does not solve the problem for unprepared applicants who are not yet able to formulate a good proposal (Staniškis et al. 2010; Staniškis and Staniškienė 2019b).

Eleventh step—monitoring: measure progress against goals, assess the degree to which goals are being met; use normalised factors; watch for shifts of wastes to other media, carefully assess toxicity, not just quantity produced; focus on social responsibility performance, including reviews at appropriate intervals, benchmarking and obtaining feedback from stakeholders.

Successfully implemented innovation project: resource-efficient and cleaner production is an ongoing effort that will be best maintained with the establishment of an awareness programme and performed by personnel in the production area. Top management have to demonstrate their continuing commitment to the RECP programme by conducting annual reviews, which should be communicated to all employees and social partners through written announcements and meetings.

The most important element of the resource efficient and cleaner production programme is training. The training programme should include all levels of personnel within the company. The main goal is to make each employee aware of waste generation, its impact on the site, regional and global environment, social responsibility and ways how negative environmental and social impact of processes, products and wastes can be reduced or prevented. More detailed training should be provided to new employees after they have been on the job for a few weeks. Specialised training sessions on environmental and social policy, procedures, and techniques should be provided to staff when their job scope is in the areas where assessment is being planned (see Fig. 2.5).

Effective communication between managers and employees is a critical requirement for maintaining a successful programme. Employee ideas should be always welcome and good suggestions should be put into practice and recognised.

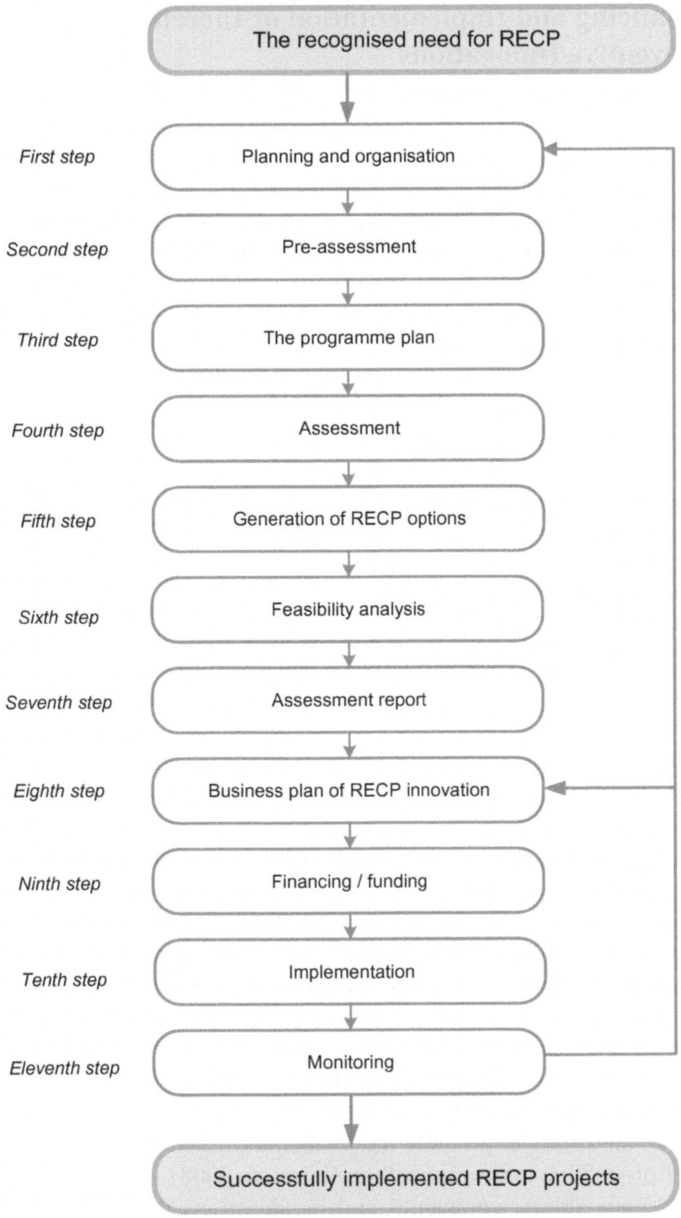

Fig. 2.5 The system for resource-efficient and cleaner production assessment

2.6 Financing and Implementation of Incremental Preventive Innovations

The best way to predict future is to create it—Peter Drucker, widely-known thinker on management, Austria

The main barrier to the sustainability of small firms is the lack of capital for investment and the difficulty in obtaining financing. From the firm's perspective, bank loans remain expensive and bureaucratic. In some countries, there are attempts to simplify credit access, along with enlargement or guarantees, improved agility in opening enterprises, financial cost reduction and expansion of access to online regulatory information. Nevertheless, strong asymmetries exist between small and big firms in term of credit access.

The financing and implementation of incremental RECP innovations mainly is based on two fundamental and sometimes conflicting approaches. The first one is indirect and generally assumes that some policy reforms and capacity building are sufficient to motivate the companies and organisations to move towards reduction of unsustainability. It focuses on tax policy reforms, use of market-based instruments, education and training, and other tools that create an external environment, which motivates top managers to choose more sustainable alternatives. The other approach is direct; it focuses on technical assistance and direct financing for companies and organisations. The best option results when indirect efforts are supplemented by direct ones.

Thus, financing of incremental innovations should focus on changing the conditions that reduce unsustainability. Business development agencies in this case can help companies and organisations by using the existing mechanisms, for instance, private and commercial banks, leasing, equity financing, etc. and/or creating special financing systems. In this case, financial institutions must be trained to understand and appreciate resource-efficient and cleaner production investments; however, this not enough. Financial institutions are hard to change, and many smaller industrial companies do not meet the basic requirements for getting a commercial loan.

From another side, before a company can apply for financing, it must know how to draw up an investment proposal professionally and a good business plan for the financial institution. This implies understanding the principles of environmental cost identification and environmental management accounting. This development step must precede all efforts to promote the financing of incremental resource-efficient and cleaner production innovations (Staniškis et al. 2008).

As experience shows, companies and organisations that have identified cost-effective and technically feasible incremental innovations to reduce unsustainability often may still be unable to make the necessary investments to realise the financial benefits and environmental advantages. There are many sources of capital, for instance, commercial or investment banks, but generally, they do not distinguish financing opportunities by project type. Furthermore, these institutions are usually not willing to take high risks, and they will secure their loans in the company itself or by guarantees from the shareholders.

There are also special banks and funds created for new industry development. These financial facilities lend money to high-risk projects or projects that improve environment and minimise pollution. Additionally, these financial sources will normally offer more favourable payback conditions or in some cases will not require a security (Fig. 2.6).

Retained profits and the shareholder capital represent an important source of capital for incremental innovations. This capital does not require interest payments, but investment decision of shareholders is rather based on the overall performance of the company and often not related to the intended use of the capital, for instance, financing of unsustainability reducing innovations. It can be expected that as the politicians and business become more aware of the sustainable economy, shareholders may consider the incremental innovations reducing unsustainability in their investment portfolio.

Environmental funds are an attractive option to facilitate the financing of incremental innovations. Usually, such funds are special-purpose oriented and limited to region, country, sector or type of investment. As long as business managers and financiers fail to understand the impact of incremental innovations based on preventive strategies on investment profitability, the end-of-pipe measures will be preferable. There is a need to use the financial and economic tools and instruments that correct this bias and address less tangible environmental and social factors and promote preventive strategies rather than end-of-pipe solutions.

The APINI experience has shown that the revolving financing facility is the most attractive source of capital for preventive innovations financing. Finances for investments could be provided by banks or donor agencies to the financial intermediaries/funds for special-purpose lending schemes. As a successful example, together with the bank APINI has developed a revolving financing facility, which has provided soft loans to the companies/organisations for preventive innovations with comparatively short payback implementation. The financing system was comprised of a

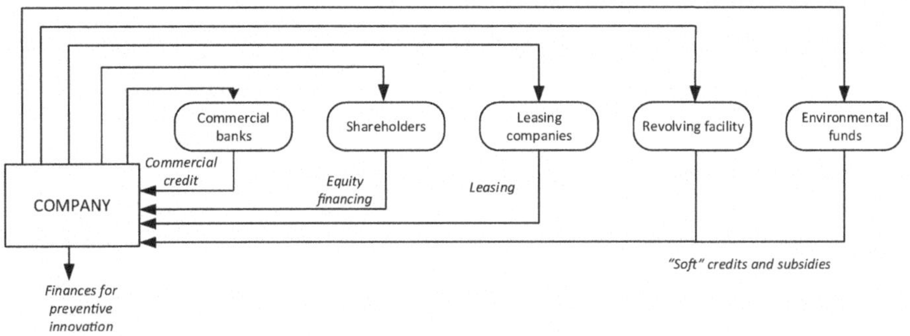

Fig. 2.6 Financial sources for the incremental innovations for unsustainability reduction implementation

pool of so called business development service (BDS) providers of preventive innovation generation, financial engineering, implementation and monitoring, whereas the financing source included a soft credit line and companies/organisations (see Fig. 2.7). Innovations identification and assessment methodology was based on company's material and energy flows, and properly evaluated environmental and social costs by using environmental management accounting. The methodology is flexible and can be applied to different types of companies to provide the decision-makers with information in terms of the intended economic, environmental and social purposes. The system was focused on:

- Ensuring that the incremental innovation is based on preventive approach and is technically feasible and cost-effective, and that no experimental technologies are being supplied;
- Confirming that the environmental and economic savings are achievable;
- Establishing that the time scales are reasonable;
- Proving that the procurement approach is acceptable and that the price estimates are realistic;
- Showing that the environmental impact of incremental innovations cannot be optimally mitigated by simple and cheap end-of-pipe measures;
- Ensuring that the materials and energy flows throughout the production/service process are correctly identified and used as a consistent basis for the development of cost-effective solutions (Staniškis and Jayaraman 2010).

For the effective realisation of preventive innovations generation, financing and implementation process, each member of the system has a specific role. For instance:

- *Business service provider*: assists the company in preventive innovation development including material and energy balance, prepares the investment project and soft loan application, assists the facility in communication with applicant, monitors innovation implementation process, develops innovation implementation progress and completion reports to be presented as a part of disbursement requests to the facility;

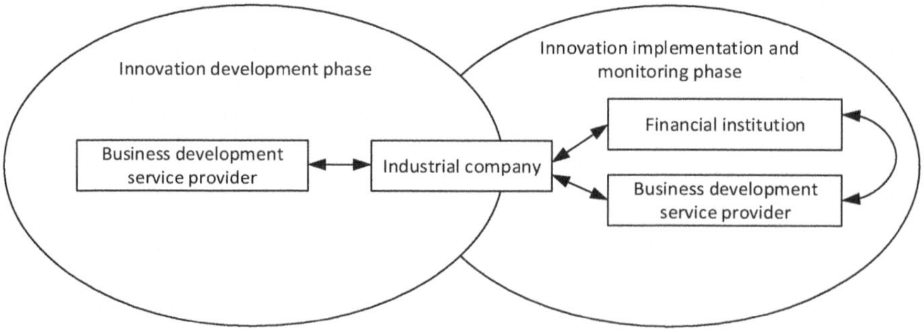

Fig. 2.7 Structure of the system for incremental innovations generation and financing

- *Company/organisation*: provides measuring, monitoring and process data, documents and technical experience to BDS, presents information and documents on collateral for loan security to BDS and facility, signs the contract with the facility, ensures innovation implementation in accordance with the loan conditions, performs monitoring on innovation implementation to BDS and facility, repays the soft loan in accordance with the loan contract;
- *Financing facility/Fund*: verifies economical, environmental, social and technical evaluation of incremental innovation, provides a loan on favourable terms for the innovation implementation, and verifies whether the investment has achieved its overall economic, environmental and social goals.

It should be stressed that the main difficulty in preparing and implementing preventive innovations lies in the identification of material flows and stocks of anthropogenic activities, for instance, insufficient depth of the process diagnosis, waste generation sources and causes. Many companies include all environmental costs into the general overhead costs together with top management salaries, advertising and any other costs not traced back to individual production processes. In such cases, non-financial support services are expected to be offered by other service providers. There is no doubt that a suitable synergy between financial services and BDS can make the credit facility system more effective and produce more successful lending programmes. The majority of the companies, especially in the developing world, need some help to convert their ideas into a technically and financially acceptable business plan and BDS can facilitate a possible partnership with investors. Financing facility may also benefit from working with BDS providers in recommending companies/organisations to get external advice when encountering problems resulting in poor repayment.

Finally, as a number of financiers and different studies have shown, there is much more capital for preventive incremental innovations than there are qualified proposals. The financing of investments through loans of a revolving financial facility heavily depends on the following (Staniškis and Jayaraman 2010):

- Creditworthiness of the company/organisation (they have to meet the conditions on liquidity, solvency, and profitability);
- Application itself—the financial costs and benefits of preventive innovation investment should be identified properly;
- The existing relationship between financing facility and a company.

The quality of environmental management and the degree of implementation of preventive innovations that reduce unsustainability are recognised as good indicators of the overall business management quality.

Reducing unsustainability will not create sustainability. The root cause of unsustainability is that we are trying to solve all apparent problems of the world, large and small, by using the modernistic frame of thinking and acting that has created the metaproblem of unsustainability. Einstein is credited with only a slightly different observation: "The significant problems we face cannot be solved at the same level of thinking we were at when we created them". It is very important that we look critically at the many programmes that have come forth in the name of sustainable development. Sustainable development in many cases is perceived as technological, technocratic programmes. Then, within the sustainable development come programmatic prescriptions like ecoefficiency, circular economy, cleaner production, and many others. All have some potential to mitigate or slow down the unsustainability trajectory of the globe, but all are only quick fixes. They are part of the problem, not the final solution and from the systems theory point of view they should be approached as the necessary condition towards sustainability. In our culture, attacking the symptoms is the underlying rationale for virtually all responses to the growing set of societal problems; yet even here, it is important to design the solutions to avoid doing even more damage. In many cases, we do know how to combat the unsustainability but will not do it because of the resulting apparent threat to one's position or economic well-being.

"Current modes of scientific research have strong historical roots in the classical experimental method that disaggregates the real world into its parts, enabling each to be studied in isolation. Yet the real-world whole is much more complex and unpredictable than the sum of its disaggregated parts, the latter being the main focus of classical experimental science. System-based analysis elucidates the internal workings of a complex system such as the climate system, and can identify the optimal intervention points for stabilisation or restoration" (McMichael et al. 2019).

2.7 Methodology for Workplace Learning on Unsustainability Reduction Innovations

> The difference between animals and humans is that animals change themselves for the environment, but humans change the environment for themselves.—Ayn Rand, writer and philosopher, US

Rapid change in the world of work and in society, fast development of information and communication technology, and changes in the organisation of work have raised the value of learning throughout the working career. There is a broad diversity of workplace learning methods. One of a few holistic models was presented by Biggs in his 3-P model of learning (Biggs 1999). This model refers to the three basic components of the learning phenomenon: presage, process, and product. There are two presage factors, one of them being prior knowledge and motivation, and the second one—teaching methods and assessment. The central part in learning is the process component that describes how students approach learning, i. e. either deep or surface level. The product part refers to learning outcomes, for instance, knowledge, skills, and attitudes. The Biggs' model was developed in the context of school learning and cannot directly be applied to workplace learning as such, but the basic model structure (presage, process and product) fits well with the concept of workplace learning. A very similar model for workplace learning was developed by Marsick et al. (2011), where the same three main components were presented, although "presage" was named as "inputs".

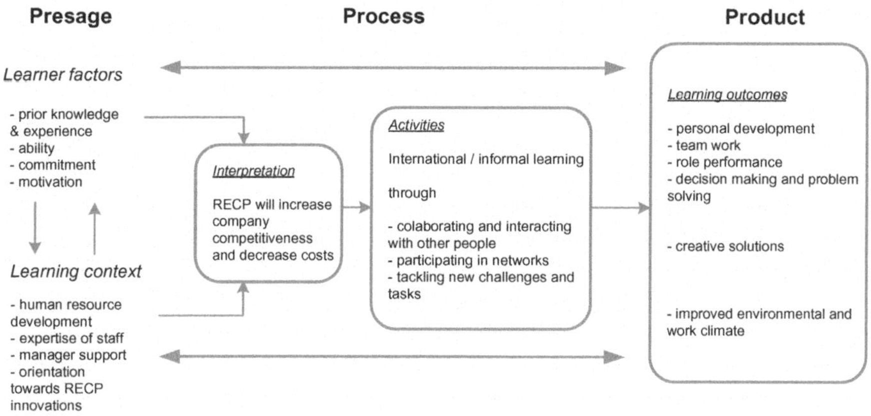

Fig. 2.8 The structure of human development and workplace learning methodology for achieving unsustainability reduction

Tynjälä adopted the 3-P model for workplace learning. First, the sociocultural environment in a wide sense, as defining the possibilities and constraints of workplace learning, was introduced. Another modification was the introduction of an additional learner's interpretation of the presage factor between the presage and process. Finally, the "student factors" label was replaced by "learner factors" because learners at the workplace are seldom in a student position. Moreover, the factors of learning context, the process and product components in the model of workplace learning are also vey different from the school learning processes (Tynjälä 2010).

The RECP assessment system (Fig. 2.5) and workplace learning methodology structure (Fig. 2.8) clearly show that the process part is the most important and at the same time—most complicated. RECP assessment system is complex, requiring technical experience, environmental and social responsibility knowledge, a basic understanding of economics and financial engineering. The learning process is interactive and combines classroom studies, group work, in-company project work, in-company advice and data search in international databases. The learning is with interaction between "out–of-company teaching" and group work and separate "in-company project". Participants from universities, consulting companies and governmental agencies take part not only in classroom group work and plenary sessions, but in the "in-company project" work as well. The objective of the 6-month learning programme that has been going on in Lithuania for more than 10 years is to train a group of 30–40 experts in RECP assessment and to create successful examples of the application of RECP on the company level. After the learning programme, the trainees should be able to execute similar RECP projects in the industry on their own. Thus, the trainees later on will become key players in disseminating and implementing RECP innovations in industry. The group of trainees usually comprises: 2–3 specialists from approximately ten companies, representatives from universities, industry associations, consultancy, local government, environmental protection departments, and NGOs. They are selected based on personal commitment to RECP as well as educational background to compose an interdisciplinary group of trainees (Fig. 2.9).

The first version of cleaner production assessment system has been developed by the Norwegian Society of Chartered Engineers (NIF) in co-operation with the Society of Engineers in Poland (Nedenes 1994). During the next fifteen years, this methodology was improved and elaborated as a RECP assessment system by many authors (Khan 2008; Environment Canada 2009; UNEP DTIE 2012). The version presented above (Fig. 2.9) was developed and implemented by experts from the Institute of Environmental Engineering at Kaunas University of Technology (Staniškis et al. 2010). During this period (1995–2015), more than 150 companies have participated in the CP/RECP programme just in Lithuania and have generated more than 250 preventive innovations with an average payback period of less than two

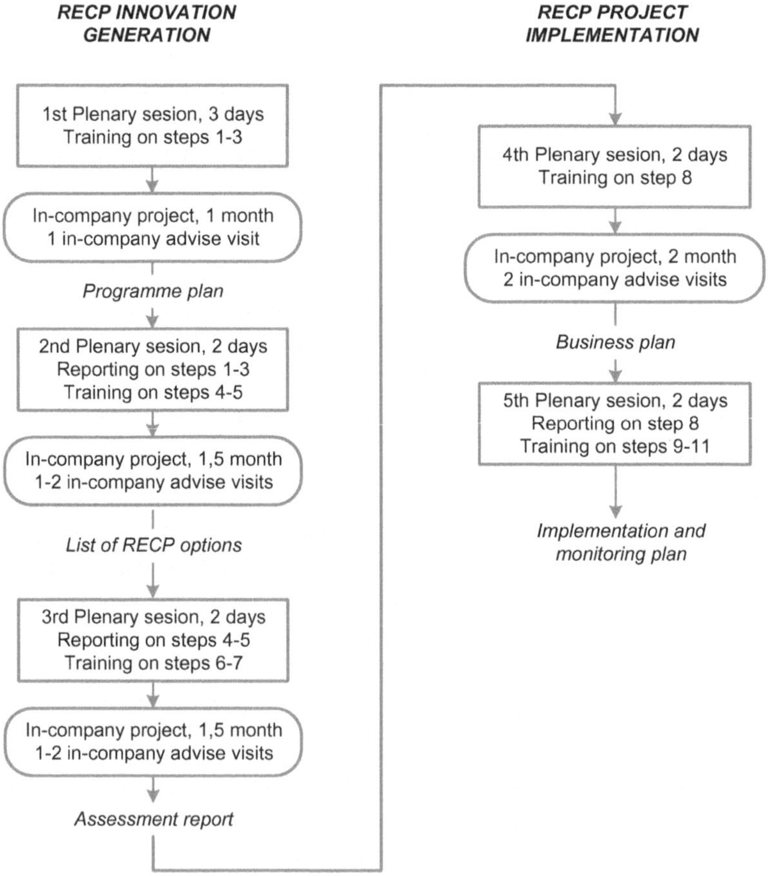

Fig. 2.9 Process algorithm for human development and workplace training

years. With increased environmental awareness, strong competition and the need to improve production efficiency, especially with regard to material efficiency, the cost of tracking and tracing material flows throughout the company are by far outweighed by the improvement potentials identified and realised.

In such cases, non-financial support services are expected to be offered by other business service providers (BDS). BDS can facilitate a possible partnership with investors but only when there is sufficient data, records and information on the business and the investment proposal. For each RECP investment project, a supervision and monitoring plan is prepared as part of the loan application The company has to provide BDS with the possibility to inspect goods, sites, factories, installations, and construction sites included in the investment project and any related documentation, and to supervise the implementation of the project. Financial and physical performance indicators are specified in the loan agreements. Environmental performance indicators (EPIs) are used to monitor the company's effectiveness and efficiency of resource management. This applies mainly to physical resources (materials, etc.).

For each project, the realised savings are verified and compared with the expected savings. A standardised reporting format is provided, with a focus on savings in energy use, water use, use of chemicals, etc. The environmental effects of each project are verified. This verification should document the reduction in emissions and wastes and the reduction in inputs (water, energy, chemicals, etc.). The mentioned standardised format ensures the management of all RECP innovation data in a special database, which is continuously updated.

More than five hundred RECP experts were trained by the methodology presented above (Figs. 2.8 and 2.9) It could be concluded that the implemented incremental preventive innovations:

- Increase profitability, lower production costs;
- Provide a rapid return on any capital or operating investments required;
- Lead to a more efficient use of energy, natural resources and raw materials;
- Increase staff motivation through reduced worker risks;
- Reduce the risk of environmental accidents;
- Reduce/avoid regulatory compliance costs significantly (Staniškis et al. 2010).

In the framework of UNEP and UNIDO projects, our experience was transferred to several countries in Africa, Central America, South East Asia, China, Central and Eastern Europe, and Russia. Some key learning points concerning RECP workplace learning and implementation:

- RECP is not a one-time event, it is a continuous process, but it is difficult to ensure the follow-up of such activities at company level after the project completion.
- Barriers to RECP are mostly related to human rather than technical factors. The end-of-pipe approach is well known and accepted by industry and engineers. The existing government policies and regulations often favour end-of-pipe solutions. There is a lack of communication between those in charge of production processes and those who manage the generated waste. Managers and workers, who

know that the factory is inefficient and wasteful, are not rewarded for suggesting improvements.

- Learning/training programmes covering social responsibility aspects play a key role in the introduction of RECP strategy in industry. Different stakeholders' involvement in the RECP learning programme is the prerequisite for local adaptation and for the continued success of the programme.
- The optimal number of companies in the programme group should be not less than ten companies from different sectors of industry. When companies are from the same sector—confidentiality issues arise; when companies represent different sectors—possibilities appear for transferring experience and ideas to different sectors of activities.
- Selecting suitable people for the learning programme has proved to be an important issue. Environmental specialists are usually very interested in the issues, but often hold a peripheral position in the companies. They are often not in the best position to carry out the changes in management and production that are looked for. Broad experience has shown that RECP project leaders in the learning programmes should preferably be recruited from production leaders, rather than environmental staff members.
- Governments should encourage the relevant non-governmental organisations (industrial and engineering associations, universities) to run modified workplace learning programmes for production managers, operators, industrial advisors, university teachers, and governmental officers.

Overall experience shows that it is often not fruitful to generalise among the workplace learning programmes or among countries. The situation in the companies/countries is rapidly changing and differs significantly, i.e. local circumstances and considerations and, most significantly, persons involved will define the outcome. Therefore, RECP as a workplace learning/training programme is more likely to be successful when carried out in companies closely linked to real issues within and outside the organisation.

2.8 Integration of Sustainability Issues into University Curriculum

The earth has its music for those who will listen—George Santayana, Spanish philospher

Higher education, especially nowadays, is subject to endless discussions and substantial reforms of performance management in universities across the world. The roots of the modern university goes back 1,000 years to the University of Paris whose unique aim was education and training. This was the first mission realised in the context of Catholic Church. The way and the form of the modern university development throughout history have always been influenced by social triggers. The trigger for the second mission of research was Alexander von Humboldt who led the establishment of the University of Berlin in 1810. As scientists are trying to understand and predict the global consequences of climate change, Humboldt's interdisciplinary approach to science and nature is more relevant than ever. His belief in the free exchange of information, in uniting scientists and in fostering communication across disciplines, is the pillar of science today. His concept of nature as one of global patterns underpins our thinking. By the beginning of the twentieth century, there was little room for a man whose knowledge had bridged a vast range of subjects. "As scientists crawled into their narrow areas of expertise, dividing and further subdividing, they lost Humboldt's interdisciplinary methods and his concept as a global force. As we have entered the Anthropocene, a new geological epoch that is shaped by the influence of human activities and one in which we will all have to deal with climate change, the acidification of oceans, glaciers melting and extreme weather patterns from droughts to floods, Humboldt's views sound alarmingly prophetic" (Wulf 2016).

The third mission for universities has emerged since 1980s as a result of global pressure on universities to play a more important role in the knowledge economy. This meant the birth of the entrepreneurial university seeking to align the creative powers of academia with economic development. This shift in focus from society to economy was promoted by the Organisation for Economic Cooperation and Development (OECD) in particular with the intention to focus the discussions concerning the third mission and societal contributions on technology and innovation transfer activities, including patenting, licensing and creation of spin-off firms and technology parks.

In the third mission approach, the entrepreneurial university is seen as an engine of economic growth with both government and academic pro-entrepreneurial discourses driven by success stories, such as the high-tech driven economic prosperity supposedly attained by linkages between Silicon Valley and Stanford. What is often forgotten is that emergence of the entrepreneurial model with active university-industry partnership and technology commercialisation is a phenomenon occurring just in a few universities, particularly in the USA (Etzkowitz 2003), although it has been framed and promoted around the globe (Trencher et al. 2014).

To capture the contemporary innovation process, a Triple Helix methodological tool has been developed, where the focus on the recursive overlay of communications among universities, industries, and governments allows for the organisation of research questions in relation to various models and metaphors (Leydesdorff and Etzkowitz 1998).

"I think there is a poison in education all across Europe which is anti-industry and anti- entrepreneurship", says Dr. Jos Peeters, managing director of Capricorn Venture Partners (ERT 1998). "A profound reform of education systems in Europe is needed. Greater emphasis must be placed on entrepreneurship at all levels of education. Despite the pressing need to manage better the transition from school to work, school-industry cooperation is still underdeveloped in Europe (ERT 1998).

Some studies have examined the contributions made by European universities to technological development in industry from the viewpoint of the recipient firm. However, a very small proactive role has been played by universities to increase the process of technology transfer from academia to local business or to regional economic development (Jones-Evans and Klofsten 1998). One of the problems is the European Research Council being run by scientists for the benefit of science on its own. While this system of self-governance has been quite successful in the past, it fails to deliver goods in terms of the contribution from the science system to the development of industry. EU science policy can be seen as an example of science-industry-government interaction at work. EU programmes tend to be designed, managed and implemented by administrators. It is only in the review and appraisal of proposals that scientists are involved (van Duinen 1998).

The push toward research and commercialisation in the United States, in Europe, and in Japan acquires greater force, but governments constantly try to cut their contribution to university budgets and therefore, require the universities to supplement them by their own earnings from research, whether through knowledge transfer, spin-offs, or equity in start-ups. According to the World Bank research, in this case, public universities are gaining more autonomy and freedom, which opens opportunities for a more aggressive pursuit of reforms to attract better students, reorganise and expand research and development (R&D), explore new sources of financing, and better understand entrepreneurship (Yusuf 2007).

The World Bank investigation also shows that in Japan, companies prefer informal ties with universities. Corporate researchers co-author papers with university faculty members, spend time working at university labs, do joint projects with university researchers, and enter into consulting arrangements with university-based researchers. At the other extreme is the United States, where university-industry relationship covers the entire spectrum, and formal contracted arrangements with universities are common. Europe falls somewhere in the middle. In the Republic of Korea and India, small firms have virtually no contact with universities as far as research is concerned, but they may seek help for the purpose of trouble-shooting from individual researchers. A similar tendency is materialising in China as a result of a determined push by governments to induce both universities and state enterprises to cooperate in developing technologies (Yusuf 2007). The same World Bank publication argues that in the pro-entrepreneurial approach there is no substantial evidence proving that the narrow economic focus of the third mission and rise in conventional university-industry links have had any 'discernible' negative impacts on universities.

However, such positive appraisals ignore an array of concerns voiced against the rise of the third mission regime. For example, in the collection of papers on the neoliberal university (Canaan and Shumar 2008), there is a general concern about the neglect of humanities at the expense of the revenue-generating fields of applied sciences. Here, Rajani Naidoo provides a grim look on the role of education in the developing world. She argues that the penetration of neoliberal policies within developing countries will not only worsen quality, purpose, and functionality of higher education in these regions, but also has little potential to contribute to development goals, which may be capable of eroding current disparities between high- and low-income countries. There is still an unanswered question how neoliberalism affects the working and personal lives, especially of other staff within higher education, such as student affairs personnel, administrative staff, custodial workers, and librarians

(Shahjahan 2012). A detailed study of Finnish academics has tried at least partially to answer these questions. They concluded that "the ethos of what it means to be academic is at stake, as a new competitive ethos is challenging the traditional collegial academic ethos. It directs those who do academic work to pursue goals that are rewarded by performance management measures and metrics, even if scholars themselves do not agree with the rationale and usefulness of these indicators" (Kallio et al. 2016). Benjamin Ginsburg in his book "The Fall of the Faculty" argues that "the problem lies in the explosive growth in administration in US universities and the concomitant decline in faculty power in influence. Put simply, 'deanlets'—administrators without doctorates or serious academic training—rule the roost, and professors do not have nearly as much institutional power as they used to" (Ginsberg 2011).

As it has been mentioned above, from the 1980s, globalisation and neoliberalism have put increasingly strong pressures on universities to behave like a business. In order to enable universities to meet these challenges, reformers started to integrate universities and tighten the links between different parts of organisation in order to make them more efficient, manageable and accountable. Although universities are still predominantly public in most countries, the way in which authorities run them has changed fundamentally, and this has been heavily influenced by the notions of 'academic capitalism' and 'entrepreneurial universities'. Such public managerialistic regimes are driven by university-state alliances, political-administrative interests and semi-competitive logic based on incentive policies, where public support depends partly on teaching and/or research performance.

In this context, the relevance of the prevailing entrepreneurial model to achieving desirable human development in universities needs to be examined in a more detailed way.

Education for sustainable development (ESD) is an increasingly important feature of higher education, which poses new challenges for teaching and learning. There are two options: ESD might focus mainly on developing new education, if this type of education is completely absent or ESD might aim to change the existing education. On the one hand, sustainable development itself is a field of study that is complex by its very nature, while on the other hand, the systems of education are also highly complex. Therefore, it is no surprise that education for sustainable development appears to be affected by many barriers and resistance. For instance, MPhil programme in Engineering for Sustainable Development at Cambridge University has been in operation since 2002, where sustainable engineering topics have moved from the "more-general" to "more-specific". Instead of broad definitions of sustainable development, students are introduced to aspects of systems theory, industrial ecology, supply-chains, and global/humanitarian engineering. According to the latest assessment of the courses and programmes at the university of Gothenburg, one third of courses and programmes deal with environment and sustainable development, while two thirds do not (Boman and Andersson 2013).

The major catalyst of the third mission and the university role of technology transfer was the emergence of an independent relationship between science, industrial innovation and government policy leading to so called "knowledge-based" economy. Yet the emergence of global entrepreneurial approach is not the last step in the ever-evolving modern university. There are many good examples, where co-creative partnership for sustainability is implemented, which is fundamentally different from conventional third mission activities. Sustainability learning challenges require new ways of transdisciplinary knowledge production with the involvement of actors from outside of academia in order to meet the goals of sustainability science as a community-based transformational scientific field. In many countries around the world, ESD is understood as integration of sustainable development in the education, research and operations as a starting point in the strategic policy of a university. In principle, university has four roles in the society. First, university acts as an educational institution and in terms of impact on society has to guide and assist the students with their learning processes and thus deliver sustainably educated professionals. Secondly, the university is a research institution and from this perspective, it delivers the results of fundamental or applied scientific research to society. Thirdly, a university is an organisation, which has all kinds of operational interactions with the outside world as any other organisation, e.g. procurement, employment of staff members, use of materials, energy and water, transport of people and goods, production of waste and air pollution. The fourth role is its direct interaction with the society, for instance, participation in the implementation of joint local Agenda 21 projects, and cultural events together with local community (Staniškis 2016).

So far, several tools have been developed to assess the universities' sustainability, for instance, the Auditing Instrument for Sustainable Higher Education (AISHE) (Roorda 2010), Graphical Assessment for Sustainability in Universities (GASU) tool (Lozano and Young 2013), Environmental Sustainability Assessment Questionnaire, and the EMS Self—Assessment (Shriberg 2000). Many of these tools are focused on improving the sustainability of campus operations. The STAUNCH system (Lozano and Young 2013) is aimed at helping universities to assess the depth and breadth of their SD-related curricula in a holistic and systematic way and to produce standardised and comparable results. A challenge that remains is how to assess the contribution and impact that curricula and university life may have on students' personal life during their studies and on their professional life in helping to make societies more sustainable.

Since the students' ability to cope with uncertainty varies both within and across courses and classes, there is no single solution for how to guide the students through the phases of sometimes very deep uncertainty and doubt without delimiting their imagination, creativity and innovation (Wangel et al. 2013). To tackle the problem of evaluation and identification of the cognitive contents associated with the concept of sustainable development, a method of analysis based on a cognitive map is proposed (Lourdel et al. 2005; Segalàs et al. 2008). Cognitive maps can be a useful tool to represent changes in the knowledge structure of students over time.

2.9 Co-Creation for Sustainability in the Universities—Beyond the Third Mission

> If an elderly but distinguished scientist says that something is possible, he is almost certainly right; but if he says that it is impossible, he is very probably wrong.—Arthur C. Clarke, British inventor and writer

Human development needs to be critically evaluated and the need for so-called social contract between academic science and society is urgently needed. "In contrast to the narrow economic scope of the third mission, for a variety of reasons, the function of co-creation for sustainability is far better equipped to bring about the sustainable transformation of a specific geographical area or societal sub-system. […] The function of co-creation for sustainability aims to address localised sustainability issues by creating socio-technical and environmental transformations with the goal of materializing sustainable development in a given geographical vicinity. Systematic comparison of the functions of technology transfer and co-creation for sustainability reveals clear differences that are so great that it is impossible to consider the role of co-creation for sustainability to be just as a mere offshoot or different enactment of the third mission" (Trencher et al. 2014).

Besides the fact that the third mission is formulated entirely in economic terms, a clash of interests could also be expected in cost-effectiveness, commercialisation of results and short- to mid-term economic gains. Nevertheless, these two approaches could be considered as two distinctly differing but compatible missions. A short presentation of three cases is provided in the next pages to show how a university might exploit the co-creation approach to bring about the sustainable transformation of a particular geographical area or societal sub-system. The second aim is to illustrate the larger point that the function of co-creation for sustainability is in fact capable of becoming an institutional priority/mission in a very different context.

There are several studies on performance management or result-based management and its implications on employees' motivation and creativity at universities. Below, selected results of two studies are presented: *Study on performance management implications for work motivation in Finnish universities* (Kallio and Kallio 2014; Kallio et al. 2016); and *Study on the effects of the transition to more quantitative performance management system in Groningen and Manchester universities* (ter Bogt and Scapens 2012).

The study in 5 Finnish universities was based on a survey questionnaire in order to gather opinions of employees related to performance management in respective universities. A sample consisting of 966 persons (response rate—33.6%) represents well the 5 universities and their 12 faculties and, with some caution, could be generalised as of Finnish university employees in general. The study results showed that universities surveyed did not succeed in developing their performance management activities, since only 15% of the respondents were satisfied with the system, 40% expressed their dissatisfaction and a similar percentage took a neutral view. What is very surprising is that with regard to monetary compensation given by the nationwide performance-related pay, numerous respondents reported that the performance management system had no effect. An explanation for this could be that numerous respondents reported that universities did not have additional financial resources necessary to reward good performance and described the system as ‘a fraud’ or ‘meaningless’.

Moreover, negative side effects of the business model based on performance management seem obvious: more than 70% of the respondents thought that their university was more interested in quantity than quality, and almost 80% agreed with the statement ‘Nowadays in universities the content of the work is secondary; what is important is to produce as much as possible’ (Kallio and Kallio 2014).

The second part of the research was to investigate how the proliferation of performance management could be seen as a catalyst for changing the very ethos of what it is to be an academic and to do academic work. The authors concluded that the performance management system significantly increased bureaucracy and at the same time did not remove subjectivity in measurement but relocated it at a greater distance from the measured subject. Survey respondents saw the adoption of performance management as a violation of academic freedom and of the traditional collegial values of university. At the same time, some scholars in the sample clearly embraced the performance management system and the competition it encouraged and saw themselves as entrepreneurs rather than members of academic communities and were self-contained and engaged in their own personal development.

The general conclusion is “that ethos of what it means to be an academic is at stake, as a new competitive ethos is challenging the traditional collegial academic ethos… A significant feature of the Finnish case is that after over a decade of more incremental change fundamental reforms are now carried out quickly and methodically, putting unprecedented pressure on scholars to reconsider their relationship to the work they do and, indeed, their academic identities. While universities in Finland have traditionally been seen in Humboldtian spirit as national cultural institutions, the new system affords them a more instrumental role and steers them towards competition with each other” (Kallio et al. 2016).

In the second study, a mixture of interviews and questionnaire surveys were employed to explore how internal changes affected the work of academics and how they perceived the effects of performance management in use at that time in accounting and finance departments of Groningen and Manchester universities. The majority of interviewees and questionnaire respondents in both universities confirmed that research output increased in the recent years, but they were uncertain whether the quality of the research increased and about the effects on teaching. The way in which the new performance management systems were applied not only put pressure on academics to meet the performance standards, but they also seemed to increase levels of anxiety and stress. A number of people indicated that the promotion criteria were not clear and that they liked their jobs less than previously.

An important conclusion is that “the claims that the new (judgemental) performance measurement systems in universities increases transparency and objectivity are debatable… The subjectivities in the previous (developmental) systems were usually located within academic departments where individuals subjected to them worked and were applied by professors who worked closely with those individuals. But now the subjectivities in the (more judgemental) systems are located at the Faculty of Business school level and applied by distant administrators—who are either not academics or if they are academics, are from different disciplines" (ter Bogt and Scapens 2012).

Studies like those mentioned above and others show that universities are now actively controlled as production companies by visible top managers who have considerable discretionary power. This has led to a new cadre of professional managers and to managerialism that is central to performance management. Although universities are somehow involved in production, the nature or essence of what they produce can be expressed metaphorically; from this, it follows that it is difficult to evaluate such an enigmatic product, and equally difficult to optimise its production, the technology of which is far from clear. This sort of investigations could be summarised by the words of the Nobel laureate Philip Sharp: "as universities become more identified with commercial wealth, they also lose their uniqueness in society. They are no longer viewed as ivory towers of intellectual pursuits and truthful thoughts, but rather enterprises driven by arrogant individuals out to capture as much money and influence as possible" (Czarniawska and Genell 2002; ter Bogt and Scapens 2012).

"Education is valuable, but its main value is not in raising productivity. It lies in its ability to help us develop our potentials and live a more fulfilling and independent life. If we expanded education in the belief that it will make our economies richer, we will be sorely disappointed, for the link between education and national productivity is rather tenuous and complicated. Our overenthusiasm with education should be tamed, and, especially in developing countries, far greater attention to be paid to the issue of establishing and upgrading productive enterprises and institutions that support them. The fact that you have graduated from university tells your potential employers that you are likely to be smarter, more self-disciplined and better organised than those who have not. By hiring you as a university graduate, your employer is then hiring you for those general qualities, not for your specialist knowledge, which is often irrelevant to the job you will be performing" (Chang 2012).

Human development needs to be critically evaluated and the need for an alternative mission and the so-called social contract between academic science and society is urgently needed. "In contrast to the narrow economic scope of the third mission, for a variety of reasons, the function of co-creation for sustainability is far better equipped to bring about the sustainable transformation of a specific geographical area or societal sub-system" (Trencher et al. 2014).

Despite the obstacles discussed above, there is an increasing call for universities to contribute to social needs, sustainability in particular. The key role in this process could be played by university-based research centres that use sustainability and/or sustainable development concepts as the main framework of their research, education and outreach activities. Such small, dynamic, attractive and independent units have scientific, educational sand organisational advantages, at the same time having the support from university for funding, infrastructure and affiliations. The exploratory study of researchers from University of Helsinki analysing a sample of 44 sustainability centres across the world shows that these "centres are not yet mainstream activity within the current regime of the universities, and therefore they seek different strategies to meet organisational and scientific challenges" (Soini et al. 2018). There are well-known challenges to fulfil these tasks, for example, sustainability research and education are strongly interdisciplinary and thus, distinct from dominating monodisciplinary approach; university administration and structure are department/faculty-based but not by interdisciplinary units; current classification, evaluation, rewarding and funding systems do not support transdisciplinary research and education. One of the successful examples of sustainability centre is the Institute of Environmental Engineering at Kaunas University of Technology, providing research, education and outreach activities locally and internationally (see the description in Sect. 2.10).

2.10 Cases of Innovative Approaches in Education

> An expert is someone who doesn't want to learn anything new, because then he would not be an expert—Harry S. Truman, US President

The first two cases clearly demonstrate the implication of global emergence of co-creation for sustainability in the mission of university—a subject that until now has been overlooked. The third one reviews some successfully designed MSc and PhD transdisciplinary engineering programmes for sustainable industries.

2,000 W Society Pilot Region Basel. In 1985, Brazilian scientist José Goldenberg posed the question: "How much energy is necessary for a good life?" Goldenberg determined that below a threshold of 1,000 watts per person, people are indeed better off if they can increase their energy consumption, however, consumption above 1,000 watts per person does not improve quality of life. Therefore, the core concept of the 2,000-W Society is sufficient energy consumption because consuming more would be undesirable, even if the energy could be provided in a completely "clean" way. This is the concept of sufficiency, also balancing higher-energy demand for heating in cold climates with the comparatively lower energy demands in warm climate zones of developing countries to establish an equitable single target for all (https://www.2000-watt-society.org/act). This programme is mainly implemented by the Swiss Federal Institutes of Technology (ETH), University of Applied Sciences of Basel (FHBB) and Novatlantis. This case seeks to foster a city-wide transition to 2,000-W per capita society, with the wider ambition of accelerating a national de-carbonisation effort. ETH established itself as a frontrunner transformative university where one of the priorities/missions became that of bringing about socio-technical transformations in view of realising a 2,000 W Society. Right from the start of the redevelopment process, particular attention was given to the requirements of a 2,000 W Society and there are plans for a proactive dialogue with both current and future users. It focused on five themes designed to promote practical and reproducible steps towards a more sustainable future:

- Energy: integrated energy services with progressive contractor models; maximising the percentage of renewable energy sources and minimising energy losses; attractive energy agreements between landlords and tenants.
- Material flows: maximum and sustained reductions in the use of resources; re-use of existing construction materials; integration of sustainable recycling systems.
- Building ecology: consistent use of environmentally sustainable building materials; healthy and motivating indoor environment; mutually beneficial materials' saving agreements between landlords and tenants.
- Mobility: targeted, attractive and reproducible transport culture; integrated public transport provision, mobility for cyclists and pedestrians; increased safety for non-motorised transport users.
- Quality of life: individual and transparent added-value by means of associated socio-economic and socio-environmental projects with performance indicators to measure the transformation (www.novatlantis.ch; Trencher et al. 2014).

The difficulty in enforcing the 2,000-W Society standards has nothing to do with technologies. "The necessary political will has to exist in order to ensure that this vision can be turned into reality, i.e., implementation plans, energy-efficiency programmes, promotion of the concept of the zero-energy house, heat pumps, biogas, low-consumption cars, hybrid vehicles, and so on. It is our political responsibility to translate this concept into practice" (Huebner 2009).

The 2,000-W Society fights climate change by focusing on the issue of energy waste and inefficient energy use. That focus has helped hundreds of cities around the world use the 2,000-W Society model to substantially lower their carbon footprint. Here is what they targeted with their carbon-reduction action plans. Since the burning of fossil fuels is a major cause of climate change, it is best to measure and benchmark CO_2 based on non-renewable source energy. That is why the 2000-W Society uses "watts" as the measuring unit. Given this perspective, one tonne of CO_2 per person per year equals a demand of about 500 watts of non-renewable source energy.

The Oberlin project. The project is an alliance between the Oberlin College (USA) and the city of Oberlin. Described as a "full-spectrum sustainability" experiment, the project functions as a decentralised system of individual sustainability initiatives, where each of them contributes to the prosperity, resilience, and sustainability of the larger community. There is a clear potential for the frontrunner institution of Oberlin College to leapfrog the widely promoted model of entrepreneurial university and become a prototype of a transformative institution—one dedicated to co-creating societal transformations with a view to sustainable development (Orr 2011). The main goals of the project included shifting the city and College to renewable energy sources, radical improvement of efficiency, sharp reduction of carbon emissions, and improvement of their economy in the process. City residents and businesses presently spend ~ $15 million each year on electricity and natural gas—twice as much as they would need to spend if they were as efficient as is now economically advantageous and technologically feasible. The project initiators proposed to reduce energy use by improving efficiency (saving millions of dollars), building a local renewable-energy economy that creates jobs and ownership, and growing the local economy while buffering Oberlin from rising energy prices and sudden cost increases.

One of the main aim was to establish a robust local foods economy to meet a growing percentage of consumption while supporting local farmers. Presently, only a minuscule fraction of what people eat is grown in northeast Ohio. As with local energy consumption, money unnecessarily flows out of the community. There was a proposal to expand the market for locally grown foods and improve the local farm economy, create new employment opportunities in farming (including summer jobs for teens) and food processing, while improving the taste and nutritional quality of food they eat.

Besides that, it was decided to create an educational alliance between the College, the Oberlin schools, a nearby vocational school, and Lorain County Community College, focused on education appropriate to the challenges and opportunities of sustainability. The transition to sustainability and a more resilient economy poses large challenges to educators at all levels. What does the rising generation need to know to live well and purposefully in the decades ahead? How do we teach them to think in systems? Broadening and deepening the conversation on sustainability to include all of the humanities, the arts, the sciences, and the social sciences was encouraged.

Collaboration with other projects and communities across the U.S. that are also developing sustainably by integrating food, agriculture, energy, sustainable economic development, education, public policy, community engagement, health, and transportation was established.

Lithuanian case of co-creation for sustainability education

> To make the world ethically better was the only justifiable purpose of economic striving—John Maynard Keynes, British economist

Sustainable development/co-creation for sustainability poses particular challenges to engineering. In the past, engineering has applied a more 'expert' approach based upon a 'control and predict' paradigm. However, sustainability issues require the participation of various stakeholder groups that might have differing perspectives, goals and paradigms. The consideration and management of differing paradigms is, thus, a critical task in engineering for sustainable development (Staniškis 2012; Gustafsson 2013).

In the work field of engineers, environment is strongly connected to the usage of raw materials and energy, possibilities for reuse and so to reduce emissions to the environment and depletion of scarce resources. In terms of sustainability, this is not only important from the anthropocentric point of view but also from the perspective of the ecosystem (Carew and Mitchell 2008). A well-known tool is Life Cycle Assessment (LCA), which not only systematically structures the material and energy flows in the system under investigation but also is a way to evaluate consequences of decisions in development planning and implementation (Thabrew et al. 2009).

Based upon this analysis, some of the authors' experiences are presented to sensitise students to different paradigms and to teach them practical approaches for community involvement. A combination of lectures, exercises and projects are proposed to introduce the tool of participatory modelling and to provide students with experiences in its application in stakeholder processes. In particular, the linking of group projects to on-going local stakeholder processes has proven to be a valuable approach for students to gain experience in the implementation and facilitation of meaningful stakeholder participation (Halbe et al. 2013). LCA directly focuses on resources and emissions and the possibility of recycling, which is of great value for engineers, as their designs usually involve large material and energy flows from and to the environment. Furthermore, the key aspect of sustainability is a long-term view for which building sets of scenarios are very useful. LCA and scenario building are generally applicable, irrespective of the engineering discipline, giving a good understanding of a sustainable engineering design (Jonker and Harmsen 2013).

A new strategy of Kaunas University of Technology was approved by the University Board in December 2011. During the development of the strategy, a solid number of meetings and discussions were held and a platform for discussions was provided for the University community and partners.

Based on the new strategy, five main strategic activities were defined. All of them emphasised the issue of social responsibility and sustainable development, i.e. the University's activities are focused on human well-being and sustainable development of the state. The University formulated its objectives:

- to reorganise the University activities and cooperation with partners for the unity of economic, environmental, social and cultural objectives and values;
- to identify—together with government authorities, municipalities, industry and business—developmental issues of the city, region and country; to actively participate in implementing the strategy based on sustainable development and knowledge-based economy;
- to organise lifelong studies which promote socially and morally correct, ethically acceptable sustainable consumption and economic development of the country;
- to develop systematic education and consulting for companies, organisations and business, using the best competencies of the University;
- to develop and support within the University such activities that are responsive to the problems of sustainable development of the city, region and country and the quality of life.

Following the strategy implementation process, from 2012, all the faculties, research institutes, and central administration departments have been shaping their activities according to the new strategy directions and developing an annual plan of activities. The annual plans of activities in the first level structural units are discussed with the sustainability management team and approved by the Rector. Within a faculty or a research institute, deans or directors set the priorities of activities on the programme level. The educational dimension is usually promoted by both institutional and divisional level. The institutional level coordinates actions related to implementation of general courses in the curriculum, while specific aspects of the programme are addressed in division and sub-division levels. The best results have been achieved at the Institute of Environmental Engineering, where the research, education and curriculum development are combined with traditional education in engineering sciences with studies in natural and social sciences and with the ultimate goal of educating scholars who are uniquely situated to undertake serious research and policy assessments to tackle sustainable development challenges. The MSc and PhD programme graduates have a unique combination of diverse skills and deep insight into the most challenging problems of future human welfare. Together with experts from industry and governmental institutions, students in the programmes conduct research in a wide variety of areas associated with sustainable development. Students also benefit from being part of APINI research programmes and projects that focus on sustainable development. Moreover, both current students and graduates are invited to conferences and other events organised by the APINI. The most motivated students are offered employment at the APINI.

Sustainable development is one of the major topics of APINI research. The monograph *Sustainable innovations in Lithuanian industry: development and implementation*, written by APINI researchers, lists more than 60 projects in the field of sustainable development and cleaner production carried out during the last decade. New projects start nearly every year, and most of them have sustainability aspects covered. All the activities of the Institute in 2013 were evaluated by QUESTE-SI (Engineering Education—for Sustainable Industries) methodology, consisting of four dimensions, and recognised to be among the best in Europe (Staniškis and Katiliūtė 2016).

Dimension 1: Institution. The Institute of Environmental Engineering systematically implements sustainable development and cleaner production programmes and projects in Lithuania and abroad. The staff of the Institute provide comprehensive assistance to business, society, non-governmental and governmental organisations in improvement of environmental performance for achieving a common goal, not a conflict, between economic growth and environmental protection now and in the future. Daily activities of the APINI are based on principles of sustainability. In 25 years, since the establishment, the APINI has increased the scope of activities and capacity to tackle key environmental problems and emerging scientific topics. While initial research activities of the Institute focused on environmental issues, social and economic topics have been gradually introduced into the research and education. The Institute of Environmental Engineering has been the initiator of the integration of sustainable development ideas at Kaunas University of Technology.

Extensive co-operation with foreign scientific institutions and universities has enabled the APINI to become an advanced and unique institution in Lithuania actively working in the area of sustainable development. The portfolio of 25-year experience and knowledge in the sustainability area has resulted in initiation and contribution to implementation of social responsibility strategy at Kaunas University of Technology.

The sustainable development strategy of the APINI is aimed at strengthening the responsibility to society and country, focusing its activities on enhancing the quality of human life and acceleration of statehood development. Thus, its principal activity is to create and translate the University's contribution into the country's viability and its sustainable economic, social and cultural knowledge-based development.

The mission of the Institute of Environmental Engineering is to disseminate sustainable development principles in Lithuania and all over the world through application of innovative sustainable solutions by means of interdisciplinary research, topical studies and continuous spread of knowledge and values.

The vision of the Institute of Environmental Engineering is a unique international leader in the field of sustainability based on interdisciplinary research and advanced studies (Katiliūtė and Staniškis 2015).

The strategic objectives for the future are as follows:

- implementation and integration of sustainable development issues in academic society;
- improvement of staff competence and working environment;
- implementation of sustainability principles in the community or community involvement.

The strategy of sustainable development includes a requirement of annual reports on the progress of defined activities. The first sustainability report was prepared at the end of 2012.

Reports on research and education activities as well as results achieved have been reported annually since 1992 by the Institute of Environmental Engineering to the Ministry of Education and Science of the Republic of Lithuania, Research Council of Lithuania, and Kaunas University of Technology.

The Institute of Environmental Engineering could be recognised as the most successful department at KTU dealing with social responsibility issues. The Institute has knowledge and capacity to provide substantial contribution in the sustainable development process at the University acting as a source of knowledge on sustainable development.

Dimension 2: Education and curriculum development. None of the central issues related to sustainable development can be understood from the sole perspective of a traditional discipline, whether in social, natural, engineering or health sciences. The MSc and PhD programmes include a set of rigorous core requirements in engineering, social and natural sciences designed to provide a deep understanding of the interaction between all three systems, and provide students with the flexibility to pursue in-depth research in a broad variety of environmental areas. Graduates of the programmes have a unique combination of diverse skills and deep insight into the most challenging problems of future human welfare. Together with experts from industry and governmental institutions, students in the programmes conduct research in a wide variety of areas, including climate change and its social consequences, causes and solutions to extreme material and energy resources inefficiency, energy systems, water resources, waste management systems, ecosystems, corporate social responsibility, environmental economics, and eco-design. Students also benefit from being part of APINI research programmes and projects that focus on sustainable development.

Many graduates pursue academic careers in interdisciplinary graduate and undergraduate programmes with the focus on industry and environment as well as in more traditional engineering disciplines. Others choose non-academic positions in governmental institutions, non-governmental organisations or private firms engaged in environmental and sustainable development projects.

During the 22 years of the PhD programme and 18 years of the MSc programme, the emphasis has been placed on research at the boundaries between social, natural and engineering sciences, and these programmes have become very popular and highly rated. Almost all PhD graduates have accepted academic positions as tenure-track professors or post-doctoral fellows, or have been employed at high-level positions in the private sector and international organisations (Staniškis and Arbačiauskas 2011).

This 2-year MSc programme equips students with the knowledge and skills necessary for managing eco-development issues at different levels and provides them with opportunities to establish ties that bridge science and industry. These qualities aid graduates in formulating policies to advance sustainable development at the company, on local, regional and global levels, to facilitate cooperation between industry, government and society in the design of a sustainable development and environment policy, and to generate strategies for environmental conflict management. To attain the MSc in engineering, students are required to:

- "demonstrate the ability to develop and design products, processes and systems while taking into account the circumstances and needs of individuals and the targets for economically, socially and ecologically sustainable development set by the community;
- demonstrate insight into the possibilities and limitations of technology, its role in the society and the responsibility of the individual for how it is used, including both social and economic aspects and also environmental and occupational health and safety considerations" (Staniškis and Katiliūtė 2016).

APINI has a double strategy regarding the implementation of education for sustainable development:

- to have educational programmes focusing on sustainable development, and
- to integrate sustainability in all engineering and architecture programmes (Staniškis and Arbačiauskas 2011; Finnveden and Strömberg 2013).

The integration of research and education activities ensures the quality and recognition of the education programme nationally and internationally, e.g. Baltic Sea Region Award, 2012 National Energy Globe award for sustainability.

The first PhD programme in Environmental Engineering and Landscape Management started in 1994. The main research and education topics of the programme initially were related to environmental technologies. Fast development of preventive concepts and the idea of sustainability has gradually led to change of priorities, contents and structure of PhD education. Therefore, a new interdisciplinary PhD programme in Environmental Engineering (in the context of sustainable development) was launched in 2012. The programme was designed to educate researchers and university teachers in engineering, and social, economic, and natural science disciplines that underpin sustainable development.

The research results presented in the programme's doctoral dissertations are based on integrative, interdisciplinary research that is needed to explore science and policy issues in the area of sustainable development. Integrated assessment methods and concepts (e.g. transitions, modelling, and scenario analysis) are instrumental in providing answers to the central questions of sustainable development.

The PhD candidates have an opportunity to choose a student-centered programme from a selection of courses offered by KTU and other involved universities. For each PhD student, a tailor-made programme is designed (Staniškis and Katiliūtė 2016).

Dimension 3*: *Student involvement. The Institute of Environmental Engineering recognises the need to provide a holistic approach to social responsibility (SR) issues for its students, including both curricular and non-curricular activities. Since the first student enrolment in the programme, the APINI has focused on involving students in project-based research activities. The APINI's research and close co-operation with industry yield knowledge, which is disseminated to students during lectures and available in the curriculum literature. Course projects are designed to give students an internal visibility of the subject by solving SR-related issues (particularly environmental) in industrial enterprises. The APINI staff support students' extra-curricular research activities and often contribute to student research by direct involvement and being co-authors of scientific publications (at least 11 publications have been prepared). Also, both current students and graduates are invited to SR-related conferences and other events organised by the APINI.

Furthermore, the APINI aims to promote sustainability issues among students, as well as staff members, through environmental non-curricular on-campus activities. For example, the APINI initiated waste recycling campaign (Green University Building) in 2010 and 'Green University' in 2012. 'Green University' was the first environmental campaign to be launched for the whole campus. The APINI has participated and invited its students to participate in the biggest environmental community campaign in Lithuania 'Let's do it' for cleaning territories from waste.

To track the information about its graduates, the APINI has established a database that enables to compile different information including graduate contacts and placement. The information is used to invite graduates to events organised by the APINI and related to sustainability. In 2012, to mark the 10th anniversary of the MSc programme, all graduates were surveyed to gather information on the impact of the master's studies on their careers and their intellectual and personal development. Graduate reflections have been used by the teaching staff for making the appropriate improvements in the MSc programme.

Dimension 4: Research, innovation, and impact on the region. Sustainable development is one of the major topics of APINI research. All the APINI members conduct research or teach subjects related to sustainability issues. Research topics, such as environmental performance, eco-design, control of chemicals, integrated waste management and others, to a great extent deal with sustainability issues.

The majority of the research projects in which APINI staff are involved are related to the sustainability issue, e.g. the recently finished FP7 project CRISP had the core objective to identify potential paths—transition pathways—to enhance behavioural and societal changes towards sustainable, low carbon Europe in 2050. FP7 project VISION RD4SD aims to ensure that Europe is able to contribute to sustainable development of the world by formulating policies and decisions based on robust, up-to-date knowledge. BSR 2007–2013 project 'Reco Baltic 21 Tech', as one of its goals, strives to help regions to reach sustainable waste management. The monograph 'Sustainable Innovations in Lithuanian Industry: Development and Implementation' written by the APINI researchers lists more than 80 projects in the field of sustainable development and cleaner production carried out during the last decade. New projects are started nearly every year, and most of them have sustainability aspects covered.

The APINI members have carried out a number of applied sustainability studies dealing directly or partly with sustainability issues. A substantial share of these studies has been commissioned by the Ministry of Economy of the Republic of Lithuania. Research of the members of the Institute of Environmental Engineering has received international recognition. In 2003 –2005, the Institute won the 5BP (Centres of Excellence) project 'Sustainable Industrial Development—Strengthening of the Competence of the Institute of Environmental Engineering' and became an EU competence centre with regard to sustainable industrial development. The Institute and its staff have received a number of awards: the National Science Award for the work on development and implementation of preventive environmental strategies in Lithuanian industry in 1993 – 2003; the Baltic Sea Award 2010, which goes annually for those putting research findings into practice, technical decisions or other activities that contribute substantially to the Baltic Sea region and its sustainable development; his Excellency the President Dr. Valdas Adamkus Award for scientific work important for Lithuania's present and future and for active ecological activities. In 2008, in the European Parliament, the Institute received the Energy Globe 2008 National Award for 'The System for Preventive Energy Saving & Waste Minimisation Innovation Development and Implementation in Industry 1997–2007' (APINI-SPIN). In 2008, APINI became a winner of the industrialists' competition 'Achievements in Environmental Protection' for successful international cooperation in development and implementation of the 'System of Generation and Implementation of Preventive Environmental Innovations'. More than 150 participants from 18 countries were invited to the Energy Trophy + competition 2007/2009. The APINI co-ordinated this project in Lithuania and 5 Lithuanian companies took part in the competition. The construction company Dzūkijos Statyba managed to save 30% of the total energy consumed at the administration building and implemented low-cost measures. Following the decision of the jury of international climate and energy experts, this company was recognised as the third place winner. The winners were ceremonially awarded the Energy Trophies at the European Prize Gala held in the course of the European Sustainable Energy Week in 2009.

As it is obvious from the lists of projects and publications of the APINI, the Institute has many linkages with industry and other organisations. The APINI staff are also involved in various associations, professional or technical bodies. This co-operation allows obtaining the latest information, getting experience and knowledge, and exchanging ideas, which contribute to the research quality and allow spreading sustainable development ideas via organisation of conferences, seminars and other events, giving trainings to society and industry on sustainable consumption and production, and teaching students (Staniškis 2016).

Discussions and conclusions. By integrating the systems theory (research) into environmental education and other programmes, students can learn more about the relationships between natural resources, environmental sustainability, and human well-being. This knowledge accompanied by appropriate action is a prerequisite in maintaining not only our natural resources but also our way of life. Thus, research and environmental education require closer co-operation and more involvement in expanding their combined effect. This will assist industrialists, on the one hand, and educators, on the other, in doing more than ever to help people to understand the environmental and social concerns and to act on them.

Many investigations and the empirical analysis point out a radical paradigm shift in the social functions of the university "via technology transfer to actually transforming and co-creating society in the pursuit of sustainable development via a much broader range of channels, approaches and actors" (Trencher et al. 2014).

Today, higher education is a short-term business system of delivering student examinations for a not well-defined global market. The co-creation for sustainability programmes as usual is started with great enthusiasm by a group of dedicated teachers. However, gradual funding squeeze, continuous internal reorganisations, and external political decision-making have led to the dismantling of the faculty organisation, which has more or less 'killed' an organised teacher influence. In the meantime, from the materials gathered and our experience, there are windows of opportunity in the existing institutional and cognitive settings. The success of these processes, however, is still to be seen, and a strategy remains to be conceived for scaling up the possible successes to other engineering programmes (Valderrama et al. 2013; Staniškis and Katiliūtė 2016).

Achieving sustainable outcomes will require engineers who can engage with diverse stakeholders, employ new tools for decision support, conduct adaptive management, and find creative solutions by integrating natural, human, and manufactured systems in novel ways (Svanström et al. 2013). There is a demand now for the services of engineers who are skilled in integrating natural, human, and manufactured systems to achieve sustainable consumption and production. That demand is likely to grow. Thus, there is likely to be a useful role for a 2-year, full-time, master's degree programme that nurtures systems-integration skills at a high level. Such a programme would draw upon existing curricula in sustainable engineering, but would probably have to break some new ground in the curriculum design (Thomson 2013). As usual, many of the original courses survive in the new programmes, and others disappear. However, with the existing recruitment situation, many courses have too few students to be economically viable. Nonetheless, many of the teaching staff have started to

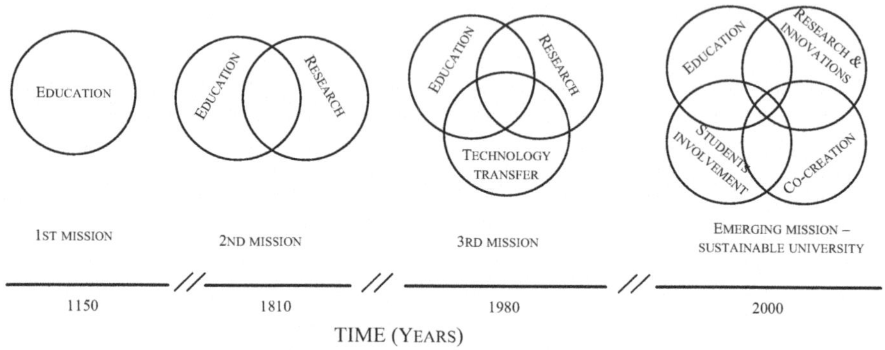

Fig. 2.10 Emergence of the 4th mission—transformative/sustainable university (Staniškis 2016)

challenge the non-inclusive and increasingly bureaucratic school system that hinders the development of high quality master's programmes.

The question that, therefore, emerges is: "How can government policy and incentive systems such as funding mechanisms acknowledge this and encourage university actors to pursue a much broader development agenda founded upon place-based sustainability needs?" (Trencher et al. 2014). "Empowering educators must be central to any professional development initiative. Educators are important agents for change within education systems. Effective educational transformation is dependent upon educators being motivated to bring about change, as well as being capable of and supported in doing so" (UNECE 2013).

The third mission of universities and co-creation for sustainability should not be viewed in isolation and emerging co-creation should not become the sole focus for a particular university. It is obvious that entrepreneurialism and technology transfer are too narrow and not much significant for many smaller and humanities-focused institutions and, at the same time, it is evident that the mission of co-creation for sustainability will not become a strategy for all the universities in the world. However, what is definitely obvious is that sustainability crisis is prompting the emergence of a new type of institutions: the transformative/sustainable university (Fig. 2.10).

2.11 The Future of Sustainability Science and Education

> Scientific research is based on the idea that everything that takes place is determined by laws of nature, and therefore this holds for the actions of people.—Albert Einstein, German-born physicist

Sustainability science is a precondition for sustainable development achievement. Boosting sustainability science and expanding access to scientific data and research globally benefit us all. Overall, insufficient mobilisation and reorientation of science more broadly—including its approaches, organisation and funding structures—threatens to derail the 2030 Agenda. Rather than standing by and allowing ourselves to come up short, the global community must enable scientific research to fulfil its transformational potential (Messerli et al. 2019).

"Seeing systems whole requires more than being interdisciplinary, if that word means, as it usually does, putting together people from different disciplines and letting them talk past each other. Interdisciplinary communication works only if there is a real problem to be solved, and if the representatives from various disciplines are more committed to solving the problem than to being academically correct. They will have to admit ignorance and be willing to be taught, by each other and by system. It can be done. It's very exiting when it happens" (Meadows 2008). It was proved by the UN Independent Group of Scientist for Global Sustainable Development Reporting when in the period of 2016–2019 a transdisciplinary team of 15 selected world scientists successfully developed the Report *The Future is Now: Science for Achieving Sustainable Development"* (IGS 2019).

"One of the goals of education should be to recognize that in the real world our problems are not demarcated by discrete disciplines. For example, climate change is neither a sociological, environmental, nor economic phenomenon. Perhaps one of our goals as scientists should be to reduce the barriers, blend the disciplines, and/or work across disciplines: that is, to be interdisciplinary. Exclusive reliance on only one discipline gives a misleading and myopic understanding" (Reardon et al. 2018).

Perhaps a more fundamental question is: can we ensure that knowledge produced now will not be a result in its use for destructive ends in the future. For example, the inventor of Freon (1930), Thomas Midgley, a chemist working for General Motors, also invented lead gas in the 1920s, which made high-compression auto and aircraft engines possible. We believe that knowledge cannot (and should not) be suppressed, but that we should engage in dialogue with each other carefully to monitor its usage and effects (particularly regarding sustainability). This can only happen with educated citizens.

Hart (1978) in his book "The 100: A Ranking of the Most Influential Persons in History" used the term "political economy" rather than economics. Indeed, economics originated as the study of both politics and economics, and was known as political economy from the time of Smith to the late nineteenth century, when economists became convinced of the superiority of their scientific credentials and divorced themselves from politics and other social scientists. Our shared belief is that it is incumbent for economics to return to its political economy roots. At the same time, we should acknowledge that "economics is a political argument. It is not—and can never be—a science; there are no objective truths in economics that can be established independently of political, and frequently moral, judgements. Therefore, when faced with economic argument, you must ask the age-old question "Cui bono" ("Who benefits") first made famous by the Roman statesman and orator Marcus Tullius Cicero" (Chang 2015). "Only pluralism is consistent with democracy and only a democracy in ideas is consistent with the ideas of education. Our world needs educated citizens and economists who understand diversity and are willing to work with others" (Reardon et al. 2018).

Anthropocene is one of the most arresting ideas to emerge from science in recent years. It could radically change the world. To do so, it must withstand intellectual scrutiny and have the capacity to alter our collective behaviour in a sustained way. Anthropocene embraces more than global climate change, encapsulating all the immense and far-reaching impacts of human actions on Earth. It says: this changes everything, forever (Lewis and Maslin 2018).

The university of the future should include a broad range of the community members, be actively engaged in issues that concern them, relatively open to commercial influence, and fundamentally transdisciplinary in its approach to both teaching and research.

To achieve similar learning outcomes, the method of future studies as a tool for ESD with some examples was proposed by Wangel et al. (2013). It was shown that future study methods work well in a problem-based learning setting in which students are to define and elaborate both problems and solutions. However, there is a need to somehow assess to what extent student perception on sustainable development has changed as a result of these methods.

Universities are beginning to recognise their unique roles as leaders in the society by demonstrating sustainability in their campus operations and, more importantly, educating the next generation on advancing sustainability in practice. In response to this recognition, a number of university sustainability assessment methods have been developed. Assessing the sustainability of a higher education institution is a complex problem; hence, the relative importance of multi-aspects should be taken into consideration when assessing the overall performance. Shi and Lai proposed a similar alternative university sustainability assessment framework and the associated rating system based on three levels: (i) formal statement, (ii) governance, and (iii) strategies for fostering sustainability, including education, research, outreach and partnership, and campus sustainability (Shi and Lai 2013).

To become a good and critical learner, one must master the following three tools: the ability to doubt and ask good questions, the willingness to imagine, and the ability to listen and engage in constructive dialogue. We do not advocate disagreeing with everything for the sake of disagreement; our objective rather is to promote a healthy scepticism, and not to doubt everything or allow our learning journey to degenerate into a ceaseless and irrelevant questioning (Reardon et al. 2018).

"The concept of sustainability implicitly recognises feedback and interdependencies among ecological, economic, and social systems; hence a systems approach to environmental management. Alas, reality is otherwise: the issues comprising sustainability have been fragmented so that isolated parts are no longer seen as one system" (Orr 2016). "Usual situation when students ask where to get it, this kind of economics. They have been studying for almost three years, and sustainable development issues have never been raised in their courses. When students directly ask teachers, there was both a logic and irony to this, because there is only one thing that economics professors ought to understand: it is the law of supply and demand" (Jackson 2017). "We, educators, have equipped our graduates with the tools and technology necessary to enlarge the human empire, but not the wisdom to understand the consequences of doing so… We have taught them how to manipulate, make, conjure, communicate worldwide, and sell everything under the sun but not how to think about the effects on themselves and others of doing such things. We have taught the future leaders of mighty corporations how to grow their companies beyond imagination, but given no guidance regarding the physical, ecological, and moral limits to the scale of the human estate or the concepts of enough and sufficiency. The students were taught to be technicians, not thinkers, in a culture that is long on know-how and short on know-why" (Orr 2016).

"In other words, if higher education is to serve the interests of humankind and life in the long emergency ahead, it must be transformed, starting with a change in our concept of education and the purposes that ideas and disciplined knowledge serve. [...] We should better understand how we came to this point and what we can do to avert the worst that could be ahead." [...] "The fast educators have neither the time nor the inclination to facilitate deep consideration of education or enable their students to think about the act of thinking or offer useful guidance on the larger issues of our time. Their model of education is the business plan, their metrics is cash flow, their students are just customers, their teachers are increasingly underpaid and overexploited adjuncts, and their pedagogy is that of the assembly line" (Orr 2016).

"Despite the growth of sustainability in the research interests, curriculum, and administration of universities, important questions still remain about the ability of higher education to promote green values. Rees, for instance, has suggested that colleges and universities often "impede sustainability" by remaining wedded to fields and methods that reproduce the habits and beliefs that generated sustainability (Rees 2003). For example, most business programs still minimize social and environmental responsibility and teach that profitability is the sole purpose of a firm. Most economics departments are still dominated by neoclassical economists who uncritically encourage growth and see the GDP as a useful measure of economic health. As a result, sustainability has often found a home in newer and less traditional fields—art and design, environmental engineering and sociology, renewable resources, international and public affairs, and so on, and because of this, there is often conflict on university campuses between sustainists, ecologists, and ecological economists, on the one hand, and those tied to business as usual, on the other" (Caradonna 2014). The ultimate goal for all of new type economists should be sustainable economy that is low-carbon, democratically decentralised and environmentally sustainable, including true cost economics, and that promotes equality, well-being, and life satisfaction. Rees offers a detailed explanation: "Neoclassical economists have traditionally been content to allow the prices of goods and services to be determined solely by the law of supply and demand. However, in unregulated markets, only direct producer costs (for rent, labor, resources and capital, for example) are reflected in consumer prices. The prevailing cost-price system does not account for the collateral damage to ecosystems, human communities, or population health caused by many production processes... Because negative externalities represent real costs, the goods and services inflicting them enter the marketplace at prices below their true costs of production. Such underpricing leads to overconsumption, inefficient resource use, and pollution—all classic symptoms of market failure" (Rees 2003).

"Beyond the primary obligation to serve students, what is the larger role of colleges and universities as "anchor institutions in their own communities? Can colleges and universities redirect their buying and investments to local and regional renewal? Can they become carbon–neutral and lead others to the same goal? Can they help restore regional agriculture, food systems, and businesses? Might they invest in restorative economic development? In short, can they operate facilities and manage their financial affairs in ways that do no harm to the world their students will inherit? And can they make these a part of their curriculum so that the young become competent in making high ideals practical and real? In response to such questions, a transformation of higher education is now under way. Construction of the Adam Joseph Lewis Center for Environmental Studies at Oberlin College in 2000 was a major catalyst for green building revolution on college and university campuses… But most important is an exercise in applied hope based on a commitment to make the world more fair and decent while preserving a beautiful and livable Earth. And if we don't stand for such things, what do we stand for?" (Orr 2016).

The study at Chalmers University of Technology, based on courses' analyses and questionnaire to students and alumni, has revealed that the courses should have an emphasis on environmental issues, but to meet alumni and industry needs the courses also must have a reasonable emphasis on economic issues, sustainable business management, social issues, and green technologies (Hanning et al. 2012). Research in Australia showed that while many universities publicly endorsed goals and values related to sustainability, the commitment was not reflected in the vision, mission and graduate attributes of business faculties/schools within the same institution. In fact, when commitment to higher education for sustainable development is not endorsed as a publicly espoused value at multiple levels of a university, the organisation's commitment to ensuring sustainable development is enacted at strategic and operational levels of the organisation may be questioned (Lee et al. 2013).

University growth and increasing specialisation have led to a neglect of matters of public interest that do not fit neatly within ever-narrower and more technical specialisation. The consequences of specialised disciplinary and sub-disciplinary "silos" have become a staple of the literature on sustainability, cited as an obstacle to university research and teaching pertaining to matters of sustainability. Scientists have understood their role in the scheme of public knowledge to be limited to doing good research within their specialisations, publishing it in peer-reviewed journals written for other scientists and training future scientists. However, they are rarely perceived as having a responsibility to communicate directly to the public and policy makers their findings suitable to guiding important decisions, nor have they been involved in interdisciplinary collaboration in order to serve the public interest. Barriers to the emergence of adequate sustainability curricula in the high education begin with the withering (shrinking) of general education in favour of fragmented, specialised, and career-focused programmes of study. Entities such as national academies have been designed to gather research that has survived peer review, identify points of consensus relevant to the public and communicate them. It is evident that wider dissemination of important findings to communities of experts is unfortunately hampered by the limitation of popular media. Coordinated efforts to inform and persuade would need to convince the public that problems of climate change and sustainability are urgent and that the affordable and attractive products marketed as lower carbon or more "green" really are what they claim to be. Since the level of misleading green claims in advertising and marketing is very high, the deficiencies of public education efforts and enforcement of truth in advertising regulation are among the obstacles towards sustainability.

"The bedrock of each university is a system of discipline-specific departments. The strength of these departments determines the success and prestige of the institution as a whole. This structure raises a few obvious questions. One is the relevance of the department-based structure to the way scientific research is done. [...] And there is a sense that, notwithstanding years of efforts to adapt to this change by encouraging interdisciplinary collaboration, the department-based structure of the university is essentially at odds with such collaboration. [...] The university of the future will be inclusive of broad swaths of the population, actively engaged in issues that concern them, relatively open to commercial influence, and fundamentally interdisciplinary in its approach to both teaching and research" (2007).

"Research institutions such as universities, academies and scientific associations should expand their evaluation systems to recognise interdisciplinary and transdisciplinary skills, and reward research that strives for societal relevance and impacts. Instituting the right incentives is crucial to fostering careers of up-and-coming sustainability scientists. Senior researchers should support and encourage their students and younger colleagues to carry out sustainability science and to engage fully in communicating it to an outside audience" (IGS 2019).

References

Aras G, Crowther D (2009) Making sustainable development sustainable. Manag Decis 47:975–988

Asif M, Searcy C, Zutshi A, Fisscher OAM (2013) An integrated management systems approach to corporate social responsibility. J Clean Prod 56:7–17

Ballou B, Heitger DL, Landes CE (2006) The future of corporate sustainability reporting: A rapidly growing assurance opportunity. J Account 202:65–74

Bass S, Dalal-Clayton B (2012) Sustainable development strategies: a resource book. Routledge

Bayulken B, Huisingh D (2015) Perceived 'Quality of Life' in eco-developments and in conventional residential settings: an explorative study. J Clean Prod 98:253–262. https://doi.org/10.1016/j.jclepro.2014.10.096

Biggs JB (1999) Teaching for quality learning at university: What the student does, 1st edn. Society for Research into Higher Education & Open University Press, Buckhingham

Bocken NMP, Short SW, Rana P, Evans S (2014) A literature and practice review to develop sustainable business model archetypes. J Clean Prod 65:42–56. https://doi.org/10.1016/j.jclepro.2013.11.039

Boman J, Andersson UP (2013) Eco-labelling of courses and programs at University of Gothenburg. J Clean Prod 48:48–53

Bonilla SH, Almeida CMVB, Giannetti BF, Huisingh D (2010) The roles of cleaner production in the sustainable development of modern societies: an introduction to this special issue. J Clean Prod 18:1–5

Caradonna JL (2014) Sustainability: A history. Oxford University Press, New York, NY

Carew AL, Mitchell CA (2008) Teaching sustainability as a contested concept: capitalizing on variation in engineering educators' conceptions of environmental, social and economic sustainability. J Clean Prod 16:105–115
Chang HJ (2012) Things they don't tell you about capitalism. Bloomsbury Press
Chang HJ (2015) Economics: the user's guide. Bloomsbury Publishing USA
Charter M (2006) Sustainable Consumption & Production, Business and Innovation. Perspect Radic Chang to Sustain Consum Prod (SCP); Sustain Consum Res Exch Copenhagen, Denmark 20:243
Czarniawska B, Genell K (2002) Gone shopping? universities on their way to the market. Scand J Manag 18:455–474
da Silva ME, de Oliveira APG, Gómez CRP (2013) Can collaboration between firms and stakeholders stimulate sustainable consumption? discussing roles in the Brazilian electricity sector. J Clean Prod 47:236–244
Diakaki C, Grigoroudis E, Stabouli M (2006) A risk assessment approach in selecting environmental performance indicators. Manag Environ Qual an Int J 17:126–139
DTIE U (2012) Understanding Resource Efficient and Cleaner Production
Eccles RG, Perkins KM, Serafeim G (2012) How to become a sustainable company. MIT Sloan Manag Rev 53:43
Elkington J (1998) Cannibals with forks: the triple bottom line of 21st century business. Br Columbia New Soc Publ, Gabriola Island
Enroth M (2006) Developing tools for sustainability management in the graphic arts industry. KTH
Environment Canada (2009) Pollution Prevention Handbook
Epstein MJ, Buhovac AR (2014) Making sustainability work: Best practices in managing and measuring corporate social, environmental and economic impacts. Greenleaf, Sheffield
Etzkowitz H (2003) Research groups as 'quasi-firms': the invention of the entrepreneurial university. Res Policy 32:109–121
European Round Table of Industrialists Competitiveness Working Group (ERT) (1998) Job creation and competitiveness through innovation. Brussels
Finnveden G, Strömberg E (2013) Developing sustainability learning outcomes for engineering. In: Proceeedings to Engineering Education for Sustainable Development 2013 (EESD13), Cambridge, England, UK, Sep 22–25 2013. Cambridge, pp 156–165
Foxon T, Pearson P (2008) Overcoming barriers to innovation and diffusion of cleaner technologies: some features of a sustainable innovation policy regime. J Clean Prod 16:148–161
Frugoli PA, Almeida C, Agostinho F et al (2015) Can measures of well-being and progress help societies to achieve sustainable development? J Clean Prod 90:370–380
Ginsberg B (2011) The fall of the faculty. Oxford University Press, New York, NY
GRI (2015) Sustainability reporting guidelines—G4
Gustafsson J-E (2013) The rise and the demise of the Environmental Engineering and Sustainable Infrastructure (EESI) master program at KTH. In: Proceedings of engineering education for sustainable development EESD13. Cambridge, pp 210–221
Halbe J, Adamovski J, Pahl-Wostl C (2013) The role of paradigms in engineering education and practice for sustainable development. In: Proceedings of engineering education for sustainable development, EESD13. Cambridge
Hanning A, Priem Abelsson A, Lundqvist U, Svanström M (2012) Are we educating engineers for sustainability? comparison between obtained competences and Swedish industry's needs. Int J Sustain High Educ 13:305-320
Hart MH (1978) The 100: a ranking of the most influential persons in history. Citadel press
Holton I, Glass J, Price ADF (2010) Managing for sustainability: findings from four company case studies in the UK precast concrete industry. J Clean Prod 18:152–160
Hörisch J, Johnson MP, Schaltegger S (2015) Implementation of sustainability management and company size: a knowledge-based view. Bus Strateg Environ 24:765–779
Huebner K (2009) 2,000 Watt Society. In: United Nations Univ. https://ourworld.unu.edu/en/2000-watt-society. Accessed 19 May 2021

Independent Group of Scientists (IGS) appointed by the Secretary-General:, Messerli P, Murningtyas E, et al (2019) Global Sustainable Development Report 2019: The Future is Now–Science for Achieving Sustainable Development. United Nations, New York

Iraldo F, Testa F, Frey M (2009) Is an environmental management system able to influence environmental and competitive performance? the case of the eco-management and audit scheme (EMAS) in the European union. J Clean Prod 17:1444–1452

ISO 26000 (2010) Guidance on social responsibility

Jabbour CJC, Santos FCA (2008) Relationships between human resource dimensions and environmental management in companies: proposal of a model. J Clean Prod 16:51–58

Jackson T (2017) Beyond consumer capitalism: foundations for a sustainable prosperity. In: Handbook on growth and sustainability. Edward Elgar Publishing

Jones-Evans D, Klofsten M (1998) Role of the university in the technology transfer process: a European view. Sci Public Policy 25:373–380

Jonker G, Harmsen J (2013) Effective teaching methods for Sustainable Design. In: Proceeedings to engineering education for sustainable development 2013 (EESD13), Cambridge, England, UK, Sep 22–25, 2013. Cambridge, pp 283–292

Kallio K-M, Kallio TJ (2014) Management-by-results and performance measurement in universities–implications for work motivation. Stud High Educ 39:574–589

Kallio K-M, Kallio TJ, Tienari J, Hyvönen T (2016) Ethos at stake: performance management and academic work in universities. Hum Relations 69:685–709. https://doi.org/10.1177/0018726715596802

Katiliūtė E, Staniškis JK (2015) Sustainable university for regional development: Quality management model that integrates employer and social partner attitudes. in: transformative approaches to sustainable development at universities. Springer, pp 75–89

Keeble JJ, Topiol S, Berkeley S (2003) Using indicators to measure sustainability performance at a corporate and project level. J Bus Ethics 44:149–158

Khan Z (2008) Cleaner production: an economical option for ISO certification in developing countries. J Clean Prod 16:22–27

Klemeš JJ, Varbanov PS, Huisingh D (2012) Recent cleaner production advances in process monitoring and optimisation. J Clean Prod 34:1–8

Krajnc D, Glavič P (2003) Indicators of sustainable production. Clean Technol Environ Policy 5:279–288

Lee K-H, Barker M, Mouasher A (2013) Is it even espoused? An exploratory study of commitment to sustainability as evidenced in vision, mission, and graduate attribute statements in Australian universities. J Clean Prod 48:20–28

Lewis SL, Maslin MA (2018) Human planet: how we created the anthropocene. Yale University Press

Leydesdorff L, Etzkowitz H (1998) Triple Helix of innovation. Sci Public Policy 25:358–364

Lorek S, Spangenberg JH (2014) Sustainable consumption within a sustainable economy–beyond green growth and green economies. J Clean Prod 63:33–44

Lourdel N, Gondran N, Laforest V, Brodhag C (2005) Introduction of sustainable development in engineers' curricula: problematic and evaluation methods. Int J Sustain High Educ 6:254–264

Lozano R (2012) Towards better embedding sustainability into companies' systems: an analysis of voluntary corporate initiatives. J Clean Prod 25:14–26

Lozano R, Young W (2013) Assessing sustainability in university curricula: exploring the influence of student numbers and course credits. J Clean Prod 49:134–141

Malinauskienė M, Kliopova I, Hugi C, Staniškis JK (2018) Geostrategic Supply risk and economic importance as drivers for implementation of industrial ecology measures in a nitrogen fertilizer production company. J Ind Ecol 22:422–433. https://doi.org/10.1111/jiec.12561

Marsick V, Watkins K, O'Connor BN (2011) Researching workplace learning in the United States. In: Malloch M, Cairns L, Evans K, O'Connor BN (eds) The SAGE handbook of workplace learning. SAGE Publications Ltd. London, Los Angeles, London, New Delhi, Singapore, Washington DC, pp 198–209

McMichael J, Woodward A, Muir C (2019) Climate change and the health of nations. Oxford University Press
Meadows DH (2008) Thinking in systems: A primer. chelsea green publishing, White River Junction
Messerli P, Kim EM, Lutz W et al (2019) Expansion of sustainability science needed for the SDGs. Nat Sustain 2:892–894
Nedenes OS (1994) The Norwegian industrial transfer of know-how programs on waste minimization/cleaner production to Central and Eastern European countries. Material of invitational expert seminar. Introducing cleaner production in Eastern Europe. Lund University, Sweden, pp 13–19
Orr DW (2011) What do we stand for now? Oberlin Alumni Mag 19–28
Orr DW (2016) Dangerous years: Climate change, the long emergency, and the way forward. Yale University Press
Pedersen C, Nielsen BB (2000) Maintaining the momentum: EMS after the certifier has left. In: R. H (ed) ISO 14001: Case studies and practical experiences. Greenleaf Publishing in association with GSE Research, pp 31–38
Perrini F, Tencati A (2006) Sustainability and stakeholder management: the need for new corporate performance evaluation and reporting systems. Bus Strateg Environ 15:296–308
Pohjola T (2005) Applications of an environmental modelling system in the graphics industry and road haulage services. In: Implementing environmental management accounting: Status and challenges. Springer, pp 169–192
Pryce V (2002) CSR–should it be the preserve of the usual suspects? Bus Ethics A Eur Rev 11:140–142
Reardon J, Madi MAC, Cato MS (2018) Introducing a New Economics. Pluto Press
Rees WE (2003) Impeding sustainability? The Ecological Footprint of Higher Education. Plan High Educ 31:88
Roorda N (2010) Sailing on the winds of change: The Odyssey to sustainability of the universities of applied sciences in the Netherlands
Salzmann O, Steger U, Ionescu-Somers A (2005) Quantifying economic effects of corporate sustainability initiatives–Activities and drivers. IMD 28(3)
Schaltegger S, Herzig C, Kleiber O, Müller J (2002) Sustainability management in business enterprises: concepts and instruments for sustainable organisation development. Federal Ministry for the Environment, Nature conservation and Nuclear Safety
Searcy C, Karapetrovic S, McCartney D (2005) Designing sustainable development indicators: analysis for a case utility. Meas Bus Excell 9:33–41
Segalàs J, Ferrer-Balas D, Mulder KF (2008) Conceptual maps: measuring learning processes of engineering students concerning sustainable development. Eur J Eng Educ 33:297-306
Seiffert MEB, Loch C (2005) Systemic thinking in environmental management: support for sustainable development. J Clean Prod 13:1197–1202
Shahjahan RA (2012) Structure and agency in the neoliberal university
Shi H, Lai E (2013) An alternative university sustainability rating framework with a structured criteria tree. J Clean Prod 61:59–69. https://doi.org/10.1016/j.jclepro.2013.09.006
Sikdar SK (2003) Sustainable development and sustainability metrics. AIChE J 49:1928–1932
Sikdar SK, Sengupta D, Harten P (2012) More on aggregating multiple indicators into a single index for sustainability analyses. Clean Technol Environ Policy 14:765–773
Silva DAL, Delai I, de Castro MAS, Ometto AR (2013) Quality tools applied to cleaner production programs: a first approach toward a new methodology. J Clean Prod 47:174–187
Soini K, Jurgilevich A, Pietikäinen J, Korhonen-Kurki K (2018) Universities responding to the call for sustainability: A typology of sustainability centres. J Clean Prod 170:1423–1432. https://doi.org/10.1016/j.jclepro.2017.08.228
Staniškis JK (2020) Future is now: science for achieving sustainable development– the global sustainable development report 2019. In: Zilahy G (ed) Sustainability in transforming societies: proceedings of the 26th annual conference of the international sustainable development research society (ISDRS 2020). Budapest University of Technology and Economics, Budapest, Hungary, pp 831–839

Staniškis JK (2016) Sustainable university: beyond the third mission. Environ Res Eng Manag Technol 72:8–20

Staniškis JK (2012) Sustainable consumption and production: how to make it possible. Clean Technol Environ Policy 14:1015–1022

Staniškis JK, Arbačiauskas V (2009) Sustainability performance indicators for industrial enterprise management. Environ Res Eng Manag 48:42–50

Staniškis JK, Arbačiauskas V (2011) Integrated theoretical and practical M.Sc. education for sustainable industrial development. In: World trends in education for sustainable development. Peter Lang Pubisher, pp 105–115

Staniškis JK, Arbačiauskas V, Varžinskas V (2012) Sustainable consumption and production as a system: experience in Lithuania. Clean Technol Environ Policy 14:1095–1105. https://doi.org/10.1007/s10098-012-0509-y

Staniškis JK, Jasch CM, Stasiškienė Ž (2005) Assessment of environmental costs for sustainable industrial development: a monograph. Technologija, Kaunas

Staniškis JK, Jayaraman C (2010) Cleaner production and energy conservation for sustainable development. Daya, Delhi 288:

Staniskis JK, Katiliute E (2017) Systems approach to resource efficient and cleaner production solutions: method and implementation. In: Sustainability Through Innovation in Product Life Cycle Design. Springer, pp 385–398

Staniškis JK, Katiliūtė E (2019) Unsustainability reduction in enterprises by incremental innovations implementation and management. J Clean Prod 236:117542. https://doi.org/10.1016/j.jclepro.2019.07.017

Staniškis JK, Katiliūtė E (2016) Complex evaluation of sustainability in engineering education: case & analysis. J Clean Prod 120:13–20

Staniškis JK, Staniškienė E (2019a) Continuous Application of Preventive Environmental Strategies as a Way to Introduce Social Responsibility in Companies. In: Leal Filho W (ed) Social Responsibility and Sustainability. World Sustainability Series. Springer, Cham, pp 189–201

Staniškis JK, Staniškienė E (2019b) Innovative approaches to learning sustainable development. Encyclopedia of sustainability in higher education. Springer International Publishing, Cham, pp 1–8

Staniškis JK, Stasiškienė Ž, Jasch C (2008) Cleaner technologies: environmental management accounting, investment appraisal and financing. Nova Science Publishers

Staniškis JK, Stasiškienė Ž, Kliopova I, Varžinskas V (2010) Sustainable innovations in Lithuanian industry: development and implementation. Technol Kaunas 458

Staniškis JK, Varžinskas V (2010) Life cycle based design and product development: application of LCA to Lithuanian Industry. Environ Res Eng Manag 54:43–53

Svanström M, Lundqvist U, Kjällstrand J, Blomqvist M (2013) Integrating Education for Sustainable Development into the Environmental Management System–Experiences from Chalmers University of Technology. In: Proceeedings to Engineering Education for Sustainable Development 2013 (EESD13), Cambridge, England, UK, Sep 22–25 2013. Cambridge, UK, pp 635–642

ter Bogt HJ, Scapens RW (2012) Performance management in universities: effects of the transition to more quantitative measurement systems. Eur Account Rev 21:451–497. https://doi.org/10.1080/09638180.2012.668323

Thabrew L, Wiek A, Ries R (2009) Environmental decision making in multi-stakeholder contexts: applicability of life cycle thinking in development planning and implementation. J Clean Prod 17:67–76

Thomson G (2013) System integration for sustainable outcomes a proposed curriculum. In: Proceeedings to engineering education for sustainable development 2013 (EESD13), Cambridge, England, UK, Sep 22–25 2013, pp 650–660

Topfer K (2000) The triple bottom line: economic, social, natural capital. UN Chron 37:39

Trencher G, Yarime M, McCormick KB et al (2014) Beyond the third mission: exploring the emerging university function of co-creation for sustainability. Sci Public Policy 41:151–179. https://doi.org/10.1093/scipol/sct044

Tynjälä P (2010) Workplace learning in transformation. Keynote presentation at the EARLI learning and professional development SIG conference. Munich, Germany, pp 25–27
United Nations Economic Commission for Europe (UNECE) (2013) Empowering educators for a sustainable future. Tools for policy and practice workshops on competences in education for sustainable development. Geneva
Valderrama A, Jorgensen U, Mathiesen B V., Remmen A (2013) How is sustainability incorporated into the engineering curriculum? The case of DTU and AAU. In: cedings of Engineering Education for Sustainable Development EESD13. pp 291–302
van Duinen RJ (1998) European research councils and the triple helix. Sci Public Policy 25:381–386. https://doi.org/10.1093/spp/25.6.381
VanDoren V (2014) Fundamentals of cascade control. Control Eng 1–4
Varžinskas V, Staniškis JK, Knašytė M (2012) Decision-making support system based on LCA for aseptic packaging recycling. Waste Manag Res 30:931–939. https://doi.org/10.1177/0734242X12448519
Veleva V, Ellenbecker M (2001) Indicators of sustainable production: framework and methodology. J Clean Prod 9:519–549
Viluksela P (2009) Environmental indicators in heatset offset printing. In: Proceedings of the 5th International Conference, EMAN
Wangel J, Höjer M, Pargman D, Svane Ö (2013) Engineers of the future: using scenarios methods in sustainable development education. In: Engineering Education for Sustainable Development, Cambridge, UK. September 22–25, 2013
Williams A, Kennedy S, Philipp F, Whiteman G (2017) Systems thinking: a review of sustainability management research. J Clean Prod 148:866–881
Wulf A (2016) The Invention of Nature. The adventures of alexander von humboldt, the lost hero of science. John Murray
Yilmaz AK, Flouris T (2010) Managing corporate sustainability: Risk management process based perspective. African J Bus Manag 4:162–171
Yusuf S (2007) University-industry links, policy dimensions. How Univ Promot Econ growth 1–25
(2007) The university of the future. Nature 446:949–949. https://doi.org/10.1038/446949a

Chapter 3
Socio-Environmental-Economic Transformations Towards Sustainable Development

Jurgis Kazimieras Staniškis

> Many people assume, wrongly, that a company exists simply to make money. While this is an important result of a company's existence, we have to go deeper and find the real reasons for our being... People get together and exist as... a company so that they are able to accomplish something collectively that they could not accomplish separately – they make a contribution to society.—David Packard, US business leader

3.1 Agenda 2030 and SDGs

> Let us take our stand here in this Assembly of Nations. And let us see if we, in our own time, can move the world to a just and lasting peace.—John F. Kenedy, former US President

Sustainable development is a way to understand the world as a complex interaction of economic, social, environmental, and political systems. It is also a normative or ethical view of the world, where the basic point of sustainable development in that normative sense is that it urges us to have a holistic vision of what a good society should be. Thus, "from a normative perspective we could say that a good society is not only prosperous society (with high per capita income) but also one that is socially inclusive, environmentally sustainable, and well governed" (Sachs 2015).

The call for SDGs is a potentially historic decision, a powerful way to move to a new global agenda that engages not only governments but also businesses, scientists, leaders of civil society, NGOs, and, of course, students everywhere. Unlike the MDGs, which apply largely to poor countries and refer to rich countries mainly as donors, the SDGs are universally applicable. "The United States, just like Mali, needs to learn to live sustainably! The rich countries like the poor have to promote more social inclusion, gender equality, and of course energy systems that are low carbon and resilient" (Sachs 2015).

The 193 countries of the UN General Assembly agreed on the 2030 Agenda, seventeen Sustainable Development Goals and adopted them on September 25, 2015. They embrace all three dimensions of sustainable development: economic, social,

J. K. Staniškis et al., *Transformation of Business Organization Towards Sustainability*, World Sustainability Series, https://doi.org/10.1007/978-3-030-93298-5_3

and environmental. In addition to the SGDs, the UN General Assembly also adopted 169 targets distributed among the seventeen goals.

Goal 1. *End extreme poverty.* The main task of the goal is to "end poverty in all its form" ensuring that by 2030 all people will be living above the line of extreme poverty.

Goal 2. *End hunger and promote sustainable agriculture.* This goal comprises the end of hunger, improved nutrition, and resilient and less destructive farming system.

Goal 3. *Ensure healthy lives for all.* By the end of 2030, all countries have to reduce children under five mortality rate to below 25 deaths per 1,000 live births, maternal mortality to below 70 per 100,000 live births and to ensure universal health coverage, including treatment of many kinds of non-communicable diseases.

Goal 4. *Ensure quality education and lifelong learning.* The goal calls countries for universal coverage of access to and quality of all levels of education starting from pre-school through at least secondary education, and then on to more advanced skills training.

Goal 5. *Achieve gender equality and empower women and girls.* The main aim of the goal is to end all forms of discrimination against women and girls globally and to ensure their equal economic and social rights.

Goal 6. *Ensure availability and sustainable management of water and sanitation.* The goal seeks that every country should ensure that every person has access to safe and affordable drinking water, sanitation, and hygiene together with sufficient water usage and reduction of water pollution.

Goal 7. *Ensure access to affordable, sustainable modern energy.* The goal encourages countries to end as soon as possible the lack of access to electricity, provide safe and efficient cooking fuels based on low-carbon power.

Goal 8. *Promote sustainable economic growth and decent work for all.* The main aim of this goal is to promote sustainable economic development ensuring the raise of incomes per person, decent work, labour rights, and end of modern slavery and human trafficking. The goal stresses the necessity to enable today's poor countries to narrow the economic and social gap compared to high-income countries.

Goal 9. *Build resilient infrastructure and promote sustainable industrialization.* The goal encourage countries to ensure that infrastructure should be sustainable, meaning resilient to environmental stresses, and also green in that it imposes minimal pressure on the natural environment. They are also urged to speed the development of new sustainable technologies and their diffusion in low-income regions.

Goal 10. *Reduce inequalities within and among countries.* The main task of this goal is significant reduction of the large income gap between the rich and the poor and thus increase of social stability, social trust, and protection of individual rights of the poor.

Goal 11. *Make cities and human settlements sustainable.* The goal encourages improved cities and settlement planning and cities partnership towards sustainable development. The management of cities should be aimed at protecting world's cultural heritage and increasing health protection and prevention of natural disasters.

Goal 12. *Ensure sustainable consumption and production.* The main task is dramatic reduction in waste generation by waste prevention and circular economy methods. This goal also includes sound management of industrial chemicals, reduction of food waste, and change of consumption and production behaviour.

Goal 13. *Take urgent action to combat climate change and its impact.* The goal calls countries for decisive actions towards reduction of greenhouse gas emissions and increase of climate resilience. High-income countries are invited to help the developing countries to confront climate change financially and technically.

Goal 14. *Conserve the oceans and marine resources.* The goal stresses urgent actions in protecting the world's marine environment from human activities like overfishing, degradation of surface waters and coastal ecosystems by preventing pollution and ocean acidification caused by fossil fuel use.

Goal 15. *Protect and restore terrestrial ecosystems, sustainably manage forests, and halt biodiversity loss.* The targets of this goal comprise the conservation of terrestrial ecosystems, protection of biodiversity and habitats, prevention of invasive species, sustainable management of forests, and combating of desertification. The goal also encourages countries to mobilise new financial resources to protect biodiversity.

Goal 16. *Promote peaceful and inclusive societies.* The main aim of this goal is to reduce the violence, to end human trafficking, to defend the rule of law, to protect the fundamental freedoms, and to promote inclusive and transparent institutions of governance.

Goal 17. *Strengthen the means of implementation and revitalize the Global Partnership for Sustainable Development.* The goal presents the main global steps needed to promote and implement the Sustainable Development Goals. It covers financing, innovations and technology development and transfer, capacity building and multi—stakeholder partnership, rule-based trading, and data and monitoring of implementation of SDGs.

Since the adoption of SDGs, there have been many positive developments; however, despite the initial efforts, the world is not on track for achieving most of the targets comprising the Goals. Limited success in progress towards goals raises strong concerns and sounds an alarm for the international community. Why do goals matter? "There are many answers to this question. First, goals are critical for social mobilization. The world needs to be oriented in a direction to fight poverty or to help achieve sustainable development but it is very hard in our noisy, disparate, divided, crowded, congested, distracted, and often overwhelmed world to mobilize any consistency of effort to achieve any of our common purposes. Stating goals helps individuals, organizations, and governments all over the world to agree on the direction" (Sachs 2015).

The UN IGS Global Sustainable Development Report states that co-benefits, trade-offs and tough choices are at the heart of sustainable development but have not always been appreciated as such. Initial interpretations that emphasised three distinct dimensions of sustainability—economic, environmental and social–tended to reinforce decision-making in thematic silos. The result, typically, was to prioritise immediate economic benefits over social and environmental costs that would materialise over the longer term (IGS 2019).

Sustainable consumption and production is reflected as a cross-cutting theme in several UN documents and is often highlighted in sustainability initiatives for a broad range of sectors and issues. In other words, the patterns of consumption and production determine the degree of sustainability in many areas, for instance, energy production is related to CO2 emissions, industrial pollution affects water, air and soil quality, wood production and mining could lead to desertification and land degradation, gender equality and education are linked to access to resources and better quality of life, etc. In this situation, the analysis based on system approach could make an important contribution to SDG implementation by emphasising that they are interlinked and that progress solely focused on one goal could have consequences for the other. One of the surprising outcomes of such analysis is the claim that reducing poverty over 2000–2015 may have come at the expense of making our economy less sustainable (Barbier and Burgess 2017).

3.2 Global Sustainable Development Report: "Future is Now–Science for Achieving Sustainable Development"

> What lies in our power to do, it lies in our power not to do—Aristotle, Greek philosopher, 350 BC

The United Nations Member States at the 2016 High-Level Political Forum took a decision to organise an Independent Group of Scientists, comprising 15 of the most world-renown experts in sustainable development science from various regions of the Planet. At the very end of 2016, the UN General Secretary Ban Ki Moon, after a careful selection procedure, personally appointed the Independent Group of Scientists (IGS) aimed at preparation of the Global Sustainable Development Report (GSDR) on fundamental transformations in order to reach the sustainable future outlined in the 2030 Agenda.

The SDGs framework addresses key systematic barriers to SD such as inequality, unsustainable consumption patterns, weak institutional capacity, and environmental degradation.

At the conceptual level, three main questions needed to be addressed in the Report:

- who are those being or at risk of being left behind?
- how can strategies and policies reach them in practice?

– what type of strategies and policies would be appropriate in order to leave no one behind?

In implementing the Agenda, countries and stakeholders will have to make choices on where, when and how to act.

The Independent Group of Scientists was supported by a task team, co-chaired by one representative each of the United Nations Secretariat, UNESCO, UNEP, UNDP, UNCTAD, and the World Bank.

The Report will provide guidance on the state of global sustainable development from a scientific perspective, which will help address the implementation of the 2030 Agenda,

– provide lessons learned, while focusing on challenges, address new and emerging issues and highlight emerging trends and actions,
– focus on an integrated approach and examine policy options
– with a view to sustaining the balance between the three dimensions of sustainable development. These policy options should be in line with the 2030 Agenda to inform its implementation
– focusing on regional dimension, as well as on countries in special situations.

In other words, the mandate of the Independent Group of Scientists was to produce the Global Sustainable Development Report, incorporating scientific evidence in a transdisciplinary manner and considering all three dimensions of sustainable development, i. e. economic, environmental and social, in order to reflect the universal, indivisible and integrated nature of the 2030 Agenda. The Group has addressed sustainable development, both as a scientific and a normative concept, using it as a guide to analyse the problem and weigh the evidence, and where needed, recommend policy-relevant solutions. The normative approach is a value-based approach to building communities, founded on the assumption that all people have a need to belong, want to have a sense of purpose, and want to experience success. This gives every individual ownership in the community. For that purpose, the "Report follows not just letter but also the spirit of the 2030 Agenda, with the overarching goal of advancing human well-being in an equitable and just fashion, and ensuring that no one left behind, while the natural systems that sustain us, are safeguarded" (IGS 2019).

The so-called developed world has seen average improvements in many areas that are important to "good life". However, along with these improvements have come worrying indications that this growth has costs we cannot continue to ignore. Something has to be done to change the development—its philosophy, structure and methods—if societies want to reverse those negative trends.

Entitled "The Future is Now: Science for Achieving Sustainable Development", the Report found that the current world development model was not sustainable, and the progress made in the last two decades was in danger of being reversed through the worsening social inequalities and potentially irreversible declines in the natural environment that sustains us. Assessments show that, under current trends, the world's social and natural biophysical systems cannot support the aspirations for universal human well-being embedded in the Sustainable Development Goals. The scientists concluded that a far more optimistic future is still attainable, but only by drastically changing the development policies, incentives, and actions. Scientific evidence is a prerequisite for designing and implementing the transformations to sustainable development. While benefiting from all inputs, the content of the Report is the sole responsibility of Independent Group of Scientists.

In the Report, the model for transformations comprising six entry points and four levers was elaborated.

The Report used the latest scientific assessments, including personal investigations, evidence bases about good practices, and scenarios that linked future trajectories to current actions to identify new urgent calls to action by a range of stakeholders, including scientists that could accelerate the progress towards achieving the Sustainable Development. The model of the transformations for country/region-specific pathways generation was elaborated. An important key for the model and system is to recognize that, while the present state of imbalance across the three dimensions of sustainable development arises from not having fully appreciated the interlinkages across them, or having unduly prioritised the short term, it is the same interlinkages that will lead to the desired transformative change, when properly taken into account. That basic understanding has guided the IGS investigation's concept and structure, leading to the identification of knowledge-based transformations for achieving sustainable development. Accordingly, six entry points that offered the most promise for achieving rebalancing at the scale and speed needed for the 2030 Agenda were identified:

- Human well-being and capabilities;
- Sustainable and just economies;
- Food systems and nutrition patterns;
- Energy decarbonisation and universal access;
- Urban and peri-urban development;
- Global environmental commons.

These entry points are the means to harness important synergies, multiplier effects and trade-offs across the Sustainable Development Goals seeking to accelerate the progress. Countries and subnational entities could then develop acceleration roadmaps/pathways based on the scientific evidence most relevant to their circumstances and context. Below, a more detailed description of entry points is presented.

Advancing *human well-being* that includes material well-being, health, education, voice, access to a clean and safe environment, and resilience, is at the core of transformations towards sustainable development. Effective action in any of those areas requires acknowledging and addressing the links among them. Pathways to advance human well-being ultimately require cooperation, collaboration, and dialogue among multiple actors and employing many actions of change. There is no single pathway, and different combinations of efforts are required across regions and for countries in special situations.

Decoupling the benefits of *economic activity* from its costs at all levels is essential and can also support the systemic transformations envisaged through the other five entry variables. Such an outcome would greatly accelerate the necessary reconfiguration and help to put people, societies, and nature on the path to sustainable development. Urgent switch from patterns of economic growth, production, and consumption that perpetuate deprivations, generate inequalities, deplete the global environmental commons and threaten irreversible damage is needed. Transitioning towards long-term decarbonised and sustainable development that maximises positive human impacts, equalises opportunities among social groups and women and men, and minimises environmental degradation is essential. A significant part of the transformation will come from changing volumes and patterns of investment—both public and private.

One of the fundamental questions is about who owns the land and whether, if land is a gift from nature, it should be allocated differently from products bought and sold on the market. The idea of land value tax was already taken up in some countries and is currently under discussion in other countries as well. So far, technological innovations in *food production* methods are prerequisites for transitioning towards environment-friendly and healthy production systems. However, technologies alone cannot deliver the transformation. Policy, institutional and cultural changes are needed to enable more equitable global access to nutritional foods and to promote agroecological practices that are deeply rooted in local and indigenous cultures and knowledge, and based on small- and medium-scale farms that have temporal and spatial diversification and locally adapted varieties and breeds that can be strongly resistant to environmental stress.

Technologies already exist *for energy decarbonisation*; however, progress has been hampered by slow progress in smart-grid management and long-term electricity storage. The potential for progress is clear. The convergence of less expensive renewable energy technologies, digital applications and the rising role of electricity is a crucial vector for transformations. Solutions need to be context-specific with energy mixes, including decentralised renewable energies, emerging from the disruptive changes in energy production and consumption and presenting significant transition risks to long-term fossil fuel infrastructure investment.

Urban and peri-urban development should proceed in a well-planned integrated and inclusive manner. Governments, businesses, civil society organisations, and individuals can use a range of policy, economic and communication tools to promote sustainable consumption and production patterns trough well-planned land use, effective urban public transport systems, rapid scale-up of renewable energy and its efficiency, and promotion of sustainable and technology-enabled businesses and jobs. A liveable city will foster "naturbanity"—a close connection between people and nature to protect biodiversity, enhance human health and well-being, and strengthen climate resilience (IGS 2019).

The *global environmental commons* comprising the atmosphere, the hydrosphere, the global oceans, the cryosphere, polar regions and natural resources systems such as forests, land, water and biodiversity, are vital for human survival and well-being. Since they are intrinsically linked to one another, achieving sustainability of the

Earth 's systems requires anticipating feedback effects among commons in order to maximise co-benefits and minimise trade-offs, both locally and globally. At all levels, it is essential to reverse the trend of overexploitation, which must be managed within boundaries that maintain the resilience and stability of natural systems, and allow for the natural renewal of resources. The management/control should avoid maldistribution and seek to repair the damage already caused by poor technical, financial and political interventions, especially where indigenous communities and other vulnerable groups are concerned with concerted efforts to leave no one behind.

However, the entry points alone will not be sufficient for the transformations system, especially if actions do not adequately address global interconnections or take full account of the non-economic, but intrinsic value of nature.

Therefore, four levers were identified:

- Governance;
- Production and consumption, finance;
- Individual and collective action;
- Science and technology.

Each lever is a powerful agent of change in its own right and impact the Goals through the identified entry points. It should be pointed out, however, that true transformation is possible only when the levers are deployed together in an integrated and intentional manner. Transformations differ from evolutionary or chaotic change, in that they are intentional changes based on societal agreement and factual understanding, and achieve outcomes at scale.

The model and system for transformations developed by the UN Independent Group of Scientists and presented in the Global Sustainable Development Report clearly show the science-based evidence that illuminate more productive sustainable paths enabling future generations to live within the limits of the Earth's ecosystem. The need for this is critical, and action must be bold and decisive, not just for change but for systemic transformations.

The Report clearly demonstrates that sustainability science can help tackle the trade-offs and contested issues involved in achieving sustainable development. The UN Independent Group of Scientists provided a successful interdisciplinary research process showing that individual researchers and research initiatives in different fields should become part of larger collective research projects and holistic programmes. Only long-term research partnerships, including financing, can identify socially relevant questions, generate meaningful insights, and bridge the gap between knowledge and action. Simultaneously, to realise sustainability science potential, significant adjustments to universities and other research and training institutions are needed.

Using this model, each organisation/country/region responding to its own conditions and priorities has a real possibility to identify specific innovative pathways towards sustainability. This is extremely important for countries in transition to rapidly implement the integrated pathways to sustainable development that correspond to the specific needs and priorities and also contribute to the necessary global transformation. Positive reaction to the Report at the UN General Assembly on

September 24–25, 2019 shows that the leaders of the UN member-states clearly understood that there was no one-size-fits-all solution, so governance approaches, as stressed and showed in the Report, needed to be diverse, tailored, innovative and adaptive, using science to support the decision–making and create institutions that deal with uncertainties and risks as well as systems for monitoring and evaluation.

The Report uses the latest scientific assessments, evidence bases of best practices, and scenarios that link the future trajectories to current actions to identify the calls for action by a range of stakeholders that can accelerate the progress towards achieving the sustainable development.

However, as it was mentioned above, despite the initial efforts, the world is not on track to achieving most of the 169 targets that comprise the Goals for Sustainable Development. Alarming is the fact that recent trends along several dimensions with cross-cutting impacts along the entire 2030 Agenda are not even moving in the right direction.

Paralel with GSDR, the Report "Transformations to Achieve the Sustainable Development Goals" was published in which six SDG transformations as modular building-blocks of SDG achievement were introduced: (1) education, gender and inequality; (2) health, well-being and demography; (3) energy decarbonisation and sustainable industry; (4) sustainable food, land, water and oceans; (5) sustainable cities and communities; (6) digital revolution for sustainable development (TWI2050–The World in 2050 2018).

The transformations for the World in 2050 are based on five principles: (1) mutually exclusive and collectively exhaustive; (2) system based; (3) aligned with government; (4) easily communicable; (5) few in number. As one of the main conclusions related to the implementation authors argue that "transformations can only succeed if they enjoy social legitimacy, so political processes should engage the public in participatory decision-making and promote transparency and accountability... they can be supported by governments, through transdisciplinary research and education" (Sachs et al. 2019).

3.3 Mathematical Formulation of the Transformations Problem

> The test of first-rate intelligence is the ability to hold two opposed ideas in the mind at the same time, and still retain the ability to function.—F. Scott Fitzgerald, writer, US

The biggest transformative potentials of the 2030 Agenda do not lie in pursuing single Goals or targets but rather in a systemic approach that manages their myriad interactions.

Progress on any Sustainable Development Goal will depend on a range of interactions with other Goals that either support achievement through benefits or hinder it through trade-offs. Progress on all the Goals will only be achieved if important trade-offs are addressed and transformed, and if co-benefits are deliberately realised,

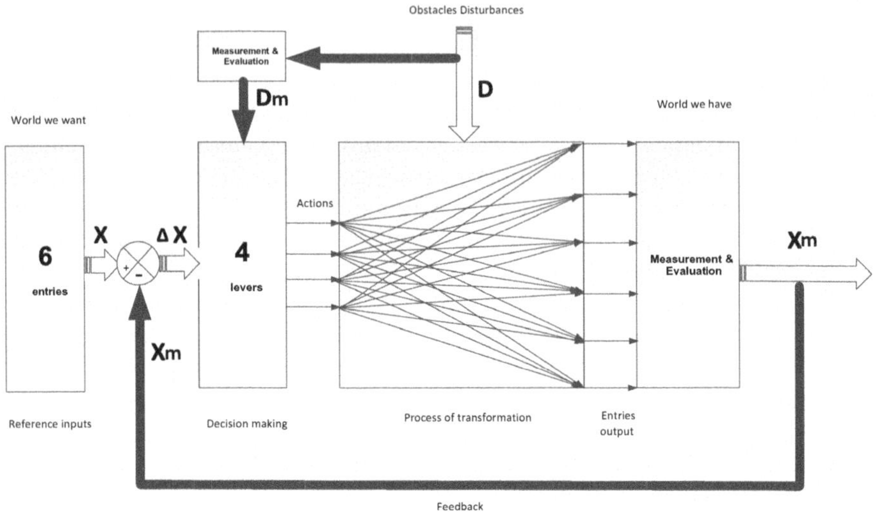

Fig. 3.1 The structure of the system for the transformations towards sustainable development (Staniškis 2020)

managing the arrows in the system that is more important than managing the boxes of individual targets (see Fig. 3.1). A system is an interconnected set of elements that is coherently organised in a way that achieves something. A system is more than the sum of the parts and may exhibit adaptive, dynamic, goal seeking, self-preserving and sometimes evolutionary behaviour.

The most efficient—or sometimes the only— way to make progress on a given target is to take advantage of positive synergies with other targets while resolving or ameliorating the negative trade-offs with yet others. As it was mentioned before, six entry points that offer the most promise for achieving the desired transformations at the necessary scale and speed were identified. These are not entry points into individual or even clusters of Goals, but rather into the underlying systems. It means that from systems point of view, the state of transformations is defined by state vector ***X*** comprising six variables ***X*** (x1, x2, x3, x4, x5, x6), where:

- x1—human well-being and capabilities,
- x2—sustainable and just economies,
- x3—food systems and nutrition patterns,
- x4—energy decarbonisation with universal access,
- x5—urban and peri-urban development.
- x6—global environmental commons.

In order to achieve successful transformations, the vector of actions/levers ***U*** should be determined. In this case, there are four types of levers/actions identified:

- u1—governance,
- u2—economy (incl. consumption and production) and finance,

- u3—individual and collective action,
- u4—science and technology.

Governance is recognised an essential action/lever in the system of transformations needed to achieve the sustainable development goals. For that, new integrated approaches that take into account systemic interactions and causal relationships between goals and policies will be needed. Effective, transparent, accessible and inclusive institutions will form the cornerstone of governance by goals. Scientific and research communities should offer evidence-based options for actions, taking advantage of the latest technologies and providing an important perspective on the potential and pitfalls of various governance perspectives. At the same time, governments need to invest in knowledge systems—indicators, data, assessments, and sharing platforms.

Economic policy and financial flows are key actions/levers for achieving transformations towards sustainable development. Economic actions typically encompass fiscal, monetary and trade policy, while financial flows include flows from public and private sources, within and across national borders. Policies that encourage trade in sustainably produced goods and services with fair prices, decent labour conditions and wages, and environmentally friendly production techniques can significantly boost progress on the implementation of sustainable development goals. Long-term decision-making combined with capital account management methods can help reduce volatility, which is important for ensuring resilience against shocks and providing consistent and predictable public spending for social welfare programmes. As it was stressed before, we are locked into unsustainable patterns of production and consumption through a complex mixture of factors, some of them institutional, some of them economic, some of them to do with infrastructure or lack of it, and some of them social or psychological in nature. Transformative change of existing patterns of consumption and production towards greener, longer-lasting and recycled goods and services will provide a better quality of life with a smaller environmental footprint.

Encouraging and enabling people to contribute *individually or collectively* expand resources for sustainable development and advance human ingenuity for innovation. Transformative change requires the reconfiguration of social norms, values, and laws that promote unsustainable or discriminatory behaviour and choices. There are many mechanisms for empowering people, changing behaviours and expanding space for collective actions. Recent research indicates that individual competencies to make qualitatively different decisions that will accelerate the transition to sustainability vary and are strengthened by the provision of supportive and stimulating environmental conditions in early childhood, formal and informal high quality education, and lifelong learning. Increasing the civic space for people to organise and participate in public dialogue and decision-making increases the likelihood of arriving at representative outcomes.

Thus, the system for transformations towards sustainable development (see Fig. 3.1) represents the most important closed loop feedback-feedforward control system that is the most widely used in current practice. The feedback-feedforward method allows for detecting serious deviations prior to their effect on the system, and feedback can counteract any imperfections in the feedforward scheme and the effects of other (unmeasured of measurable) disturbances. In practice, it is usually necessary to build up a complex set of feedback, feedforward adaptive control systems capable of representing the whole range of system disturbances and changes, which may occur. Transformations control systems of this type are capable of using past behaviour to modify future response and/or the performance measure. It is important to understand that whilst feedback control systems are very useful in industrial transformations, they are far less appropriate to economical, environmental, and social systems since the fundamental human ability to anticipate is ignored. Hence, it is preferable in most cases to attempt at detecting disturbances ***D*** before they occur and before deviation from the performance index/target value can happen. Thus, the main advantage of the feedforward action is that it uses the human ability to anticipate the course of events and to **prevent** disturbances ***(for the business organisation, the main disturbances D are obstacles and drivers that are explored in detail by the authors and presented in Part 2 of the book).*** The most serious disadvantage of feedforward control is that there is no comparison of the system output with the reference performance index to determine how far the control is effective. Therefore, the feedback-feedforward system for transformations towards sustainability control was developed where two actions are considered: vector of uncontrolled inputs ***D*** and vector of controlled actions ***U***. Actions vector represents a set of actions/levers that can be manipulated by the decision-maker to influence the system of transformations in order to approach the entries output more closely to reference settings. Mathematically, the dynamic model of transformations could be expressed by the system of the first-order differential equations:

$$
\begin{aligned}
dx1/dt &= f1(x1, \ldots, x6,\ u1, \ldots, u4, t)\\
dx2/dt &= f2(x1, \ldots, x6,\ u1, \ldots, u4, t)\\
dx3/dt &= f3(x1, \ldots, x6,\ u1, \ldots, u4)\, t)\\
dx4/dt &= f4(x1, \ldots, x6,\ u1, \ldots, u4, t)\\
dx5/dt &+ f5(x1, \ldots, x6,\ u1, \ldots, u4, t)\\
dx6/dt &= f6(x1, \ldots, x6,\ u1, \ldots, u4,\ t)
\end{aligned}
$$

In many real cases, it is essential to construct a model of the real object and, on the basis of this model and of the existing information, monitoring and estimation systems, to generate forecasts of the future behaviour of the object. From these forecasts, by using a feedforward loop, it is possible to anticipate future object deviations before they occur. This forecast is of greatest utility for example, for environmental and social objects, with long lag, reaction and relaxation times and where there are significant delays in the provision of information on transformation system behaviour and performance. Without forecast by feedforward loop, the transformation system can act only as a feedback control system which is capable of counteracting deviations from goal only after they have occurred.

Performance criteria or control targets. As usual, these are quantifiable policy objectives which can be applied to the transformations systems to achieve sustainable development. There are many different performance indexes and the most common of those are:

Static economic efficiency criterion—the goal is to achieve the proper level of utilisation of resources and effective allocation of resources between competing users. It means that marginal rates of substitution in consumption should be equal for all users; marginal rates of transformation and substitution in production must be equal for all firms; and the marginal rate of substitution in consumption must equal the marginal rate of transformation in production. Full employment is a special case of maintaining supply–demand equilibrium.

Growth criterion—the goal expressing extensive and/or intensive growth. Intensive growth generally is more preferable in which productivity is increased by economies of scale through managerial and technological measures. A special feature of the achievement of economic growth is the accumulation of capital, because only by accumulation of capital the technical progress and increase in productivity can be achieved. It should be stressed that economical growth could destroy independent small business and cause negative environmental consequences.

Social welfare criterion—the goal is designated to increase the general well-being of the society. In most cases, the welfare is associated with economic growth assuming that the high level of national income provides more opportunities for social well–being. In the most cases, this is not true, because increased incomes could go to other expenditures or be spent for elimination/treatment of environmental damage. Secondly, it is very much dependent upon whose welfare is introduced into the growth index. Theoretically, improved social welfare could be achieved by redistribution of incomes, by political and institutional reforms, or by changing the pattern of consumption and production.

Equity criterion—the goal is to improve the distribution of the final output of economy. The equity index could be expressed differently. For instance, in a capitalist system state, income is an equivalent of society's value of contribution an individual makes to the economy and is determined by the market mechanism; in socialism, income is proportional to labour input, whereas with a communist state, the aim is to make income proportional to need (Bennett and Chorley 1978).

Biological criterion—it is designed to protect nature (flora, fauna, waters, air and earth resources) and counteract the damaging effect of human intervention. The biological goal is often ignored in socio-economic systems and in decision-making accordingly. However, it cannot be ignored and should serve as a constraint on growth, welfare, and productive efficiency.

Sustainability criterion should comprise all three dimensions of sustainable development: economical, environmental and social. The basic structure of the criterion is:

$max\,\mathbf{J} = economic\,benefit\ +/-\ environmental\,benefit/damage\ +/-\ social\,benefit/damage.$

Mathematically, it could be expressed in many different ways.

Constraints. There are two types of constraints: cost constraints and system behaviour constraints. The second type of constrains should be developed considering the essential relationship between human and natural systems. There are two broad types of this relationship: (I) the man stands apart from the rest of nature in which he operates; (II) man operates as a part of nature. This gives the rise of the two types of systems. Firstly, the intervention and secondly, the symbiosis. In case of intervention, a socio-economic system operates apart from physico–ecological systems to cause them react in a predetermined and advantageous manner. Such systems involve a considerable amount of monitoring and a rather analytical view of environmental system. Figure 3.1 demonstrates a real example of intervention system where monitoring and feedback loop adjustments related to the control of system elements exposed to environmental effects. It is clear that in such type of systems, the link between the elements and the action is involved first with sensors and then by means of complicated negative feedback loops to generate responses which provide the required control. The desired control/optimisation task can now be stated as maximising the performance criteria

$$J(x1, x2, x3, x4, x5, x6, u1, u2\,u3\,u4)\ -\mathbf{max}$$

subject to constraints

$$C(u1, u2, u3, u4).$$

The constraints $\boldsymbol{C}$, as it was mentioned above, will be either cost constraints or constraints coming from physico-ecological and socio-economic systems.

Each stage of decision-making process comprising information gathering, estimation, monitoring, forecasting, evaluation, and correction with respect to objectives has presupposed that the manager has the power to act. In practice, however, control of economic, environmental and social goals is highly constrained by the administrative, legal and political issues in which they are imbedded. In most complicated cases towards sustainability in which the executives have power over resources and environment or society itself, management must use a trade-offs strategy over a range of conflict issues.

Information monitoring and estimation. The aim of the monitoring system is to detect deviations from target values of the performance indicator in the control system. Therefore, we cannot collect information until we know the nature of the system and the information required. The type of system goals and methods by which control can be achieved are also important, especially the policy instruments that are going to be used.

The essential roles of any monitoring and assessment in control or decision making support systems are:

- to detect important economic, environmental and social problems at early stage in order to effectively use this information in the feedforward loop;
- to determine how far a given control/policy output is succeeding in meeting objectives in order to use it in the feedback loop;
- to determine what corrections/control actions are necessary to improve the policy or control system performance, and
- to provide the assessment of directly unmeasurable system outputs based on indirect measurements and indirect assessment model.

In unsustainability reduction systems, the decision making support systems require to quantify energy and material flow balances in order to identify and assess the source of waste or pollution generation.

The special importance of modelling, assessment and forecasting systems is simplification of the real world system in order to aid the decision-making and to allow the anticipation of changes and deviations within the system behaviour. A model as usual is a simplification of reality, as well as an abstraction; it omits a lot in order to distil its essence. Given the complexity of economy, a model is a necessary aid to understanding. Consequently, to criticise a model for its simplification is a mistake. We cannot expect a model to cover all aspects of reality, but nevertheless we expect it to capture its essence.

It is important to point out that system for the management of unsustainability reduction (see Fig. 2.1) and system for transformations (see Fig. 3.1) are interconnected in a hierarchical way, i.e. the system for transformations generates the reference input for the unsustainability reduction system.

Corporate systems, military systems, ecological systems, and living organisms are arranged in hierarchies. Hierarchies represent a brilliant systems invention, not only because they reduce the amount of information that any part of the system has to keep track off. In hierarchical systems, relationships within each subsystem are denser and stronger than relationships between subsystems. Everything is still connected to everything else, but not equally between subsystems. If these differential information links within and between each level of the hierarchy are designed right, feedback delays are minimised. No level is overwhelmed with information; thus, the system works with efficiency and resilience. Besides that, systems often have the property of self-organisation—the ability to structure themselves, to create new structure, to learn, diversify, and complexify. Resilience, self-organisation, and hierarchy are three of the reasons a dynamic system can work so well. Promoting or managing these properties of a system can improve its ability to function well over the long term—to be sustainable.

Transformations are not simple or painless, but rapid change can happen when actors work in an integrated way towards the agreed goals, sharing and applying available scientific technological and policymaking knowledge. Positive results of actions are further amplified when multiple outcomes are considered and evaluated by decision-makers, as opposed to working in silos in which only a single goal or outcome is considered. There is a rich store of scientific evidence, technologies, and knowledge-based solutions across disciplines and regions that must be mobilised to shape actions (IGES 2010).

The transformation process requires rethinking all concepts and ways of socio-environmental–economic life. Therefore, the discussions and some conclusions on socio-economic models, responsible consumption and sustainable production, role of markets, role of state, transformations in society, corporate social responsibility, nations' health, globalisation and localisation are presented below. The description of the topics mentioned above is made having in mind the possible structural transformations of our life and activities towards sustainability.

3.4 Review of Socio-Environmental-Economic Models and Systems

> Virtually every bad social trend from crime to obesity is strongly linked to the unequal distribution of income and opportunity, risk and reward. The fact is that the global capitalist economy is trending toward greater concentration of wealth and is therefore increasingly prone to lurch from crisis to crisis and to foster a growingly public disaffection.—Thomas Piketty, French economist

Eco-innovations, eco-efficiency and cleaner production are important for unsustainability reduction in enterprises; however, sustainable consumption and production, corporate social responsibility and socio–economic systems in general require transformative changes to achieve long-term social, environmental and economic sustainability.

Features of a route towards environmental and economic sustainability might be:

- A system that encourages minimising consumption, or imposes personal and institutional caps or quotas on energy, goods, water, etc.;
- A system designed to maximise societal and environmental benefit, rather than prioritising economic growth;
- A closed-loop system where nothing is allowed to be wasted or discarded into environment, which reuses, and remakes in preference to recycling;
- A system that emphasises delivery of functionality and experience, rather than product ownership;
- A system designated to provide fulfilling, rewarding work experiences for all that enhances human creativity/skills;
- A system built on collaboration and sharing, rather than aggressive competition (Bocken et al. 2014).

"In the beginning of 1980, a fundamental change in perspective started to emerge about what had enabled post-war economic growth. During the immediate post-war years, it was believed that increased economic prosperity was something that everyone had contributed to, and so it had to be shared by all. It was an industrial model of progress built on partnership between company owners and their workforces. By contrast, the growth phase of the 1980s was based more on market fundamentalism and individualism and less on state intervention or the building of a social contract. I think this was mistake" (Schwab 2021).

Looking at the way the world works through the lens of the new economics, there are some bizarre questions that seem to fly in the face of orthodoxy (generally accepted theory, doctrine, or practice):

- why do we work longer hours than some medieval peasants?
- why are the best mechanics in the world—Cubans?
- why do we export as many chocolate waffles out of the UK as we import?

Although GDP in the UK has doubled over the last 30 years, most measures of well-being remained steady or dipped down. The winners of the system are suffering from rising debt, rising stress, rising depression, and ill mental health. Twelve million people in Europe are involved some way in downshifting – earning less money for greater well-being—then you know the mainstream, which demands we should accelerate our earning and spending, has a problem (Boyle and Simms 2009).

Cuba demonstrated that it is possible to feed the population under extreme economic stress with very little fossil fuel input. As calorie intake fell by more than a third, and fuel was unavailable, the proportion of physically active adults more than doubled and obesity halved. Between 1997 and 2002, deaths attributed to diabetes fell by half, coronary disease by 35%, strokes and all other causes by one fifth. This required between 15 and 24% of the labour force to get involved in growing food, while in the UK, this number comprises less than 1% (Boyle and Simms 2009).

Old economics confines its interest to the point where money becomes involved and to the point when a product is thrown away, i.e. the narrow definitions of economics exclude the beginning and end of the product life style. Conventional economics assumes that the resources of the planet are infinite and nature's capacity to absorb waste is also infinite.

In other words, capitalism needs growth in order to survive, but economic growth is closely associated with an increase in the use of material resources, which has enabled us to live beyond our ecological means. However, many businesses and a number of sustainability advocates suggest that scientific and technological innovation, new modes of doing business and different ways of running economy can lead to a virtuous (high moral standards) circle of increased productivity, increased resource efficiency, and green growth (Rifkin 2014). Some argue that in order to create a more sustainable world we must abandon the ideology of continuing economic growth—whether in capitalism or socialism. Moreover, we must severely curtail the operation of the profit motive and the private accumulation of wealth, while seeking forms of economic ownership and control that respect ecological limits, valuing the significance of the commons while retaining the benefits of individual and collective enterprise and innovation (Reardon et al. 2018).

Jackson argues that "one thing is pretty clear. The fundamental changes to the economy that I have advocated (prosperity without growth) are totally incompatible with the "casino capitalism" or the "consumer capitalism" that has characterized the richest economies in recent decades. But this is not the same thing as saying that we have arrived at the end of capitalism entirely" (Jackson 2017). *("Casino capitalism" is an economic system in which private businesses, especially banks, risk large amounts of money on investments, shares,* etc.*)*

Standard models take no account of the use of finite resources and environmental constraints, and are blind to social outcomes in terms of equity and, of course, human well-being. Macroeconomic models are open-ended by nature, with growth being the primary output of interest. We need to reverse this. That is, to start with the hard outcomes we need: environmental sustainability, reverse engineering, equitable social and economic justice, and high levels of human well-being. We then propose to link these to relevant economic determinants within the model (aggregate output, income distribution and working hours, respectively, for example) and to 'reverse engineer' what this would imply for the levels and types of differing inputs. The role of enterprise, the quality of work, the structure of investment and the role of money supply taken together are elements, which hold the potential for radical transformation of consumer economy and offer solid formation for the economy of tomorrow (Jackson 2016). *(Reverse engineering, shortly, is the reproduction of another manufacturer's product following detailed examination of its construction or composition.)*

Norgaard sees the failure of the Soviet Union as significant because it was a major power to collapse while diligently pursuing modernity. Its version of progress ended up destroying people's creativity and initiative in bureaucracy, wasted soil in an effort to modernise agriculture, and polluted air and water to accelerate industrial development. "Democratic capitalist nations are wrong to interpret its failure as a victory for their version of modernity, not realizing that the differences in versions only amount to how quickly the breaking point is reached. Capitalism and communism both were the forms of "industrialism" different only as how best to divide the proceeds from exploiting the Earth" (Norgaard 1994).

"Communism and capitalism have many similarities, including their dependence on economic growth and industrial paradigm rooted in a rationalistic philosophy that regards the world as so much dead material waiting to be transformed in ways suitable for human use and then discarded without consequence. Both ideologies worship at the same altar, even if they argue who owns the church" (Piketty 2014).

Marx did not make a fetish of material production either. On the contrary, "Marx thought it should be done away with as far as possible. His ideal was leisure, not labour. If he paid such unflagging attention to the economic, it was in order to diminish its power over humanity. His materialism was fully compatible with deeply held moral and spiritual convictions. He lavished praise on the middle class, and saw socialism as inheritor of its great legacies of liberty, civil rights and material prosperity. His views on Nature were for the most part startlingly in advance of his time" (Eagleton 2018).

Karl Polanyi argued that the dehumanisation of capitalism was a result of the particular institutional construct of a market society, which of course is not natural, because land, labour and money are "fictitious commodities" and they were never intended to be produced and sold in market. Now, selling our labour for a wage in order to buy goods and services is a second nature for many of us, and we may think it normal and even natural. Life is not sustained by market forces but by a process of transferring aspects of society from generation to generation: in households, in communities, in society. Land is not simply a commodity, because it is part of nature (Polanyi and MacIver 1944). This is important since it suggests that we can develop new institutions to push the economy in a more human and sustainable direction (Standing 2016).

The decoupling of economic growth from resource use and environmental degradation has often been promoted as one of the key means for achieving sustainable consumption and production. But in reality, after more than two decades of international policy discussions on sustainability, there have been only a few examples achieving relative decoupling, when material consumption and associated environmental grew at less rapid rates than the economies. At the same time, there are no examples of absolute decoupling which is necessary to achieve sustainability. Thus, "decoupling approach remains largely theoretical, based on questionable assumptions, for example of rapid technological progress with limited undesirable side effects in the form of rebound mentioned above. Besides that, decoupling gives primacy to the economic dimension, because it is based on the assumption that economic growth can and should continue, seemingly ad infinitum, and it does therefore in practice assign a secondary role to other sustainability objectives" (Akenji and Bengtsson 2014).

Sharing economy or crowd capitalism is one of the new versions towards unsustainability reduction. "There are at least four key drivers in shifting to crowd–based capitalism impacts that shift to crowd-based capitalism is likely to have: (I) the changes in the impact of capital, (II) changes, in consumption driven by greater variety and different models of access, (III) changes in economies of scale and network effects, and (IV) the promise of inclusive growth. The motives for people engaging in crowd-based capitalism are not always purely monetary. This suggests a need to focus not just on economic growth as measured by income and dollar value output, but on economic development broadly defined. The key question that remains to be answered is whether or not these changes will ultimately create a better world of work, and what can we do to nudge things in the right direction (Sundararajan 2017).

David W. Orr suggested economics and resulting business practices oriented around four well-known principles. First, *the economy is a subsystem of the ecosphere* and is thereby bound by its limits and is subordinate to the bio-geo-chemical cycles, energy flows, and ecological functions that govern the Earth and the health of its constituent parts. The demands of the economy for resources and energy and for absorbing its waste products, including all chemicals and all of their various combinations, must not exceed what the larger system or its components' ecosystems can provide in perpetuity. From the other side, the prices we have been paying do not reflect the true costs of things purchased. If we finally and fully assessed the costs, for instance, of fishery collapse, soil erosion, desertification, loss of tropical and other forests, waste treatment, toxic releases, and mass extinction of species, the price tag would be completely different. It means that in a decent economy prices would tell the truth by including all costs and externalities. Second, *the economy is a means, not an end*. The purpose of a good economy is to provide and fairly distribute sustenance for living such as food, water, shelter, and healthcare. Beyond basic needs, it would further encourage the arts, beauty, kindness, solace, and conviviality for everyone, not just the well-off. There is evidence that capitalists will sell anything for a profit, but some things such as climate stability, human health and dignity, sacred groves, children, and grandmothers should not be for sale at any price. Good economy would not grow merely for its own growth. Third, *the economy must be non-violent*. Decent economy does not employ violence in its various forms without harming not only the people but also the social and ecological conditions that underwrite their well-being. We glorify violence in our movies, advertisements, politics, and sports, and all too easily overlook the violence happening in our names in different places around the world. British economist E. F. Schumacher wrote that "wisdom demands a new orientation towards the organic, the gentle, the non-violent, the elegant and beautiful" (Schumacher 1973). Fourth, *politics are more basic than the economy*. Neoclassical economists argue that economics is more fundamental than politics and that economics mostly determines our political reality. However, if democracy really matters, the rules that govern the economy should be made public by properly elected representatives serving an informed electorate, not by oligarchy meeting behind closed door. The public should participate in decisions having to do with the distribution of wealth, risk, reward, and the sustainability of the entire human enterprise. They should be made aware of the social, political, and ecological consequences of economic and fiscal decisions, taxation, purchasing, and investment. Unrestrained and minimally regulated capitalism has led relentlessly to greater concentration of wealth, economic monopoly, and ever larger ecological, social, and political crisis (Orr 2016).

Recently, Klaus Schwab (2021) introduced the concept of stakeholder capitalism instead of the shareholder capitalism. The well-being of people and the planet are natural stakeholders with people being simply all human individuals and planet being the natural environment we all share. Four key stakeholders can optimise the well-being of people and the planet. They are: governments, civil society, companies, and the international community. Each of these stakeholders has their own primary objectives:

- Companies pursue profits and seek long-term value creation;
- Civil society's primary aim is each organisation's purpose or mission;
- Governments pursue equitable prosperity; and
- The international community works for peace.

In a stakeholder model, all of these groups and their goals are interconnected. One cannot succeed if the others fail. The model is simple, but it immediately reveals why shareholder primacy and state capitalism lead to suboptimal outcomes: they focus on the more granular and exclusive objectives of profits or prosperity in a particular company or country rather than well-being of all people and the planet as a whole. By contrast, in the stakeholder model, neither of more granular objectives is set aside, but interconnectivity and the overarching well-being of people and the planet are central, ensuring a more harmonious outcome over time.

It looks like an approach and there are many questions on the decision-making process: the model, performance criteria, system boundaries and constrains, etc. Schwab explained it this way: "Let us consider how the decision-making process among stakeholders can function. It is easy to imagine this becoming a hodgepodge if no clear process and guidelines exist... It may well be that the interest of various stakeholders diverge or... most vocal stakeholders would try to monopolize or block decision–making, leading to a standstill or lopsided outcomes... The solution, I believe, lies in separating the consultative process from the decision-making one. In the consultative stage, all stakeholders should be included, and their concerns should be heard. In the decision-making stage, by contrast, only those mandated to make decisions should be able to do so, which means in the case of companies, respectively the board or executive management" (Schwab 2021).

Conclusion. "So capitalism, yes, but we need to end our love affair with unrestrained free-market capitalism, which has served humanity so poorly, and install a better-regulated variety. What that variety would be depends on our goals, values and beliefs" (Chang 2012). Development of productive capabilities, especially in the manufacturing sector, is crucial if we are to deal with the greatest challenge of our time—sustainable development, including climate change. However, the structure, performance index, and boundary conditions for consumption and production should be transformed according to the model based on systems theory and presented above. It is important to mention that economy's productive capabilities include the capabilities of non-enterprise actors such as the government, universities, research institutions, industry associations, training centres and, certainly, the economic institutions, which also facilitate production and improve the effectiveness. These organisations should be involved in the transformations as well and discussions on that follow in the next subchapters.

The quality of employment. In 1930, John Maynard Keynes predicted that, by the end of the century, technology would have advanced sufficiently so that countries like Great Britain or the United States would have achieved a 15-h work week. There is every reason to believe he was right. In technological terms, we are quite capable of this. And yet it did not happen. Instead, technology has been marshalled, if anything, to figure out ways to make us all work more. In order to achieve this, jobs have had to be created that are, effectively, pointless. "Huge swathes of people, in Europe and North America in particular, spend their entire working lives performing tasks they secretly believe do not really need to be performed. The moral and spiritual damage that comes from this situation is profound. It is a scar across our collective soul. Yet virtually no one talks about it" (Graeber 2013).

For employees, it is very important to reap the rewards of productivity growth in terms of reduced hours worked per person, sharing the available work amongst the workforce and shifting economic activity to more labour-intensive sectors. As a practical policy tool, post-Keynesians have suggested a so-called job guarantee, which would ensure that all people capable and willing to work would be able to get a permanent, state-funded, and locally administered job. The most suitable jobs for the programme would be those that almost anyone can do with short training. The jobs could be modelled after the needs of sustainability transition and adapted to climate change: for example, installing decentralised energy solutions and preparing for floods. In addition to getting the transition going, the job guarantee would ensure full employment. It would lessen insecurity and the need to compete for environmentally destructive jobs—on the individual and the collective level (Järvensivu et al. 2018). Living wage is a wage that enables an individual to live, or, in other words, a wage that allows one to adequately provision. A living wage attempts to connect the means (income) with the ends (capability). The right to earn a living age is a human right, included in the Declaration on Fundamental Principles and Rights at Work (International Labour Organization 1998). Of course, what is considered an acceptable quality of life is time- and place-specific.

Over the course of the last century, the number of workers employed as domestic servants, in industry and in the farm sector has collapsed dramatically. At the same time, "professional, managerial, clerical, sales, and service workers" tripled, growing "from one-quarter to three-quarters of total employment". But rather than "allowing a massive reduction of working hours to free the world's population to pursue their own projects, pleasures, visions, and ideas, we have seen the ballooning of not even so much of the 'service' sector as of the administrative sector, up to and including the creation of whole new industries like financial services or telemarketing, or the unprecedented expansion of sectors like corporate law, academic and health administration, human resources, public relations, and advertising. While corporations may engage in ruthless downsizing, the layoffs and speed-ups invariably fall on that class of people who are actually making, moving, fixing and maintaining things; through some strange alchemy no one can quite explain, the number of salaried paper-pushers ultimately seems to expand, and more and more employees find themselves effectively working 15 h just as Keynes predicted, since the rest of their time is spent organising or attending motivational seminars, updating their Facebook profiles or downloading TV box-sets. Once, when contemplating the apparently endless growth of administrative responsibilities in British academic departments, I came up with one possible vision of hell. Hell is a collection of individuals who are spending the bulk of their time working on a task they don't like and are not especially good at" (Graeber 2013).

There is a lot of questions one could ask here, starting with, what does it say about our society that it seems to generate an extremely limited demand for talented poet-musicians, teachers, scientists but an apparently infinite demand for specialists in corporate law, communication, marketing and advertising? Why, despite our technological capacities, we are not all working 3–4 h days?

3.5 Sustainable Consumption and Production. Industrial Ecology. Product–Service Systems

> Our enormously productive economy demands that we make consumption our way of life, that we convert the buying and use of goods into rituals, that we seek our spiritual satisfactions, our ego satisfaction, in consumption… We need things consumed, burned, worn out, replaced, and discarded at an ever-increasing rate.—Victor Lebow, journalist and economist, Canada

Irresponsible **consumption** problems are not new. Indeed, the accumulation of a big number of bad habits and unsustainable practices has led to critical stresses on societies and the environment. In spite of fast economic growth, the world has been on a course leading to resource depletion and serious social crises, and old ways of problem–solving have proven inadequate.

Consumption in economic jargon is related to the equipment, tools, products, services, resources, etc. that are used in the process of providing some sort of satisfaction. Economists refer to these objects by counting up all the money we spend in the market place in certain categories. Modern forms of consumption have become addictive, and they produce serious unintended consequences. First of all, in the world with limits, this model of consumption underlies unsustainability. The fact that critical natural resources may be depleted is of no consequence. Technology and innovation will always show up just in time and offer substitutes.

The overarching hypothesis, which was emphasised at the Rio +10 Summit in Johannesburg, is that sustainable consumption can drive sustainable production and lead to structural changes in the world economy. Sustainable consumption is defined by UNEP as "the use of services and related products which respond to basic needs and bring a better quality of life while minimizing the use of natural resources and toxic materials as well as the emissions of waste and pollutants over the life-cycle so as not to jeopardize the needs of future generations". In theory, sustainable consumption, by definition should lead to sustainable production, as someone concerned with sustainable consumption would not willingly consume a product that was produced unsustainably, or service that was delivered in an unsustainable way.

At the same time, Lowell Centerfor Sustainable Production defined sustainable production as "*the creation of goods and services using processes and systems that are nonpolluting, conserve energy and natural resources, are economically efficient, are safe and healthy for workers, communities, and consumers, and are socially and creatively rewarding for all working people*".

There is an overarching hypothesis that sustainable consumption drives production towards sustainable development and leads to structural changes and transformations of economical system. Therefore, human population cannot continue to expand indefinitely at the levels of consumption and production practised by the developed countries. Over time, industry, heavily supported by advertising, has permanently delivered the wrong message for happiness: the more you consume, the more attractive and happier you will become with luxury equated to liberty. Due to absolute limits of our natural environment and climate constrains, unsustainable consumption and production patterns have to be altered and, in some cases, reduced. This requires decision-makers to do necessary policy shifts and transformations leading towards sustainability. Essential elements of effective policy approaches to achieving sustainable consumption and production are summarised by the 6 re philosophy. The basic principles of this approach are: (I) re-thinking the product and its functions (replacing goods with services, (II) making products that are easy to repair and ensuring that skilled tradesmen are trained in repairs, (III) replacing harmful substances in products with safer alternatives, (IV) designing products for disassembly so that parts can be easily reused or recycled, (V) reducing energy, material consumption and impacts throughout the product's cycle, and (VI) recycling (IGES 2010).

"It is important to note that recycling in many waste, especially disposable packaging, management hierarchies, including circular economy, should not be regarded as an effective solution. Our experience in waste reduction innovations implementation clearly shows that priority should be given to waste prevention, avoidance, industrial ecology, and product modification by design for unsustainability reduction. The main reasons confirming that are:

- The whole recycling chain including collection, transportation and material processing consumes a lot of energy and thus generates high costs and extra pollution;
- Most wastes are composite or mixed and technically challenging or very costly to recycle;
- The quality of recycled materials in most cases is worse than that of virgin materials and they can only be used for certain applications with low technical requirements;
- Recycling of plastic wastes requires that different kinds are separated and this can be difficult to achieve due to the households' lack of knowledge and low motivation to sort" (Hawken et al. 1999).

It is still more profitable to make a washing machine that breaks down after five years, than to make one that will last indefinitely, partly because that there is a whole industry that has grown up to service that obsolescence, including extended warranties that rely on products being unreliable. When labour costs are very high, as they are in most developed countries, repair becomes unfeasibly expensive. That is why conventional electronics favours built-in obsolescence. Although according to some directives, for example the WEEE (Waste of Electrical and Electronic Equipment) Directive, manufacturers are now liable for the disposal of their old products, unfortunately they still end up polluting poor communities in the developing countries. Therefore, it is strongly recommended to end the export of e-waste and hazardous chemicals to countries that do not have the advanced infrastructure to manage them (IGS 2019).

"Economies that are driven by the growth imperative are always looking for ways to increase the volume and speed of production and consumption. This can involve conscious efforts to reduce the lifespan of a product—otherwise known as "planned obsolescence"—so that they need to be replaced. The lifespan of most equipment and motor vehicles has been reduced and repairs have become more costly or impossible and it has become easier and often cheaper to replace than repair. New-generation electronic devices are rendered obsolete by even newer versions and generational lifespan of such devices has been reduced. Market-driven efficiency reduces the resources and labour used to create goods. This lowers costs, thereby increasing demand and enabling people to buy greater quantities of the more efficiently produced goods" (Mulligan 2014). This is well understood in the experience of recent decades. The money gained through energy savings, for example, is often used to buy more energy-using appliances and services. Dramatically more efficient car engines in the 1970s to 1990s led to heavier cars laden with new features and accessories.

"The new economics criticises how this style of shopping offers superficially attractive apparently cheap products, which are in reality bought at high cost to the local economy, community life, local distinctiveness, and the environment. The retail model depends on enormous hidden subsidies, like transport infrastructure and cheap fuel to bring the customers in and to operate its just-in-time distribution systems, without which these technocratic retailers could not survive. It also relies on a wholly unbalanced and frequently abusive power relationship between retailers and producers and suppliers. It is important to redefine work more radically, so that we can shift the mainstream economy into supporting "core economy" of family and community, on which we rely. The growing wealth and power of multinational corporations have given new force to advertising and we are witness of a new wave of globally recognised brand names" (Boyle and Simms 2009).

Collaborative consumption is a new approach to consumer access of goods and services based on an interdependent peer-to-peer model. Elements of the collaborative model include bartering, sharing, gifting, lending and leasing or renting. One of the best-known examples of collaborative consumption refers to the emergence of companies aimed at very flexible car hiring or car share schemes that enabled some city dwellers to go without the car of their own. It is important to note that new communication technologies have given collaborative consumption schemes a greater chance of success.

"Sustainable consumption and production are reflected as a cross-cutting theme in several UN documents and are often highlighted in sustainability initiatives for a broad range of sectors and issues. In other words, the patterns of consumption and production determine the degree of sustainability in many areas, for instance, energy production is related to CO2 emissions, industrial pollution affects water, air and soil quality, wood production and mining could lead to desertification and land degradation, gender equality and education are linked to access to resources and better quality of life, etc. In this situation, the analysis based on system approach could make an important contribution to SDGs by emphasising that they are interlinked and that progress solely focused on one goal could have consequences for the other. One of the surprising outcomes of such analysis is that reducing poverty over 2000–2015 may have come at the expense of making our economy less sustainable" (Barbier and Burgess 2017).

There are also interesting results of a new comprehensive analysis by a team of scientists from the Potsdam Institute for Climate Impact Research (PIK). According to the study shared to Databank, responsible consumption and production seem to be a bottleneck, as data from the past shows. Improvements in well-being, economic prosperity, and lifestyles currently still come to a large extent through an increase in consumption and, therefore, with the growing environmental and material footprints. To successfully implement the 2030 Agenda for Sustainable Development, such conflicts in objectives need to be identified, governed and tackled (Arillo 2017).

We are often locked into unsustainable patterns of production and consumption through a complex mixture of factors, some of them institutional, some of them economic, some of them to do with infrastructure or lack of it, and some of them social or psychological in nature. As far back as Adam Smith, economists have struggled with the paradox of value, where diamonds are socially useless, yet expensive and water—essential to our lives—is infinitely cheaper. If we rely on the unregulated market to price everything, then we will find ourselves in the ultimately self-defeating situation where we only value life-giving water once it has become scarce and polluted. The effects of this can be seen on the non-market economy, or the core economy. "The core economy is made up of all the unpaid labour that is crucial to keeping society and communities functioning—the time that is put into caring for older people, volunteering for the community, exercising democratic rights to protest, and raising children. These tasks (disproportionately carried out by women) have also been systematically devalued as the market economy encroaches on the time that can be committed to them. Where markets have emerged in these activities they are poorly rewarded, making up the notorious '5Cs'—cleaning, cooking, care, clerical, and cashiering" (New Economics Foundation 2010).

"The most well-known instances deliberately involving manipulation are the mass media and advertising. Some economists, following the works of George Stigler, a leading free-market economist of the 1960s and the 1970s, have argued that advertising is basically about providing information about existence, prices, and attributes of various products, rather than manipulation of preferences. However, most economists agree that much of advertising is about making potential consumers want the product more eagerly than they would otherwise do—or even want things that they did not know they needed. Once poor people are persuaded that their poverty is their own fault, that whoever has made a lot of money must deserve it and that they too could get rich if they tried hard enough, life becomes easier for the rich. The poor, often against their own interests, begin to demand fewer redistributive taxes, less welfare spending, less regulation on business, and fewer worker rights" (Chang 2015).

"The problem itself is, simply put, that human beings are not rational—or that they possess only bounded rationality. We are too easily swayed by instincts and emotion in our decisions—wishful thinking, panic, herd instinct and others. Our decisions are heavily affected by the "framing" of the question when they shouldn't be depending on the way it is presented. And we tend to over-react to new information and under-react to existing information; this is frequently observed in the financial market. We normally operate with an intuitive, short-cut system of thinking, which results in poor logical thinking. Above all, we are over-confident about our own rationality" (Chang 2015).

Greenwashing is a way of exaggerating or fabricating the environmental benefits of a product, practice, or service—and indeed many corporations have greenwashed their own products through buzzwords and misinformation by advertising in an effort to tap into the growing market of 'green products'. "Greenwashing and the denial industry serve as potent reminders that those who profit from unsustainable industrial practices will use underhanded tactics to further their business interests—even if those interests stand in direct contrast to the long-term interests of the global community" (Caradonna 2014).

The problem with these exchanges is when they undermine the aspects of planet or communities, either because they have to come by airfreight or lorry—undermining the climate as they come—or because they replace productive capacity nearer home. "That is because narrow economics ignores these externalities—and the far greater costs of global warming that we now understand—is that it distorts what economy is doing. We also might have a better chance of feeding the world if people's basic needs were met as close to home as possible, rather than forcing them to rely on unstable delivery systems. It is important to end subsidies to unsustainable energy production, fisheries, agriculture, and transport" (Boyle and Simms 2009).

In SD approach, this socially defined concept of value is placed at the centre of decision-making and progress towards it is measured. In public policy, achievement of such value would be instituted as the central goal. Instead of living with the consequences of spiralling inequality and climate change whilst politicians are applauded for achieving economic growth, politicians would be held to account on the social and environmental as well as economic outcomes that they achieve. Policy decisions would be audited across the 'triple bottom line' using such approaches as Social Return on Investment (SROI).

However, it is not just in the public sector that this new socially defined concept of value would have an effect. Involving businesses makes this potentially transformative. A 'great repricing' that aligns prices with value would be instigated. This would require businesses to take full account of the costs of any (unintended) negative consequences. Through intelligent use of the tax system, the price paid by the final consumer would be aligned with real value (New Economics Foundation 2010).

"Take the example of wood furniture manufacturing. At the moment, manufacturers are not penalised for sourcing wood unsustainably. They pay for the wood, make their chairs and put a price on them based on whatever their costs were and the margin they want to make. If we value forests and require anyone that buys unsustainable wood to pay for the environmental costs of deforestation, then the price of the unsustainable wood and/or the furniture made from it would go up relative to the sustainably sourced wood. Similarly, food grown locally is often more expensive than that shipped around the world, not least as the environmental costs of the transport involved are not reflected in the price. Repricing in line with what we value would fundamentally change incentives for consumers and so for business there would be a significant disincentive to engage in activities that lead to social or environmental harm and an incentive to do good as this would lead to greater price competitiveness. Successful companies would be those that build the most social and environmental value. Those that are destructive of this value would not be able to compete, forcing them to change their ways or to go out of business" (New Economics Foundation 2010).

Researchers and practitioners have explored and proposed imperatives and implementation mechanisms for fostering the sustainability of production-consumption-systems (Lorek and Spangenberg 2014):

- Innovations in production processes significantly reducing the environmental impact per unit made;
- Producers are responsible for waste generation and for waste from the disposal of products at the end of their life. Production waste prevention and waste as a resource (industrial ecology) are the highest priority of waste management;
- Consumers, researchers, and authorities are involved in the design and redesign of products to meet functions with significantly less environmental impact;
- Producers provide services rather than sell products; this reduces the number and increases the life-time of products made while still providing to consumers the functions they need (product-service systems);
- Production companies impose the same standards on their suppliers and partners internationally to improve environmental and social performance;
- Consumers buy labelled products, based on independent certification. Unethical practices in marketing and advertising are restricted and based on product information only;
- Behaviour change by education of consumers about impacts of individual products, classes of products and consumption patterns. Any change to be significant must include a shift of institutional settings (rules, laws, habit preferences). Otherwise it can only call upon people to behave "right" within "wrong" structures, resulting in a rather limited potential for change.

To reach a society where all—or at least most—of the elements listed above can be established, some fundamental transformations in societies have to take place, for instance, (I) support from powerful actors, like WTO (World Trade Organization) and big business organisations, influencing the structures which are dedicated to make sustainable consumption an issue of public debate, (II) introducing scenarios for consumers, policy and business decision makers as a mean to assess the social, economic and environmental impacts, risks and costs and their trade—offs, (III) increasing the demanding responsibility of governments to implement, monitor and enforce sustainable consumption strategies and action plans because delegating such responsibilities to social actors for them means a "participation overkill" rather than achieving real-world changes, (IV) appreciation of the potential of social innovation which still remains insufficiently explored, the efforts of scaling up extremely limited and the political macro structures to foster this process undeveloped, (V) NGOs could play a catalyst's and an initiator's role through closer cooperation with academia because often grassroot networks stimulate and catalyse, inform, motivate and link their members and supporters.

"As the necessary reduction of environmental impacts calls for shrinking instead of growing markets, sustainable economies revalue the contribution of non-market activities for human well-being like sharing or voluntary work. Thus, the already existing enabling mechanisms for sustainable consumption have to be consequently developed to meet a strong sustainable consumption approach" (Lorek and Spangenberg 2014).

A modus vivendi of sustainable development means being more concerned with economic well-being than simply producing more stuff, using more of the world's natural resources, creating more and more waste, etc. A new model of sustainable consumption and production management of the company was elaborated to incorporate a set of widely accepted sustainable development measures and tools, such as resource-efficient and cleaner production, industrial ecology, life-cycle assessment, eco-design, eco-labelling, environmental product declaration, corporate social responsibility, stakeholder engagement, and sustainability reporting, that can be applied to the three main aspects of the company activities, namely manufacturing processes, products/services, and stakeholders. The consumer was for the first time included in the general consumption and production system as an active participant, who can influence the producer through the feedback mechanism (Jonkutė and Staniškis 2016).

"Hence, capitalism cannot function without scarcity; it is created artificially by advertising industry and thus undermines attempts to achieve sustainable development and must be ended. One principle offered to change the existing patterns of consumption is "sufficiency", where the economic system removes incentives for over-production and over-consumption above the level necessary for a needs-based economy. Sufficiency economics would remove the impetus for environmentally damaging growth and create the potential for socially just de-growth. Degrowth is a planned reduction of energy and resource use designed to bring the economy back into balance with the living world in a way that reduces inequality and improves human well-being" (Reardon et al. 2018)..

Industrial symbiosis

> Nothing was made by God for Man to spoil or destroy.—John Locke, English philosopher and physician

The proponents of intensified recycling have argued for the co-location of industries with eco-industrial parks in order to maximise the transformation of waste into resource in a new kind of industrial symbiosis, but global waste flows are making this harder to achieve. A successful case was presented in one Lithuanian fertiliser company. Fertiliser production is one of the most important industries for meeting a growing demand for food resources. Regrettably, this industry is characterised by energy high intensity. Even today, fossil fuel such as natural gas is the main raw material and the main source of primary energy in nitrogen fertiliser production. Resource-efficient and cleaner production, industrial ecology, material flow analysis, environmental impact assessment, and evaluation of environmental efficiency were integrated into a special methodology. The initial environmental analysis of nitrogen fertiliser production processes revealed that despite the fact that company had already implemented several RECP projects seeking to minimise energy intensity and emissions, an enormous amount of energy was still wasted in the main and additional processes (for example, ammonia production, cooling process, and heat energy production). By using the methodology, several alternatives for resource efficiency and energy saving have been developed and one of them in which pollution prevention was introduced into the industrial symbiosis method was successfully applied (Kliopova et al. 2016; Malinauskienė et al. 2018).

While new and efficient industrial precincts have been established in some countries, especially in Northern Europe, consumption is favouring cheap goods produced by cheap labour in relatively remote locations. Electronic goods and materials make-up the fastest growing waste stream in the world. Electronic goods and materials discarded in the developed world are more likely to be reprocessed in nations with low-paid workers and most of these nations are not signatories to the Basel Convention. The expanded industry is raising concern about its impact on the environment and the health of those who work in largely unregulated reprocessing sites (IGS 2019).

While there is now widespread acceptance of the severity of our environmental position, there is a presumption that growth can be decoupled from environmental impact through more efficient use of resources and through the dematerialisation of economies. But the environmental economist Herman Daly put it like this: "The notion that we can save the "growth forever" paradigm by dematerialising the economy, or "decoupling" it from resources, or substituting information for resources, is fantasy. We can surely eat lower down the food chain, but we cannot eat recipes…" (Daly 1997).

Product–service systems

Consume less; share better.—Hervé Kempf, French journalist and author

The key idea behind product-service systems (PSS) is that consumers do not specifically demand products, per se, but rather are seeking the utility these products and services provide. By using a service to meet some needs rather than a physical object, more needs can be met with lower material and energy requirements. A product-service system is a competitive system of products, services, supporting networks and infrastructure. The system includes product maintenance, part recycling, and eventual product replacement, which satisfy customer needs competitively and with lower environmental impact over the life cycle. Industry is particularly well positioned to take the lead and implement new strategies by using 'out of the box' thinking. A PSS is designed reflecting the strengths of a manufacturing company to provide added values to the customers.

A rational systems approach was elaborated by Friedrich "Bio" Schmidt-Bleek for improving the chances for a long-term human existence on earth. The approach is based on a simple idea:

"*To create* our wealth and well-being with far less natural resources than hitherto the case, on the average to dematerialise Western economy by at least a factor 10. Technically, this is achievable through systems innovation without losing end-use satisfaction.

To organise a cost-neutral shift of overheads, charges and taxes from income to natural resources, in order to internalise the costs of using nature, stimulate eco-innovation, give incentives to producers for dematerialising goods and services, create a price structure on the market that rewards purchasing and using eco-efficient goods and services, and make labour less costly and thus create new jobs.

To use any other reasonable and cost-effective options available in order to lessen the use of natural resources, including through education, elimination of perverse subsidies, review of norms and standards, and on occasion new control legislation.

In brief, we need to add a new dimension to the traditional environmental protection: *rather than just fighting* one symptom of our impact on the ecosphere after another, we must get to the root cause of our problems.

We must understand the currently fundamental mismatch between generating welfare for people and the stability of the carrier system earth" (Schmidt-Bleek 2008).

There are three main uncertainties regarding the applicability and feasibility of PSSs: the readiness of companies to adopt them, the readiness of consumers to accept them, and their environmental implications. The main finding is that successful PSSs will require different societal infrastructure, human structures and organisational layouts in order to function in a sustainable manner (Mont 2002).

Research findings were accumulated from the field to present a framework supporting the implementation of well-established categories of PSS business models, i.e. product-oriented, use-oriented, and result-oriented business models. Each business model category is linked to five operational-level tactics that ensure the model can be implemented successfully and subsequently generates value. These tactical sets include (1) contracts, (2) marketing, (3) networks, (4) product and service design, and (5) sustainability operational practices.

Chemical Leasing

> The greatest threat to our planet is the belief that someone else will save it.—Robert Swan, UK, author, the first person to walk to both Poles

One of the good examples could be the Chemical Leasing concept, developed by UNIDO. Chemical Leasing is a performance-based business model for sustainable chemicals management. Chemical Leasing aims at a more efficient use of chemicals in the production process by redefining the business relationship between the user and the supplier of chemicals.

The conventional business model assumes that the more you sell, the more you earn. However, in Chemical Leasing, the supplier does not sell quantities. The supplier sells the performance, the function of the chemical. Built on strong cooperation between partners and based on mutual trust, it increases the efficient use of chemicals, reduces the risks to human health brought about by their use, improves the economic and environmental performance of participating companies, and ultimately enhances business performance. Chemical Leasing projects show that basing business relations between producers and consumers of chemicals on the chemical product performance results not only in environmental benefits, such as reducing harmful emissions and decreasing the amount of waste. It also boosts companies' fitness by streamlining their inventories, improving their working conditions, and establishing long-term partnerships based on mutual trust, while at the same time enhancing customer satisfaction and innovation.

Since the first book "Chemical Leasing Goes Global" was published in 2008 and the first Global Chemical Leasing Award was held in 2010, this performance-based business model has become ever more relevant for national and international policy frameworks and initiatives, while at the same time being more systematically applied by the private sector and academia. Chemical Leasing, as a holistic and systematic approach, is not only a proven performance-based business model but has also been shown to be an innovative policy tool for sustainable management of chemicals. It facilitates partnerships with and among the private sector and its business associations, academia, and government, and contributes to the Sustainable Development Goals (Jakl and Schwager 2008; UNIDO 2020).

3.6 The Role of Markets

> Nature provides a free lunch, but only if we control our appetites.—William Ruckelshaus, American attorney

The market economy remains one of the humankind's most awesome inventions, a powerful machine for the creation of wealth. This notwithstanding, the measure of the society is the quality of life at the bottom of the pyramid, not the top. At the same time, markets are constructed by people, for purposes chosen by people—and they can be changed and rebuilt by people too.

The wider the range of activities covered by the market, the fewer are the issues on which explicitly political decisions are required. However, reliance on the market grants priority to people who have money. This misapprehends the human nature. Relationships are like muscles rather than textiles. They are strengthened by use. In the depths of the Great Recession, only the most foolhardy purists continued to insist that markets should be left to their own devices. Milton Friedman said of John Maynard Keynes that if he had lived long enough, he would have been at the forefront of the free market counterrevolution. Perhaps if Friedman had lived a few years longer, even he would have recognised that the counterrevolution had gone too far. 'The idea that the markets were always right was mad,' said French president Nicolas Sarkozy, 'Laissez-faire is finished' (Applebaum 2019). However, it was not obvious what should come next.

The existence of equilibrium and tendency of market systems towards equilibrium represent a central element of neoclassical economics. "There are two reasons: (I) the nineteenth century founders of neoclassical economics were heavily influenced by the Isaac Newton's three laws of motion, which assume a body is at rest (i.e. in equilibrium) until acted upon by a force), (II) which in turn generates an equal but opposing force, and (III) with motion continuing until the force ceases, resulting in a new equilibrium. From sustainability perspective, much more relevant and practical than the concept of equilibrium is the carrying capacity of our planet and of ensuring that our level of economic activity is comparable with available natural resources and our ecosystems" (Reardon et al. 2018).

"The collapse of the Soviet Union solidified the political consensus on the free market ideology. The division of the world between communist and capitalist societies was one of history's great natural experiments, and the results seemed clear. Capitalism became a self-satisfied monopolist in the marketplace of ideas with predictable consequences: in the absence of alternatives, it was difficult to muster the will to deal with its evident shortcomings. The importance of economic growth became the nearest thing: as President George W. Bush told the nation after the September 11 attacks: "we must stand against terror by going back to work" (Applebaum 2019).

Classically, the divide in terms of economics is between central planning at one extreme and free markets, or laissez faire, at the other point. Socialists argued that a central planner with sufficient information about the needs and wants in a society and the resources available to achieve these was best placed to do this optimally. In fact, central planners cannot get this information from inside people's heads. The result of a centrally orchestrated allocation process will thus be the interpretation of a small group of people of what an entire population *might* want. At best this might cover basic needs, but even this is not guaranteed, and it cannot but be strongly influenced by the whims of the planner.

In principle, at least, the price of a thing should reflect its scarcity, the demand for it (i.e., how much people value it and can afford it) and its attractiveness relative to other goods. Through the interaction of countless separate transactions, markets should allow the 'discovery' of a price that balances all of these components. The way that prices change over time then reflects changes in the balance of these aggregated individual decisions, reflected in price levels.

There are clear problems with markets, which have long been recognised. Managed or 'social' markets have sought to regulate, temper and steer the market so as to reduce negative social impacts and prevent market economies becoming 'market societies'. In the UK, the 'third way' attempted to combine the logic of free markets with the public service ethos and the energy of civil society. None of these often laudable attempts have really succeeded, and we would argue that they could never do so until we 'get the prices right'.

"This is simply unsustainable, and no amount of tinkering with regulation or appeals to business leaders to take a triple bottom-line approach has been sufficient to offset the logic of the market expressed through the single bottom line of profitability. In the SD economy, this should be decisively resolved. By using the tax system so that 'goods' become cheap and so plentiful, and 'bads' become very expensive and so very rare indeed, the strength of the market is harnessed towards positive goals, rather than constantly undermining them" (New Economics Foundation 2010).

"According to the IMF Fiscal Monitor, public debt in advanced economies reached more than 120% of GDP and in emerging economies shot up to over 60% of GDP (in the wake of COVID-19). Our ecological footprint is greater than ever, as we use up more than 1.75 times the resources the world can replenish. Global debt, including public, corporate and household debt, at the end of 2020 amounted to $ 281 trillion, or more than three times the global GDP. During the COVID-19 crisis, many governments have effectively used "helicopter money" to sustain the economy; they printed money, creating an even higher debt with their central banks, and handed it to citizens and business in the form of one-off subsidies and consumption checks so that they could get through the crisis unscathed. In the short term, this approach was necessary to prevent an even worse economic collapse. Nevertheless, in the long run, this debt too will need to be repaid. The goal is to, ultimately, restore higher growth as people consume more, companies invest more, and governments spend more" (Schwab 2021).

This is not the only problem, however. The argument set out above might suggest that the things people want is a sort of natural phenomenon just waiting to be uncovered by the market. If this was the case, why would companies bother spending billions on advertising products? Clearly, 'wants' can be and are manufactured and manipulated. Consumption has been stoked to unsustainable levels, becoming an end in itself for many. The SD economy would inevitably see overall consumption reduced, and strict restrictions placed on advertising, particularly to children. At the very least, advertisers would be required to prominently feature the social and environmental costs and benefits of their products, but the fact that these would also be incorporated into market prices would limit the extent to which the advertising of 'bads' could succeed anyway (New Economics Foundation 2010).

"Branding and advertising not only aim to sell products, but also mould cultural values, desires, and frames. Worldwide, companies with global brands are spending around $ 500 billion a year on advertising. They bombard TV viewers, social media users, and radio listeners; they slip messages into songs, museum displays, and video games; they embed subliminal images into movies, teenage fiction, and cartoons" (Dauvergne 2018).

Subsidies in the most cases introduce economic, environmental and social distortions with unintended consequences. They are expensive for governments and may not achieve their objectives while also inducing harmful environmental and social outcomes. Interestingly, for many sustainable development issues there is a very strong argument to be made in favour of using taxation and other market-based mechanisms instead of subsidies: there are the chances that policy-makers will identify every initiative worthy of support and make the appropriate subsidy, without accidentally supporting some initiatives, which turn out to have negative effects.

According to Chang smetimes even Nobel Prize-winners in financial economics, top bankers, high-flying fund managers, prestigious colleges, and the smartest celebrities have shown that they do not understand what they are doing, how can we accept economic theories that work only because they assume that people are fully rational? "For example, Nobel prize winners in 1997, Robert Merton and Myron Scholes, were the chief science officers of two funds: LTCM (Long-Term Capital Management) and PGAM (Platinum Grove Asset Management) that went bankrupt. There is a saying in Korea that even a monkey can fall from a tree. Yes, we all make mistakes, and one failure—even if it is gigantic like LTCM—we can accept as a mistake. But the same mistake twice? Then you know that the first mistake was not really a mistake. Merton and Scholes did not know what they were doing. The upshot is that we are simply not smart enough to leave the market alone. But where do we go from there? Very often we need regulation exactly because we are not smart enough to leave things to the market" (Chang 2012).

In more recent years, there has been a second round of privatisation of infrastructure, primarily in response to an increase in public debt and, in some countries, the public desire to reduce tax commitments in an era where wages, especially among the middle and working classes, have not kept up with the cost of living. It is not surprising, then, that local and state government have looked to privatising more and more public infrastructure. However, private companies overseeing infrastructure are often far more aggressive in squeezing profits out of what they regard as more as of a business than a service, which often leads to what industry watchers call "asset stripping". This is a common problem, experienced over and over again, with privately run prisons, toll roads, schools, and the like (Rifkin 2019).

3.7 Is GDP a Wrong Indicator? the Economy Expressed by GDP

The corporations don't have to lobby the government any more. They are the government.—Jim Hightower, agriculture commissioner and journalist columnist, US

The way we construct economic indicators has huge consequences for how we organise our economy, what kind of policies we implement and ultimately how we live our lives. There are various ways of measuring GDP, but the most common one is based on expenditure approach:

GDP = Consumption + Government expenditure + Private Investment + Exports - Imports.

We believe that it is not possible to have an economy without society, without the people who provide the ideas and the labour-power to make the economy function. GDP is the most basic measure that governments have traditionally used to measure our economy. This is a rather strange measure to set at the heart of policy-making, since it excludes many things we value that are never bought and sold (most importantly, environment):

- GDP does not distinguish between beneficial and harmful activity,
- GDP focuses on flows rather than stocks of wealth,
- The focus on market value means that activities not bought and sold in the market are not considered to have economic value,
- GDP is a total or average measure and does not give information about the distribution of wealth.

Last but not least, "GDP is built on great lie. The lie says that markets are only producers of wealth… Nature, the ultimate provider of all richness, is enslaved and devalued. GDP gives mankind the illusion that growth is about production, when it actually should be viewed as a transfer. Mankind does not produce anything. It simply turns natural wealth into money. Price tags are the ultimate symbol of GDP. Continuous production and endless consumption are its underlying values. Durability, reusability and self- production are its worst enemies… Things that we produce for ourselves are even worse, because they are not priced at all" (Fioramonti 2013).

"In 1999, the US Commerce Department declared GDP to be the most important invention of the twentieth century" (Reardon et al. 2018). GDP tells us about consumption, but it does not tell us about well-being; it tells us about production, but not pollution or the resource use; it tells us about government expenditure and private investments, but not about quality of life.

The Kuznets curve implies a sort of economic law. Inequality worsens as a nation begins to develop, but as development continues, inequality subsides. In other words, the price of inequality societies pay for development early on is offset by higher development and lower inequality later. Branko Milanovic, a former lead economist at the World Bank, showed that inequality fluctuates, as waves of technological progress and policy responses to them take hold. His point is clear: technology has a tendency to increase inequality. It should be stressed that Simon Kuznets never subscribed to the notion of any "Environmental curve", which hypothesised that harm to the environment would decline as an economy developed.

"The gross national product does not allow for the health of our children, the quality of their education or the joy of their play. It does not include the beauty of our poetry or the strength of our marriages, the intelligence of our public debate or the integrity of our officials. It measures neither our wit, nor our courage, neither our wisdom nor our learning, neither our compassion nor our devotion to our country, it measures everything in short, except that which makes life worthwhile" (Meadows 2008).

President Jimmy Carter was trying to deal with flood of illegal immigrants from Mexico. He suggested that nothing could be done about that immigration as long as there was a great gap in opportunity and living standards between the United States and Mexico. Rather than spending money on border guards and barriers, he said, we should spend money helping to build the Mexican economy, and we should continue to do so until the immigration stopped. Although that never happened, it was the right decision based on preventive principle that always pays off.

On the top 100 economics based on GDP and revenues, 60 are countries and the rest companies. Such vast economic influence gives companies power and leverage. Oil companies have lobbied governments against policies geared to combat global warming because of the threat posed to their business. Many corporations pay professional lobbyists to influence the decisions of politicians. For example, only in the US in 2014, almost 12,000 registered lobbyists were working to influence 535 members of Congress with total lobbying spending of $1.5 bn.

"Despite the fact that global GDP is increasing and few countries have negative growth rates, the gap between rich and poor countries is growing. In 1990–2014, the most dramatic growth came in the emerging economies of China, Vietnam, and Qatar. The global middle class, those with a daily spending power of $10–100, is expanding. Around 1.8 billion people were categorised as a middle class in 2009 and this number is expected to rise to 4.9 billion. The influence of middle-class consumers in the developing world is growing too. By 2030, it is estimated that around 35% of the global middle-class consumption will come from India and China" (Juniper 2016).

The economic power shift away from the established advanced economies in North America, Western Europe, and Japan is predicted to continue. By 2050, the combined GDP of the E7 nations (China, India, Turkey, Mexico, Indonesia, Brazil, and Russia) is expected to be double that of the G7 (USA, UK, Japan, Germany, France, Italy, and Canada).

In 2019, global public debt reached more than $70 trillion. Countries with large debts in relation to their national income face greater difficulties than those with a smaller financial burden in proportion to their GDP. Countries with stable government, low corruption, and economic strength, like Japan, are able to borrow even when heavily indebted (Japan—227% of GDP, US—71%, China—22%, Russia—13%.)

There is an urgent need to introduce a new indicator that would be more inclusive than GDP—one that incorporates social and environmental issues. This is especially important given global challenges such as climate change, global poverty and inequality, pressure on resources and their potential impact on societies. Today, already there are some serious elaborations: the New Economics Foundation derived the "Happy Planet Index", which uses survey data about a country's happiness and divides it by the amount of resources consumed to produce the level of well-being. The UN proposed "The Human Development Index", which measures country performance on a range of social measures, including life expectancy and education, while also incorporating GDP that will skew it in favour of wealthy countries, and "The Genuine Progress Indicator", which measures whether the environmental impact and social costs of economic production and consumption in a country are negative or positive factors in overall health and well-being. Recently, the new social indicator for sustainability was developed by Lutz el al. "The indicator, years of good life (YoGL), reflects the evident fact that in order to be able to enjoy any quality of life, one has to be alive and thus is primarily based on life expectancy. However, since mere survival is not considered good enough, life years are counted conditional on meeting minimum standards in two dimensions: the objective dimension of capable longevity (consisting of being out of absolute poverty and enjoying minimal levels of physical and cognitive health) and the subjective dimension of overall life satisfaction. The idea behind YoGL, on the other hand, is to study sustainability by focusing explicitly on the constituents of well-being and its change over time" (Lutz et al. 2021).

3.8 The Role of State

> We have allowed the interests of capital to outweigh the interests of human—Archbishop Desmond Tutu, South Africa

There is a huge spectrum of opinions on the appropriate role of the state including two extreme views. At the one end of extreme, there is a free market view, which needs no more than the minimal state, at the other end—the Marxist view, which believes that the market should be marginalised, and the whole economy must be coordinated through central planning by the state. For instance, according to the Libertarians, the only justified actions of the state are things like provision of law, especially including the protection of property rights, national defence and supply infrastructure. Anything beyond these minimal functions is seen as violating the sovereignty of individuals.

However these days the state produces a huge range of goods and provides services—law and order, infrastructure, education, research, health protection, pensions, and childcare and care for elderly, culture services and a lot more. Besides that, most countries also own companies that produce goods and services, for instance, electricity, oil, steel, banking, airline services and so on. Successful state intervention does not, of course, mean that more government is always better. In the world there are cases where governments have harmed the economy, sometimes even disastrously. But the fact is that the state still remains the most powerful organisation that the humankind has invented and thus essential transformations, for instance towards sustainability, are impossible to achieve without it. Below there are different views on state and government role in achieving sustainable development.

What we get from nature is fundamental to our economy, and without these inputs, we would in fact produce nothing. Man is a part of nature and his war against nature is inevitably a war against himself.

The state is ‘us’ not ‘them’—or at least it should be. It is not a malign force or an alien entity, but the product of collective decision-making and an expression of shared interests.

In terms of the state’s role in the economy, however, it could be argued that getting the prices right would greatly reduce if not eliminate it. Much of the justification for state intervention in the economy comes from the fact that the market does not take account of externalities. Because of this, the state has long been seen as having a duty to prevent public ‘bads’ – like pollution – and to produce public ‘goods.’ But if market prices reflect social and environmental costs and benefits accurately, ‘bads’ become expensive and are produced less while ‘goods’ become cheap and are produced more. Would we still need the state at all then?

We can think of the rule of law as a public good, for example. Similarly, maintaining broad economic and financial stability is a public good that is essential to underpin local, regional and national economies. Everyone benefits from this, but no single individual or group of individuals could or would maintain the rule of law or macroeconomic stability across a society or an economy. In a related way, maintaining reasonable levels of individual and regional equality can be seen as a public good—the negative effects of high levels of inequality extend beyond the affected individuals to society as a whole. "We have this crisis caused by people using money they don't have, to buy stuff they don't need, and in doing so helping to drive the planet to the edge of collapse. Now the best our government can do in response is to give us more money that they haven't got, so we can do more of the same. Is that really the best we can do? Is that how far we've come" (Gilding 2011).

International financing institutions claim on eradication of extreme poverty as measured by changes in the number of people who find themselves below poverty line. However, the very idea of the international poverty line has many critics, both because monetary income level has been set so low and because monetary income does not reflect numerous causes of poverty and disadvantage, such as access to adequate food and clean water supplies, access to adequate health care, and access to education. Entrenched poverty is spread very unevenly across the world and there is a lot of data suggesting that entrenched poverty is getting worse and that many people are struggling with the rising costs of living.

Hungarian economist Karl Polanyi argued that economics must be embedded in culture and society. The social framing of economics helps create a stronger link between economic development and the social equity outcomes of that development and this, in turn, links back to the consideration of both intragenerational and intergenerational equity. Focusing on the social outcomes of economic development also brings in play the principle that prevention is better than cure because it can be difficult and expensive to wind back economic development that is having harmful social consequences for a significant number of people. A classic argument for the role of the state is thus to fill gaps and produce the public goods that we would otherwise lack, and which in many ways are the hallmark of a real society that is more than the sum of its cities, towns, villages, and neighbourhood.

"Unemployment is not just a lack of money; it is also a lack of purpose and of opportunity. Therefore, economic growth cannot sensibly be treated as an end in itself. Development has to be more concerned with enhancing the lives we lead and the freedom we enjoy" (Amartya 1999).

In July 2008, in London, a group of green economists and other activists published in the Guardian the Green New Deal, urging the governments to embrace it as comprehensive, self-reinforcing programme. It involved policies and new funding mechanisms enabling to reduce emissions and allowing to cope better with the coming energy shortages. What the promoters of the Green New Deal understood that there was no point in tackling the crisis alone but alongside redistributing the wealth of the planet. In February 2019, the mission of Green New Deal for America was launched by creation of a selected committee, which would develop an industrial plan to address climate change, decarbonise the economic infrastructure within ten years, create new business opportunities, and employ millions of disadvantaged workers in an emerging green economy.

"The US Green New Deal is not just about mobilising the public to pressure governments to loosen the purse strings, pass legislation, and incentivise green initiatives. Rather it is the first call for a new kind of peer political movement and commons governance that can empower entire communities to take direct change of their futures at very dark moment in the history of life on Earth" (Rifkin 2019).

In December 2019, the European Commission launched an ambitious roadmap termed the European Green Deal that aims to make Europe the first carbon–neutral continent by 2050, whilst promising to help companies to become world leaders in clean products and green technologies.

The European Green Deal roadmap seeks to design a set of "deeply transformative policies" at regional and national level across eight key areas: increased climate ambition for 2030 and 2050; clean, affordable and secure energy; a clean and circular economy; energy and resource-efficient buildings; sustainable and smart mobility; a fair, healthy and environmentally-friendly "farm to fork" food system; preserving and restoring ecosystems and biodiversity; and zero pollution for a toxic-free environment. One of the tools to achieve this as well as the other SDGs will be a new circular economy action plan. The circular economy package will among other actions include a 'sustainable products' policy to support the circular design of all products with a focus on reducing primary resource use and promote further the reuse of materials, thus keeping the value embedded in materials and products in the economic cycle as long as possible.

In a recent report "Widening the European Green Deal" (Neßhöver et al. 2021), the German Environment Agency analyses the European Green Deal from its own perspective and places it in the context of the global challenge of achieving the United Nations' sustainable development goals. In addition to necessary measures in the thematic fields of the European Green Deal, the report also addresses the structural adjustments needed as key levers to achieve the desired goals.

The report concludes that the European Green Deal is an important step forward, but that further efforts beyond those described there are still needed in order to achieve a sustainable Europe.

"However, for the UN's Sustainable Development Goals to be achieved, it is crucial that citizens are involved in the process, providing input and oversight to policy decisions and working alongside both the business sector and the European institutions.

In addition to necessary measures in the fields of human well-being, sustainable economic activity, sustainable food systems, climate and energy, urban development and digitalisation, the paper also clearly states that structural adjustments are needed as key levers to achieve the desired goals. New approaches are needed in the governance of sustainability policy, in the economic and financial sector, in civil society involvement, and in science and innovation." (Neßhöver et al. 2021).

It is difficult not to agree with the conclusion, because the European Green Deal, although very important, looks like a re-worded and re-structured version of SDGs with the emphasis on climate change, which is in fact just one of the Goals. At the same time, the UN Independent group of Scientists in their Report on global sustainable development "Future is Now: Science for Achieving Sustainable Development" explicitly state that the actions by a range of stakeholders that can accelerate progress towards achieving the SDGs should "derive from knowledge about interconnections across individual Goals and targets, recognizing that the true transformative potential of the 2030 Agenda can be realized only through a systemic approach that helps identify and manage trade-offs while maximizing co-benefits" (IGS 2019).

"It is obvious, that green economy cannot provide the solution, although a substantial greening of the economy would be one of the corner stones for the transformation to a sustainable society, but this transformation requires not only sustainable production and responsible consumption but more justice and democracy, empowerment of marginalised groups, and international collaboration to solve the global environmental and social problems" (Lorek and Spangenberg 2014).

Effective action in any of the SDGs implementation requires acknowledgement and addressing the links among them – close ties between climate change and human health and poverty, for instance, or the ways in which biodiversity loss and deterioration of ecosystem services exacerbate inequalities. Pathways to advance human well-being ultimately require cooperation, collaboration and dialogue among multiple actors, and employing many levers of change (IGS 2019).

It is evident that there is a need for good governance for promoting peaceful and inclusive societies for sustainable development, providing access to justice for all, and building effective, accountable and inclusive institutions at all levels (Goal 16). It means that governance is an essential lever of the systemic transformations in order to achieve all 17 Sustainable Development Goals and adequate state capacity is among the key factors for successful sustainable development policies.

3.9 Society, Social Responsibility, Climate Change, and Nations' Health

> We destroy the beauty of the countryside because the unappropriated splendours of nature have not economic value. We are capable of shutting off the sun and the stars because they do not pay a dividend.—John Maynard Keynes, British economist
>
> Few trends could so thoroughly undermine the very foundations of our free society as the acceptance by corporate officials of a social responsibility other than to make as much money for their stockholders as possible. This is a fundamentally subversive doctrine.—Milton Friedman, US economist
>
> 'Who is society', once asked British Prime Minister Margaret Thatcher in 1987, justifying her relentless attacks on social services. 'There is no such thing. There are individual men and women and there are families.'

The London School of Economics defines the civil society as "the arena of uncoerced collective action around shared interests, purposes, and values. Indeed, people are social animals, and their absolute well-being is less important than their relative well-being. During most of the world's history, the reference point was a local one. In the Industrial Revolution—a national one, in the post-world war decades—West or Soviet Union. Now—the connective technologies provide people's reference point anywhere in the world (Schwab 2021).

"It is fundamental, perhaps even biologically based, sense that part of what is to be human means re-creating ourselves at every moment— on and on until we die. But because we can be blind in certain areas, this fundamental, positive engine for unfolding human potential has become buried so deeply that it comes to the surface only when our lives are imminently threatened" (Ehrenfeld 2009).

"When John Maynard Keynes made his famous distinction between art and ideas, which should be international, and goods, which should be primarily local, he was setting out a truth that new economists have been developing ever since. When Sir William Beveridge urged that voluntary action was a vital ingredient of the new welfare state that he had outlined, he was providing a glimpse of the ideas about social capital that would also be so central to the new economics" (Boyle and Simms 2009).

What is social capital:

- Social capital is a set of shared values that allows individuals to work together in a group to effectively achieve a common purpose.
- The idea is generally used to describe how members are able to band together in society to live harmoniously.
- In business, social capital can contribute to a company's success by building a sense of shared values and mutual respect.

The social glue that holds our lives together, and makes the economy possible, is also unravelling: families, neighbourhoods and relationships are fracturing under the pressure of high mortgages, benefit regulations and the kind of monoculture that drives out local enterprise, institutions, and community life from many areas in the name of efficiency, centralisation and corporate success.

AT&T's dramatic approach to human capital highlights some important questions that many companies are grappling with today:

- What is your responsibility for retraining employees whose skills become obsolete as a result of compressive or big bang disruption?
- How much should you invest to develop workers who are highly skilled or unique in their capabilities, knowing they may leave or, if they stay long enough, require reskilling?
- Should you instead minimise uncertain and unpredictable skills investment and just buy the talent you need, understanding, however, that the freelance market suffers from a short supply of the most current technical skills?

"No matter how you balance your financial portfolio in the old, the now, and the new, the successful execution of a wise pivot requires a thoughtful and strategic management of one of your most value assets: your people" (Abbosh et al. 2019).

Puzzlingly, many of the people who suffer the most under free-market policies support these approaches, because they have come to believe that small governments, fiscal rectitude, business freedom, and self-help are the foundations of a dynamic economy and a decent society. Such is the power of free-market ideology that people vote for politicians who do things to hurt them, often electing leaders who are even more committed to free-market ideology than those that ruled before. Thus, it is increasingly urgent that people understand what is really going on with the economy. Voters and consumers need to realise that there was nothing "natural" about free market order that the financial and economic crisis could have been prevented, and that there are alternatives to the unhappy present.

"We should build a system where material enrichment is taken seriously but it not allowed to become the only goal. Organizations – be they corporations or government departments – should be designed to reward trust, solidarity, honesty and corporations among their members. The financial systems need to be reformed to reduce the influence of short-term shareholders so that companies can afford to pursue goals other than short-term profit maximization. We should better reward behavior with public benefits (e.g. reducing energy consumption, investment in training), not simply through government subsidies, but also by bestowing it with a higher social status. People should be given a real, not superficial, second chance through unemployment benefits and publicly subsidised retraining" (Chang 2012).

The powerful testimony to the limitation of standard economic theory is the assumption that self-interest is the only human motivation that counts. Adam Smith maintained that "the beauty of the market system is that it channels what seems to be the worst aspect of human nature- self-seeking, or greed, if you like—into something productive and socially beneficial". Some—the so-called "New Public Management" school—went even further and recommended that the management of governments and public organisations itself should be exposed to greater market forces: a more aggressive use of performance—related pay and short term contracts for employees; more frequent contracting—out of government services; a more active exchange of personnel between the public and the private sectors. If we did that, people would feel that they are not trusted and refuse to act in moral ways, making it necessary for us to spend a huge amount of resources monitoring, judging and punishing employees. If we assume the worst about people, we will get the worst out of it. Certainly, self-interest is one of the most important, but we have many other motives, for example, honesty, self-respect, altruism, love, sympathy, faith, sense of duty, solidarity, loyalty, public-spiritedness, patriotism, and so on, which are in many cases more important than self-seeking as the driver of our behaviours.

Instead, a narrower shareholder paradigm was put forth by the University of Chicago economist and Nobel Prize winner Milton Friedman starting in 1970. He held that free markets were what mattered above else and that the "only social responsibility of business is to increase its profits".

Milton Friedman said "the business of business is business". Thus, shareholder capitalism became the norm across the West as companies globalised, loosening their ties with local communities and national governments, and focusing instead on maximising short-term profits for shareholders in competitive global markets. At the same time, labour unions, governments, and other civil society stakeholders lost much of their power and influence, further weakening the fabric in which a stakeholder model could prosper. Together with financial industry pressures to boost short-term results, the single-minded focus on profits caused shareholder capitalism to become increasingly disconnected from the real economy.

Libertarianism is an even more extreme philosophy developed by Charles and David Koch, based on three kinds of arguments. The first is the moral assertion that every individual has the overriding right to liberty, that is, the right to be left alone, free from taxes, regulations or other demands of the state. The second is political and pragmatic: that only free markets protect democracy from the despotism of the government. The third is economic: free markets alone are enough to ensure prosperity.

For the SD economy, we shall need the state not only to enable individuals to make choices that serve their own long-term interests, but also to promote and protect the long-term interests that we share. "A state that gives priority to prevention will have a different framework for decision-making—one that recognises the value of investing in upstream measures, where benefits accrue across sectors and over the longer term. It will seek to prevent 'ill-being' by addressing the underlying causes of unequal opportunity and tackling avoidable risks to physical and mental health. It will invest to prevent the waste of human potential by fighting inequality and entrenched patterns of unemployment. It will act to prevent wealth escaping from poor neighbourhoods by keeping resources circulating locally, through support for local businesses and trading within communities. And of course it must prevent damage to the environment—most urgently, climate change. There is a double prize for getting it right—more well-being for all, while public services are safeguarded for the future" (Coote and Franklin 2009).

Boulding argues that "there is a great deal of historical evidence to suggest that a society which loses its identity with posterity, and which losses its positive image of the future, loses also its capacity to deal with present problems and soon falls apart" (Boulding 1966). Such evidence suggests that "any society that fails to extend brother- and sisterhood to future generations, particularly by way of invoking the principle of intergenerational equity, is likely to rapidly disintegrate. Because human well-being cannot possibly be sustained within the confines of a self-destructing society, the goal of sustainability emerges as a necessary condition for achieving any form of human development" (Lawn 2000).

In her book "The Value of Everything", Mariana Mazzucato wrote that the dominant belief in the recent years has been created mostly by and in companies, and within that, to a large degree by financial institutions. In fact, the value in society is not only or primarily created by companies and their employees but also by educators, scientists, cultural actors, government institutions, and above all, society and the natural environment itself. Mazzucato pointed out that "the different view on value created, and the practices that went along with it, created a system in which value extraction became normalised, as many of the most productive members of the economic system, including those active state-funded scientific research, education, and social services, were undervalued. It also led to a financialisation of the economic system, in which revenues and profits were confounded with true value creation. And it created a cult of CEOs and tech start-up founders, in which private innovations generated more praise and protection than the fundamental breakthroughs achieved through public funding and institutions" (Mazzucato 2018).

Human societies are complex and adaptive system and in case of the shocks, negative feedbacks and feedforward loops should return the society to its central functions. This will happen in case of a resilient system. But, according to anthropologists Simon Lewis and Mark Maslin, there are three major reasons why from a complex adaptive systems perspective, the current mode of living appears unlikely to continue into long term: (I) positive feedback loops drive the current system, and these usually end in fundamental changes; (II) the energy, information and collective human agency factors that underline all human societies are changing increasingly faster; (III) there are some core challenges, including environmental impacts, which may cause collapse. "The capitalist development system relies on two endlessly cycling positive feedback loops: the solving of problems via the scientific method, which improves technology, thereby allowing greater numbers of problem to be solved; and the investment of profits into the production of ever more profits, which requires ever more energy and resources. The results of this dynamic process include social and environmental destruction, extreme inequality, reduction of people and their potential to 'human capital', and creation of atomised societies where people compete instead of forming cooperative communities. It should be noted that regardless of which view is taken, these positive feedback loops have led to five centuries of increasingly fast changes, which must end by the system either settling into a stable state, collapsing or moving to a new mode of living. Such rapid, radical changes suggest that a collapse or a switch to a new mode of living is more likely" (Lewis and Maslin 2018).

Corporate social responsibility

> Never doubt that a small group of thoughtful, committed citizens can change the world. Indeed, it is the only thing that ever has.—Margaret Mead, American cultural anthropologist

The idea behind corporate social responsibility—that corporations have an obligation to consider the impact of their activities on the environment, economy and society—is an old one. How companies treat their employees and what kind of product they submit to their customers have been the topics of discussion for centuries. The problem with corporate social responsibility is that it is voluntary. However, time has already come when companies by government requirements must fulfil their environmental and social obligations both at home and abroad according to sustainable development principles.

"Many refer to the creation of profit as the purpose along the lines of economist and Nobel laureate Milton Friedman's famous statement, "The social responsibility of business is to increase its profit" (Ehrenfeld 2009).

The concept of "natural capitalism" proposes four basic principles for SCP: (I) increasing resource productivity (by at least a factor 10), (II) eliminating the concept of waste entirely, by redesigning an economy based on closing the loops of material flows, (III) shifting from processing materials and making "stuff" to creating services and qualitative improvements in the standard of living, and (IV) rehabilitating the planet by investing in natural capital instead of financing capital divorced from the physical world (Hawken et al. 1999).

The anthropologists and proponents of "human economy refer to well-being, to the satisfaction of all human needs – not just those that can be met through private market transactions, but also the need for public goods, such an education, security and a healthy environment, and for intangible qualities such as dignity that cannot be reduced to dollars spent per capita. We live in an era when market mechanisms (always the result of social construction and never "free") have been extended into new sectors, with the aim of increasing "economic efficiency". But people are beginning to realize that making a market for a valued good such as education is not morally neutral. It often gives to misleading statistics which obscure reduction in quality that sets in when teachers and professors are treated like any other provider of commercial services. We may agree that economy does indeed shape kinship and religious institutions in the long run. But we are sceptical of evolutionary models grounded in notions of efficiency and abstract individual rationality, and argue instead for a more rounded approach to economic organization that does justice to the material, historical and ethnographic record" (Hann and Hart 2011).

"Humans are clearly a resilient species, but there is no guarantee that we will ever learn lessons from our mistakes". Therefore, "forget about vanity: instead, enjoy any present fame and family that we have, and admit that within generation or so we will be lucky to be remembered by anything more than a name on an ancestry tree or on gravestone. Our world is guided by leaders, and the rest mostly follow without serious dissent. Therefore, to make a change we need far better educated majority and leadership who all recognise not just local but also global issues. The reality is that we already have a global economy, and therefore all solutions must be considered in terms of global survival, global sustainability, and a global welfare. At the same time, "Humans are quite arrogant and self-centred, so we tend to assume that because we have a highly developed society, and have survived for a few ten thousand of years, we will always overcome natural disasters. Optimism is great, but pragmatic view is very different as we realize that, precisely because of our dependence on advanced technologies, our civilization is far more likely to be destroyed by catastrophes, that would have been only minor features for earlier generations" (Townsend 2018).

Analysts tend to measure a society by classical economics and social statistics: its deficit to GDP ratio, or its unemployment rate, or the rate of literacy among its adult women. Such statistics are important and revealing. However, there is another statistics, much harder to measure, that "I think is even more important and revealing. Does your society have more memories or more dreams than memories? By dreams I mean, life–affirming variety. When memories exceed dreams, the end is near. The hallmark of a truly successful organization is the willingness to abandon what made it successful and start fresh" (Friedman 2006).

Many degrowth and economic justice thinkers call for the introduction of a basic annual income, a wage given to every person, regardless of income, as a recognition that the system cannot provide jobs for everyone and that it is a counter productive to force people to work in jobs that simply fuel consumption.

"Unlike encouraging energy efficiency, the measures we must take to secure just, equitable, and inspiring transition away from fossil fuels clash directly with our reigning economic orthodoxy at every level… It requires visionary long-term planning, tough regulation of business, highest levels of taxation for the affluent, big public sector expenditure, and in many cases reversals of core privatisations in order to give communities the power to make the changes they desire. In short, it means changing everything about how we think about the economy so that our pollution does not change everything about our physical world" (Klein 2015).

However, more often changes in climate have been detrimental, contributing to food shortages, famines, infectious disease outbreaks, weather disasters, and conflicts over resources that foment social disorder and topple regimes. The Maya, for example, became obsessed with building ever-grander edifices, just like leaders of today's Dubai. "Now humankind is on a treadmill, attempting to produce more food, electricity, safe drinking water, houses, and consumer goods for an ever-growing population, and doing so within a largely deregulated market-based economic system that routinely discounts long-term environmental damage" (McMichael et al. 2019). Effective stewardship requires appropriate response on many fronts. There is old sailors' saying that "we do not control the wind, but we can set our sails".

The detailed theoretical and practical insights on organisational transition towards corporate social responsibility are presented in Chap. 4.

Climate change

> Climate change is destroying our path to sustainability. Ours is a world of looming challenges and increasingly limited resources. Sustainable development offers the best chance to adjust our course—Ban Ki-moon, former UN Secretary-General
>
> It becomes increasingly evident that neither famine, nor earthquakes, or microbes, nor cancer, but man, is the greatest danger to man…—Carl Gustav Jung, Swiss psychologist

While the dangers of climate change are real, other social and ecological issues that threaten the creation of a sustainable society may also occur. The climate change problem has to be challenged for two reasons: it encourages the restriction of proposed solutions to the technical realm and it detracts attention from the planet's ecological predicament as a whole, by virtue of claiming the limelight or the one issue that trumps all others. Issues such as "species extinction, endocrine disruption endless development, and biodiversity crisis" are all problems that exist independently of the climate crisis, even if the latter tends to make them worse (Crist 2007).

Gilding asks where will we go with technology. "I certainly believe technology will play a critical role in the way forward. However, I am not a techno-optimist who believes technology will fix it all… Climate change is just one issue in sustainability, and there is no way we can address all our challenges with technology and keep growing the economy. This is made even stronger by the rebound effect, where technology drives efficiency and efficiency drives more technology use. This means behavior change and shifts in how we organize our lives will in the end be essential. But technology will be an important enabler… In other words, remember we are designing a society, not a gadget" (Gilding 2011).

"Greta Thunberg spoke in Davos: "Adults keep saying: 'We owe it to the young people to give them hope', but I don't want your hope. I don't want you to be hopeful.

I want you to panic. I want you to feel the fear I feel every day. And then I want you to act. I want you to act as you would in a crisis. I want you to act as if our house is on fire. Because it is" (Schwab 2021).

Trying to soften the accusations, Klaus Schwab stated: "The emissions that lead to climate change are not just the result of a selfish generation of industrialists or Western baby boomers. They are the consequences of the human desire to create a better future for oneself… It is how countries around the world have gone from rural and agricultural societies to urban and industrial ones. It is a process full of trial and error, growing pains, and trade-offs, but to this day, it is still the most successful development model the world has ever known" (Schwab 2021). The model he mentioned was made at the World Economic Forum (WEF) by the Alliance of CEO Climate Leaders with consulting from Boston Consulting Group, based of their action focus centred on three domains:

- Reduce the greenhouse gas intensity of own operations and that of activities in the supply chain. Often reductions can be made by simply using energy more efficiency;
- Refocus investments in other companies, to include only those that are clean, and apply internal carbon prices to reveal the true cost of certain operations, and
- Innovate business models, by transforming the existing ones and pursuing new green opportunities (Schwab 2021).

The broad range of consequences of climate change must include implications for loss of life and population. Unfortunately, many, particularly economists, are greatly tempted to dodge the ethical issues. "Unless they think in this way, the evidence they offer is likely to be ignored in ways a decision-maker can use as useful. Or the conclusions and analysis may be missed or distorted because ethical positions are buried below the surface" (Stern 2016).

When the science of climate change tries to generate actions, they meet difficulties that arise from four key elements of the processes at work: (I) scale; (II) risk and uncertainty; (III) lags and delays in consequences; and (IV) the "publicness" of greenhouse gas emissions. An understanding of these four features is crucial to an understanding of the obstacles to, and thus how to create, the necessary political will for accelerating action (Stern 2016). Economists often are tempted to force everything into a simplistic cost–benefit analysis in which changes are marginal and all relevant effects can be expressed in terms of a single denominator, such as money. The method and system based on system analysis theory and presented above could help find a solution identifying the possible alternatives that could be used for decision–making process. This could facilitate the examination where we may be going under different assumptions about policy. Businesses have to play a powerful and constructive role in what can be done and how to combine the development and environmental responsibility. There are some examples around the world showing that they understand the risks of climate change and their crucial impact to this phenomenon and vice versa. At the same time, they should recognise the market opportunities of a low–carbon world and seek clear signals from governments so that they could invest with confidence in the sustainable economy.

"Climate change not only intensifies the existing global asymmetries that may result in wars and outbreaks of violence; it may also strengthen environmental trends that originally had nothing to do with climate change. Other environmental issues that must be described as "classical"—sea pollution, soil contamination and declining of biodiversity, burning rainforests, drying up the rivers or disappearance of lakes—have taken a back seat, although the cause of them too is the insatiable hunger for resources in societies fixated on growth. Whichever classical theme of the environmental movement one takes—the problem is becoming worse as a result of globalization, yet it appears largely remote from everyday consciousness" (Welzer 2017).

Decision-makers should replace the frequently rigid and sequential development paths that place climate change or economic growth ahead of social equity and environmental protection. Instead, policy makers should adopt systemic approaches, following different pathways to sustainable development that offer multiple solutions and drivers, across different sectors and jurisdictions. Effective action in different systems will require that the links among them be acknowledged and addressed—the connection between climate change and human health, for instance, or between climate change and inequalities (IGS 2019). Indeed, fighting inequality on every front and through multiple means must be understood as a central strategy in the battle against climate change.

Nations' health

> During my nearly five years as director-general of WHO, high-level policymakers have increasingly recognized that health is central to sustainable development.—Gro Harlem Brundtland, former Prime Minister of Norway

"Our current model of economic growth has led to increasing deforestation and loss of biodiversity, accelerated urbanisation, intensive animal farming, global travel—all factors known to increase the risk of zoonotic viruses jumping to human hosts and spreading with alarming ease. Scientists and public health experts around the world have been warning us for many years that a pandemic caused by an unknown virus was not a matter of "if" but of "when", and that we needed to prepare. However, as German virologist Christian Dorsten pointed out, "there is no glory in prevention", and pandemic preparedness has not received the necessary resources or attention. Hopefully, this will change. Ultimately, the best way to reduce the risk of future pandemics is by avoiding a return to "business as usual". As economist Mariana Mazzucato and others have been saying, "it is time to rethink capitalism" (Yong 2020).

Poor and vulnerable communities are at greatest overall risk to health and physical safety from climate change and weather extremes. We also see clearly the extent of negative externalities, environmental, ecological, and social damages that market-driven extraction, production, consumption, and waste disposal cause.

In fact, every type of inequality—gender inequality, racial inequality, income inequality—has been exposed and amplified during this pandemic. To cite some examples: in the US and the UK, black, Asian, Latin and other minority ethnic groups were two–four times more likely to die from COVID-19. People who lived in deprived areas had higher diagnosis rates and death rates than those living in less deprived areas. In Spain, recent seroprevalence results showed that women who worked as cleaners or carers, as well as migrants, were more exposed to the virus than the general population. Borrowing a line by British writer Damian Barr: "we may be in the same storm, but we are in different boats" (Ford 2020).

The consumption of wildlife per se does not increase the risk of disease transmission. Freshly killed deer in the Scottish highlands can provide venison that is less risky than intensively farmed chickens, which are routinely infected with human pathogens. When wildlife are stressed, farmed in small cages and kept in close contact with humans during the entire rearing and slaughtering process, including in wet markets, the risk of disease transmission rises (Phillips 2020).

International expert investigations concluded the virus was most likely of animal origin. It probably crossed over to humans from bats, via an as-yet-unknown intermediary animal, at an unknown location. Such "zoonotic" diseases have triggered pandemics before. But they are still working to confirm the exact chain of events that led to the current pandemic. Sampling of bats in Hubei province and wildlife across China has revealed no SARS-CoV-2 to date (Phillips 2020).

"As well as importing more meat, the Chinese government has rapidly changed production systems from "peasant-style" agriculture to intensive animal production systems. Recent urban expansion has also put more pressure on agricultural land. Furthermore, there is a big opportunity to bring "clean meat" into the Chinese diet. Clean meat is grown synthetically from muscle cells, without the massive land and water resources required of traditional meat production in China, without the emissions of pollutants and, most importantly, without the risk of transmission of novel diseases" (Phillips 2020).

Bat populations, for example, often change their roosting colony sites because of changes in climatic and feeding conditions. Bats harbour a myriad of coevolved hosts for their replication, leading to diseases. For example in Malaysia in 1998, forest-feeding bats sought alternative sustenance from fruit orchards adjoining commercial pig farms, where they infected the omnivorous pigs via half-eaten fruits. The pigs then infected their human handlers, resulting in deaths (McMichael et al. 2019).

COVID-19 is the most wide-reaching public health crisis our world has faced in a long time. It has significantly impacted practically every aspect of business operations. COVID-19 has made three aspects very clear: first, the centrality of sustainability for business is resilience. "Companies that integrated sustainability and transparency strategically into their business operations prior to the COVID-19 crisis have put an even stronger focus on it during the crisis. More importantly, they were much more agile in responding to unexpected events. Second, health, safety and wellbeing became central to the resiliency and sustainability discussion in a way that it has never occurred before. Third, the role of partnerships; nobody can deal with a pandemic on their own" (Viliani 2021).

"The response to the pandemic has transformed our economies, our lifestyles, how our public and private sectors interact, and it has done so overnight. Millions and millions of people are sheltering in place, working from home, and changing their habits of consumption and travel. We are collectively capable of the kind of change we will need in order to address our larger climate crisis" (Lovins 2020).

Quammen states: "We invade tropical forests and other wild landscapes, which harbor so many species of animals and plants — and within those creatures, so many unknown viruses. We cut the trees; we kill the animals or cage them and send them to markets. We disrupt ecosystems, and we shake viruses loose from their natural hosts. When that happens, they need a new host. Often, we are it. This virus comes from our failure to respect these places and ecosystems that not only need our protection, but that also threaten us if not cared for properly. What is the lesson here? Human activity and ecological interruption is the cause of our current pandemic and the cause of our ongoing climate crisis. The path forward must include greater care for the integrity of ecosystems" (Quammen 2020).

Hunter Lovins points to the lessons emerging from this crisis: "Our reliance on neoliberalism and market forces to achieve the common good does not work. We need to embrace our common humanity and the reality that we survive best when we care for the group as a whole. The goal of sustainability efforts is clearer than ever: shared prosperity on a healthy planet. We know what the solutions are: investing in renewable energy instead of fossil fuels; investing in nature and reforestation; investing in sustainable food systems and regenerative agriculture; and, shifting to a more local, circular and low carbon economy. These positive actions can also be a much-needed source of collective hope and optimism for life regeneration in these uncertain times" (Lovins 2020).

Inequity per se is a pandemic, and tackling it should become a priority at the national and global level. The question is whether we are ready to take the measures (including tax reforms and investment in public goods) that are required. In Pope Francis's words, "while the earnings of a minority are growing exponentially, so too is the gap separating the majority from the prosperity enjoyed by happy few. This imbalance is the result of ideologies that defined the absolute autonomy of the marketplace and financial speculation" (McMichael et al. 2019).

The vaccine alone will not get us out of this crisis. We need solidarity (within and between countries) and we need trust (in the vaccine, in science, in institutions). 'Without equity, we cannot end COVID-19, HIV or any other pandemic', wrote Peter Sands from the Global Fund.

Prior to COVID-19, the world was highly focused on increasing environmental problems and social issues that many believe arose from controversial economic policies and the global trade. Sustainability has been a term and concept used to bring balance and create responsibility for economic activity and development. The general and most used definition of sustainability is "meeting the needs of the present without compromising the ability of future generations to meet their needs". Hakovirta and Denuvara propose to rethink and redefine sustainability as the intersection of the economy, environment, society, and human health. "The significance of adding human health as one of the sustainability development goals can be seen through the results of the current COVID-19 pandemic. It is practically impossible for life to go on as usual when the health of the society quite dramatically crumbles down due to a human health risk of global magnitude" (Hakovirta and Denuwara 2020).

Those directly responsible for pollution have rarely, if ever, paid for the costs. Moreover, they have created a lot of pollution: greenhouse gas emissions, oil spills, toxic waste, every other kind of waste, chemical pollution, and on and on forever. We are the ones who have paid for it, and we will continue to pay if we do not demand that they change their bad behaviour. We suffer the effects of bad air and dirty water. We pay for the cost of the illnesses associated with exposure to toxic pollutants (Schlossberg 2019).

The COVID-19 pandemic is causing a growing financial burden on all countries, disrupting economies and causing hundreds of thousands of deaths globally. Low- and middle-income economies will additionally suffer from the lack of international funding available for achieving the 17 Sustainable Development Goals, climate change mitigation and adaptation, and biodiversity conservation. The pandemic is likely to further undermine progress towards the SDGs by 2030, which was already faltering even before the outbreak.

'Ultimately, it's our choice as a global community whether we face another lost decade or not. We know what can be done to restart vaccination campaigns. We know what it takes to maintain essential health services during a crisis. We know a lot about the kind of inclusive policies needed to help people withstand the economic shock. It comes down to a question of political will and working together beyond our own borders. Covid-19 doesn't determine that. We do,' said Melinda Gates (Gates and Gates 2020).

"If sustainability is to be revived as a development objective, then low and middle-income economies will need to come up with policies that are affordable and achieve multiple SDGs simultaneously. We have identified three policies that meet these criteria: a fossil fuel subsidy swap to fund clean energy investments and dissemination of renewable energy in rural areas; reallocating irrigation subsidies to improve water supply, sanitation and wastewater infrastructure; and a tropical carbon tax, which is a levy on fossil fuels that funds natural climate solutions. Through such interventions, developing countries can foster greater progress towards achieving the SDGs through cost-effective and innovative policy mechanisms that do not rely on external funding to implement" (Barbier and Burgess 2020).

3.10 The Great Localisation and Engagement

Don't judge each day by the harvest you reap but by the seeds you plant—Robert Louis Stevenson, Scottish novelist

I sympathize with those who would minimize, rather than those who would maximize economic entanglements among nations. Ideas, knowledge, science, hospitality, travel—these are things that of their nature should be international. But let goods be homespun wherever it is reasonable and conveniently possible, and above all, let finance be primarily national.—John Maynard Keynes, British economist

The SD economy will involve a shift to much more local food production. Partly, this should happen naturally, because repricing will see the cost of much of the non-local food we buy today from supermarkets significantly rising in price as social and environmental costs are factored in.

Much as with food, sustainable development will make renewable energy much more affordable (Rifkin 2011). In addition, the emergence of co-operative community renewable energy schemes will give people a secure and reliable place in which to invest their savings. People will have a genuine economic stake in creating and investing in energy production.

"The lesson is that globalisation is, theoretically, a force for good, but in practice, it can be a positive force only if guardrails ensure that it benefits and ensures resilience and sovereignty. Generally, economic globalisation works best for everyone when at least three conditions are met. First, globalisation can take off only if a social compact is in place. The long term of the social compact means that individuals are willing to leave short-term or selfish considerations aside, knowing that they would gain in the long run, just as the other stakeholders contributing in the past. Second, globalisation thrives when political leaders find a balance between providing direction to the economy and caring for their people on the one hand and opening up to the world in terms of trade and investment on the other. And third, societies benefit from globalisation when the reigning technology of the era is congruent with the comparative advantages an economy and society have. Finally, globalisation's adverse effects can get amplified by technology: if people are not well skilled or educated to make the best use of the latest technologies, others in other countries will take their place in globalised economy (Schwab 2021).

Business moves in an elevator lifted by the force of creativity; government and regulatory agencies take the stairs of incremental learning. This situation poses a particular responsibility to companies in ensuring that all technological advances are well understood, not only in terms of their functionality for individual users but also what their mean for society more broadly.

It is incorrect to cast enterprises solely in terms of the throughput of material products. Rather, in sustainable economy they should deliver "human services" that improve the quality of lives.

A lot has been written in recent years about the perils of automation with predicted mass unemployment, declining wages, and increasing inequality. Rutger Bregman proposed a solution to this phenomenon—universal basic income and a 15-h-week—which has been tested in some countries or regions. "I firmly believe that a universal basic income and 15-h–week are the most effective answer to the dilemma of advancing robotisation. Not because robots will take over all the purposeful jobs, but because a basic income would give everybody the chance to do work that is meaningful. I believe in a future where the value of your work is not determined by the size of your paycheck, but by the amount of happiness you spread and the amount of meaning you give. I believe in a future where the point of education is not to prepare you for another useless job, but for a life well lived. I believe in a future where "jobs are for robots and life is for people" (Bregman 2017). Incidentally, a famous British economist John Maynard Keynes (1930) in his essay "Economic Possibility for our

Grandchildren" was predicting that due to labour-saving technologies we would all be working just 15-h weeks by the year 2030.

It is possible to identify some clear operational principles that enterprises should fulfill in sustainable development: (i) provide an equitable distribution of the goods and services needed for prosperity, (ii) use as little as possible in a way of materials and energy and minimal damage to the environment; and (iii) offer people meaningful employment and the opportunity to participate in society and contribute to the vitality of the community. People often achieve a greater sense of well-being and fulfilment, both as producers and as consumers of these activities, than they do in time-poor materialistic, supermarket economy in which much of our lives is spent (Castel et al. 2011; Victor and Jackson 2016).

"Humans have never faced a more vexing and dire convergence of problems caused not so much by our failures as by our successes. We are in a progress trap where we attempt to solve the problems caused by progress with the same methods and mind-set we used to progress. Material progress creates problems that are—or seem to be—soluble only by further progress. We are wealthier than ever, but the gap between the super wealthy and the rest of us continues to widen and the collateral effects of inequality and demoralisation infiltrate every sector of modern society. As a result, a small group of oligarchs hold our common future hostage" (Orr 2016).

If we are to survive, then we need to be aware of possible dangers in the world driven by greed, profit, and power. "We have not to forget that in spite of the fact that in many parts of the world progress has been successful, we have to remember how it was achieved. Prior to the mechanical items that we put under the umbrella term of technology, a highly profitable way to greater productivity was to use slaves. Slaves brought immense wealth to those countries; without it, they might never have risen to the world richest nations of the nineteenth century, or indeed have present status. According to evaluations of this process, more than 12 million people have been shipped to America, Britain and other countries to work as a slaves. Many informed estimates say the current world slave population is around 30 million and increasing. Therefore, implementing the sustainable development goals, we have to plan ahead and to be able to cooperate across the world stage for the benefit of all mankind, not just for local political, racial, or religion reasons. "If we are unable to change our greed and exploitation to obliterate such a cancer in society, because of the inherent profits, then no matter how vulnerable we are likely to become, because of natural events and dependence on technology, I do not believe we will ever manage to change our behavior to plan ahead to safeguard future generations" (Townsend 2018).

Indeed, "although the stakes of the economic system are more global than ever, the implementation of any approach by multinational companies will mostly be done at a more local level. Communities are locally embedded, and people know and trust those they live in close proximity to. Consider just for once the alternative: a global government regulates multinational companies in global markets, and people gather in a global democracy and global unions. It is an unrealistic and undesirable goal, as it increases the distance between individuals and the immediate social ecosystems they are part of. When the centre of power is too far removed from people's realities, neither political governance nor economic decision-making would have popular support" (Schwab 2021).

The shift from globalisation to glocalisation is transforming the relationship between national governments and local communities, in a sense, reversing the locus of responsibility for the workings of the economy and the affairs of governance from the nation state to region. This change in governance presages a revolution in the way humanity organises its common and social life. This shift to glocalisation will transform the sphere of actions for federal government. They will be a key player in the infrastructure build-out in the country, establishing new codes, regulations, standards, tax incentives, and other means for the transition towards sustainability.

"Private markets work well when there are many suppliers and consumers, but they begin to misfire when economic logic calls for a single supplier, for example, to operate the police force, fire department, army, court system, highway network, etc. […] Free markets fail also when producers cause adverse spillovers to the rest of society, for example, by polluting the rivers with toxic chemicals." In such cases, there is a need for "corrective pricing" such as tax levied on the pollutant, in order to reduce negative spillovers. Private markets fall short in the case of scientific research as well, where spillovers of knowledge occur. […] The implication is that one of humanity's most important activities –scientific discovery - needs to be promoted in ways other than the pure profit motive." None of the great promoters of market economy denied a major role for government in a market system (especially Keynes, Samuelson), only the present-day free-market followers of Hayek and Friedman who neglect the key role of government in ensuring the efficiency and fairness of the market system (Sachs 2015).

3.11 Decision–Making Support Systems

> We need to learn how to work with nature, rather than against it.—Sir David Attenborough, natural historian, UK

A decision-maker represents one essential link between the control targets, management costs and the individuals, groups and components, which make up the system being controlled. For this reason, the decision-maker and the decision-making process must usually be viewed as political (Bennett and Chorley 1978). For this, the decision-maker needs the mathematical description of the problem complemented by the human capacities of anticipation, learning, innovation, adaptation, and thought. The world in the twenty first century is marked by close systemic interlinkages, by positive synergies and by negative interactions and externalities that imply complicated trade-offs between different dimensions and processes of development—sectorial, local, regional, global and temporal. At the same time, the possibilities of producing potentially disastrous side effects from any control action together with the demand upon governments to satisfy an increasing proportion of societies needs, desires and aspirations, have stimulated the creation of so-called science-policy interface, which by large-scale decision-making support systems can provide actions that are in some sense rational and optimal. Science for sustainable development must provide the evidence to support decision-making process through the current social, economic and, especially, political decisions to enable creative and transformative solutions that bring far-reaching changes. Often driven by pressure to produce quick results, many scientists, engineers, and development practitioners continue to rely on simple framing and research, or intervention methods, even for difficult problems, such as the transition to decarbonised energy systems. Instead, there should be innovative partnership between science, technology, policy, and society. Scientific research focused on sustainable development has to uphold the highest standards of scientific rigour, in particular transparency, reproducibility, falsifiability, and compliance with specific standards of the discipline, but it should also consider relevant social norms and objectives, as well as people's and communities' aspirations and preferences, and explicitly address these as part of the research (IGS 2019).

The structure of a very simplified deterministic decision-making support system is presented in Fig. 3.2. The main role of a decision-maker in the constrained world is to make a disposition of resources between policy instruments and other system variables so that the control objectives are achieved as closely as possible within costs or other constraints, i.e. to rationalise and optimise the transformation system performance to achieve a given set of goals. Therefore, the result of the decision-maker's response must be transferred to those managers or groups who are responsible for the actual implementation of transformation control process (see system for transformations in Fig. 3.1).

"Unlike encouraging energy efficiency, the measures we must take to secure just, equitable, and inspiring transition away from fossil fuels clash directly with our reigning economic orthodoxy at every level. It requires visionary long-term planning,

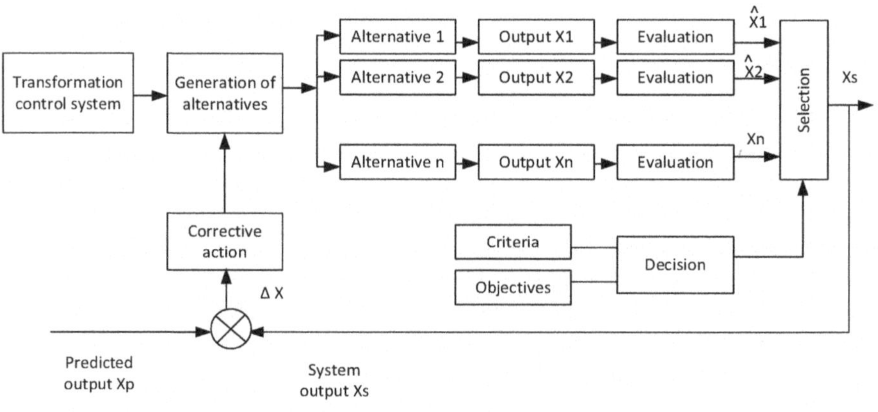

Fig. 3.2 The structure of simplified decision-making support system

tough regulation of business, highest levels of taxation for the affluent, big public sector expenditure, and in many cases reversals of core privatisations in order to give communities the power to make the changes they desire. In short, it means changing everything about how we think about the economy so that our pollution does not change everything about our physical world. It must always be remembered that the greatest barrier to humanity rising to meet crisis is not that it is too late or that we don't know what to do. There is just enough time, and we are swamped with green tech and green plans. And yet the reason is we are afraid—with good reason—that our political class is wholly incapable of seizing those tools and implementing those plans, since doing so involves unlearning the core tenets of the stifling free-market ideology that governed every stage of their rise to power " (Klein 2014).

Herbert Simon, the winner of prize in memory of Alfred Nobel (1978) argued that our rationality is "bounded". He did not believe that we are entirely irrational, although he himself and many other economists of the behaviour school have convincingly documented how much of our behaviour is irrational. The world is too complex for our limited intelligence to understand fully. This means that very often the main problem we face in making a good decision is not a lack of information but our limited capability to process that information. In other words, the world is full of uncertainties. In this situation, companies operate with "productive routines", which simplify their options and search paths (Chang 2012).

But where do we go from there? Very often, we need regulation exactly because we are not smart enough to leave things to the market (Chang 2012). The key question is about whether capitalism is actually compatible with creating a sustainable world?

It is evident that business must turn to more holistic scientific models, which account for the connectedness between a business organisation and the natural environment, because the current paradigms that underlie the economic myth do not adequately guide to their relationships with the Planet. "Organisations need a new scientific paradigm which allows them to understand that they are integral parts of interconnected, holistic, entropic planetary processes. Organisations also need a new, community-based economics paradigm which allows them to recognise that there is such a thing as enough" (Stead and Garner Stead 1994).

3.12 Transformations in Financial-Monetary System and Structure of Sustainable Investment

> Only when the tide goes out do you discover who has been swimming naked.—Warren Buffet, businessman, US

Capitalism would not have developed in the way it has without the development of the financial system. As a result, the financial sector has become much more profitable and this has enabled to offer salaries and bonuses that are much higher than those offered in other sectors, attracting the brightest students, regardless of the subject they studied in universities. It means that a lot of higher educational spending has been wasted, as many people are not using the skills they were originally trained. "Unfortunately, following the ride of "new finance" in the last three decades, our financial system has become a negative force. Financial firms have become very good at generating high profits for themselves at the cost of creating asset bubbles whose unsustainability they obscure through pooling, structuring, and other techniques. When the bubble bursts, these firms deftly use their economic weight and political influence to secure rescue money and subsidies from the public purse, which then has to be refilled by the general public through tax hikes and spending cuts" (Chang 2015).

The 'financialisation' that has accompanied marketisation creates particular governance challenges in the context of transitions to sustainability. These need to be addressed using a variety of approaches and instruments in different transitions arenas. In terms of mainstream markets, sustainability and environmental improvement projects face fundamental obstacles concerning the availability of financial resources. Such projects are often long-term undertakings with high up-front costs and significant market uncertainties (Criscuolo and Menon 2015). They are difficult to finance when investors are risk-averse and expect short-term returns on investment. Environmental projects face the specific challenge that common resources and ecosystem services do not feature in orthodox accounting frameworks, with values assigned only to what is immediately in demand. Speculation—and current opportunities for speculative behaviours to manipulate and distort currency and equity markets—also represent a barrier to more 'responsible' investment behaviours (Staniškis et al. 2005, 2008).

"The global financial system has a fundamental problem: it accelerates 3 trillion dollars through the system every day, nearly 90% of which is speculation, mostly speculation in the foreign exchange markets. We find ourselves colluding in that system through our savings, pensions and credit card debts, but it has nothing to do with the jobs that the financial sector is supposed to do—to facilitate the exchange of goods and services, to make capital available so that they can create productive business in the future. In fact, the global money system no longer meets the needs of vast majority of people on the planet, partly because it is out of control, partly because it is focused on speculation rather than production, and partly because it is pricing production out of existence" (Boyle and Simms 2009).

Banks do not really make loans, because a loan implies that the lender gives up what the borrower receives. When the bank lends money it gives up nothing, creating deposits *ex nihilo* up to the limit set by reserve requirements. The real "lender" is the community at large, whose money balances lose in purchasing power with the issue of new money. Prices are bid up since *ex nihilo* creation of money (demand) can increase much more rapidly than can the *ex materia* creation of new physical wealth (supply). "Banks are like counterfeiters who lend false money, accept their own false money in repayment, but receive the interest in real money transferred to them by the rest of the community and which is not destroyed. Also, by continually changing the value of money as they create and destroy it, the banking system converts the money into a rubber yardstick, in effect making a mockery of all physical measurements standards. Thus, debts are subject to the laws of mathematics rather than physics" (Soddy 1926).

Thanks to the deregulation of capital control, the state itself is also increasingly subordinate to the needs of business and finance, and openly so. This is a partly logical consequence of the focus on income and profit growth: the business and financial sector are seen as the means through which this can be delivered, and so their needs are given priority. This means that the purpose of social, health or education spending, for example, is not to make people healthy and well for their own sake, but to increase their productivity. "The education is not seen as a good thing in itself, but is valuable because it equips people with the skills needed for business to compete in a global world. In fact, everything becomes important and defined according to the ways in which it can increase this narrow profitability, primarily for corporations. Policy regards people simply as passive consumers of goods or services, which are "delivered" by service providers. Health care consumption is profitable in the short term for providers, preventative health is not, so we become consumers of health "solutions". This "profit growth" is facilitated by financial institutions, which have grown themselves to an enormous scale, but which also profit from arranging mergers and acquisitions in the business world" (Boyle and Simms, 2009).

Jeffrey Sachs, along with climate and environmental scientists, has shown that the world's ecological crisis affects the entire globe. There is no escape and no hiding, even if the coming changes will not affect all parts of the world in the same ways or at the same time. Yet global change, loss of biodiversity, impairment of ecosystems, and a growing population are threats to all human societies. We are all downstream from one to another. Therefore, the wealthier countries should give to the developing world the financial and logistical aid. The logic is that helping others is not only ethical and altruistically good but creates global stability that is ultimately beneficial for the industrialised world too (Sachs 2015).

There are numerous ways to align investment incentives with environmental conservation goals. These include developing proper markets for natural capital and ecosystem services, using combined public–private financing programmes and instruments to support investment in green projects, and developing 'green' and 'responsible' investment funds. A potentially important innovation involves extending the concept of offsetting by requiring project developers to compensate for any loss of natural capital and ecosystem services resulting from their project. This is being explored through the development of new governance principles, such as No Net Loss (NNL). Another possible intervention is a tax on financial transactions to deter short-term speculation and to promote longer-term investments in 'real' projects. Although referred to as a 'Tobin tax' (after James Tobin, who suggested taxing all spot conversions of one currency into another), the financial transaction tax was actually suggested by Keynes in 1936 as a means to constrain speculation and reduce or eliminate 'noise' transactions. The technical feasibility of deploying such a tax is an unsettled issue, though several proponents suggest that this is made easier by ICT (information and communications technology) advances. The 2007 inquiry of the All-Party Parliamentary Group on Debt, Aid and Trade suggested that revenues from a financial transaction tax could be dedicated to achieving the

Millennium Development Goals and to investment in climate change mitigation and adaptation (All Party Parliamentary Group 2007). Alternatively, revenues could be used to insure the global tax payers against a future banking crisis.

New forms of financing and financial governance are also emerging in the complementary economy transition arena. An example is crowdfunding using internet platforms to raise capital for promising initiatives and projects (Vulkan et al. 2016). Performance assurance bonds are another important instrument that could bridge mainstream and complementary markets, by helping social innovations organisations take over some functions of service delivery now handled directly by central and local government or their agencies. These instruments could also overcome sectorial and departmental divisions in government and its agencies, and enable public money to be pooled and directed toward cross-cutting initiatives and projects that address more than one area of need (Weaver et al. 2015). Global transformations of the economy towards sustainability urgently need changes in the investment strategies:

- *The allocation of capital (investing in the future).* Investment in the sustainable/green economy needs to be focused on the protection and maintenance of the assets on which future prosperity depends.
- *The nature of money economy.* Money economy (creation, maintenance, and stability of the money supply) is a vital component of sustainability. Unconstrained creation of money through commercial debt stimulates unsustainability in investment and instability in financial markets. The information of national accounts, which includes vital information about financial flows and provides full account of the balance sheets of each major sector, should be taken seriously. Prosperity in one region at the expense of high levels of indebtedness in another region flies in the face of social justice, and has no place in a sustainable/green economy.

Wolfgang Merkel challenges nothing less than the fundamental assumption of post-war political science that capitalism and democracy are birds of a feather: that just as capitalism needs as well as supports democracy, democracy needs as well as supports capitalism, the two flocking together in everlasting pre-established harmony. Is capitalism compatible with democracy?, Merkel argues that according to him, the historical developments of the past three or four decades have deeply diminished the efficacy of democracy in capitalist world, so the answer to this question must be increasingly not. Rightly, Merkel designates capitalism as the challenger, the independent variable, while democracy functions as the dependent variable. In particular, Merkel focuses on the transformation of the post-war "social market economy" of what he calls "embedded" capitalism, in the course of its financialisation, a process that began in 1980 and reached its climax, for the time being, in the financial crisis in 2008. Post-war democratic capitalism was not a selection by skilful social engineers or concerned citizens from a range of less optimal alternatives, but a historical compromise between a then uniquely powerful working class and an equally uniquely weakened capitalist class that was as never before on the political and economic defensive – which was true in all capitalist countries at the time, among the winners of the war as well as the losers (Streeck 2016).

Many Post-Keynesians, working through the framework of modern monetary theory, emphasise the economic role of states or state unions with their own currency and central bank claim of these scholars is that states can never run out of their own currency. A sovereign currency is not a limiting factor in collective action such as a sustainability transition, unlike natural, social and technological resources. This has been the case since the dropping of the gold standard and the adoption of fiat money in the 1970s. The state can always spend and invest in its own currency. It also does not have to hold on to particular jobs or industries for the sake of tax revenues. In other words, from this perspective, collective action, organised at least partly through the state, should be guided not on the basis of securing public funds but on the basis of social goals and material boundary conditions (Järvensivu et al. 2018). Key international institutions, such as the International Monetary Fund, which has been known for its policies of privatisation and export-led industrialisation, will need to be reconfigured accordingly.

"Financial system needs to be regulated because of its power and importance. It is not simply because of sector's lobbying power that most politicians and regulators have been reluctant to radically reform, despite the incompetence, recklessness and cynicism in the industry which it has revealed and also because of their ideological conviction that maximum freedom for the financial industry is in the national interest" (Townsend 2018).

Transformations towards sustainability and sustainable development projects need long-term investments, but the rise of the "shareholder value maximisation" model in the era of new finance has dramatically reduced the resources available for long-term investments in non-financial corporations.

The regionalisation and localisation of production and consumption could be assisted by the emergence of electronic regional and local currencies, complementing those operating at national and international levels. These currencies could provide effective economic feedback loops and further assist regional economic development in a dynamic fashion. So if a region suffers from a downturn, its regional currency will naturally depreciate against the national currency and other regional currencies, increasing within country 'export' demand and encouraging import substitution and more local production of services (Greco 2009).

Governance of transformation

> It is horrifying that we have to fight our own government to save the environment—Ansel Adams, environmentalist, USA
>
> A politician thinks of the next election. A statesman of the next generation—James Freeman Clarke, US theologian, and author

Sustainability represents a hugely complex challenge, requiring coherent activities across multiple sectors and scales. New technologies alone will not deliver transformations to sustainability. As Mazzucato and Perez (2015) note, "renewables alone do not constitute a synergetic technology system that results in a long wave for the economy". A green long wave "golden age" would need to include "conservation, pollution control, reduction of material content per product; designing for durability;

replacing products, possession and waste with services, rental and maintenance and recycling, respectively; promoting the flourishing of the creative economy; making cities more liveable and less polluting; revamping transport systems and the built environment; promoting collaborative and sharing economies; focusing on health (including preventive and personalised medicine); and promoting all forms of education, in and out of schools" (Mazzucato and Perez, 2015). Strict environmental regulation, introduction of sustainability and welfare indicators, scepticism about the GDP indicator as a relevant guide for public policy, acceptance of lower material consumption levels without severe losses of happiness, reduction of status competition in consumption, and more cooperation at the community level are the main multipurpose instruments. It is important to note that one of the mentioned issues can translate into problems associated with another and the timing of the application of different strategies is very important (Antal and Van den Bergh 2013). The SDGs need to act as a system of interacting components that together move the world into a safe and just operating space. Studies show that the SDGs are much more than just a collection of targets, but a system of synergistic reinforcement. While no single SDG has the power to transform the world alone, the whole set of SDGs together does (Arillo 2017).

The proper balance between markets and government has been a centre of debates for generations, going back to Adam Smith's explanation on self-organising markets. Below there are five of Sachs' conclusions regarding this debate:

- In productive sectors of many producers and consumers, where strong market competition applies, we should rely on market forces. They are decentralised, voluntary, and do not require the very difficult work of creating cooperation among a large number of people.
- We should turn to government to ensure the fairness and sustainability of market outcomes, including the broad distribution of income in the society.
- We should recognise that the knowledge of science and technology is a public good that should be promoted actively by government alongside the private sector. Markets alone will not create the twenty first-century knowledge society.
- As economic life becomes more complex, we should expect the role of government to become more extensive. We need fresh thinking about our circumstances, especially at a time of rapid globalisation, environmental threats, and a knowledge-based economy.
- We should appreciate that circumstances as to the appropriate role of markets and government differ across countries. History has shown that the emerging economies should devote special government resources and policies to closing the technology gap, while countries in the lead should devote special government resources to cutting-edge research and development (Sachs 2012).

"Sustainable development never criticised the markets in themselves. On the contrary, markets with clear social and environmental parameters can be effective ways for human beings to interact and get what they need. Nobel Prize winning economist Amartya Sen said that "to be generically against markets is like being generically against conversation". Markets are part of life, can be vibrant and bustling at street level in towns and villages binding communities together. This notwithstanding, on the larger scale, markets can be faceless, bland and destructive, the economic equivalent of aerial bombing, in which the pilot never gets to see the damage they cause on the ground "(Boyle and Simms, 2009).

When prices "tell the truth" about costs, consumers adjust their consumption patterns accordingly, then markets would operate more efficiently, producers would innovate and adopt resource-efficient and cleaner production processes, and total production and consumption would decline.

"I believe in my kind of economics", Samuelson said, adding how the theories that had preceded his work, written by his own teachers, had been imperfect. "During the Great Depression, they said that government should do nothing about it because they would only make things worse." Samuelson had a different belief. He felt he was not only a researcher but an educator too and wanted to introduce students to a different approach. This included the ideas of British economist John Maynard Keynes, who advocated that markets needed active government participation and controls.

Sustainability economists tend to favour greater government regulation of markets and arrange extra-market mechanisms: firstly, pollution taxes that make wastes and ecological destruction costlier and thus less attractive; secondly, shift of government subsidies from polluting to green industries; thirdly, promotion of sustainable advertising and marketing practices and consumer education to reduce environmental impact; fourthly, support to cap-and-trade system, that imposes upon industries fixed emission levels with tradable pollution rights, and finally, natural capital – the goods and services that come from the natural environment – need to be protected, restored and valued.

The upshot is that despite a great deal of talk about systems, we continue to administer, organise, analyse, manage, and govern complex ecological and economical systems as if they were a collection of isolated parts and not an indissoluble union of energy, water, soils, microbes, land, forests, biota, and air. Humankind has the capacity to create far more information than anyone can absorb, to foster far greater interdependency than anyone can manage, and to accelerate change far faster than anyone's ability to keep pace. The managers of environmental/urban systems require the capacity to shift from seeing parts to seeing wholes, from seeing people as helpless reactors to the present to creating the future (Senge 2006).

Governance is not only about our politics and government, but also about major organisations that are key actors in sustainable development, including our private sector. There are certain shared principles of governance for the public and private sectors (Sachs 2015):

- Accountability,
- Transparency,
- Participation,
- The polluter-pays principle,
- Corporate responsibility, and
- Clear affirmative commitment to sustainable development.

"My view is closely related to an ancient and very important doctrine known as *primum non nocere,* which means, "First, do no harm". A principle of good governance in my view is first, do no harm. Even if the law for whatever reason allows an enterprise to impose costs on others, it is the company's responsibility not to do so, because our higher responsibility is an ethical responsibility to do no harm" (Sachs 2015).

Companies will need to carry out a new materiality assessment to ensure they have captured what is really material to them. They will then need to address these issues in their sustainability strategy and report about it. Addressing the interlinkages between environment, social, economics, and wellbeing are the initial building blocks. However, a new and more transparent governance which uses adaptive management to deal with crisis is needed (Viliani 2021).

"It must always be remembered that the greatest barrier to humanity rising to meet crisis is not that is too late or that we do not know what to do. There is just enough time, and we are swamped with green tech and green plans. And yet the reason is we are afraid – with good reason – that our political class is wholly incapable of seizing those tools and implementing those plans, since doing so involves unlearning the core tenets of the stifling free-market ideology that governed every stage of their rise to power" (Klein 2015).

The common cliché is that "knowledge is power", but political power comes from an electorate that only has carefully selected information (definitely not knowledge). Townsend argues that our world is guided by leaders, and the rest mostly follow without serious dissent. Therefore, to make a transformation towards sustainability we need a far better majority and leadership who will recognize not just local but also global issues. This means that we need to move away from the strict divisions of party politics, and the simplistic "tribal thinking". Instead, we need a common aim for our own nation, and for the world as a whole. (Think globally, act locally). "Reality is that we already have a global economy, and therefore all solutions must be considered in terms of global survival, global sustainability, and global welfare… therefore, a radical thought is that we reorganize how we operate such as local councils, Parliaments, the Senate and Congress, and the United Nations ". Otherwise, we then become irrational in our behaviour if we are with a pack of people with similar objectives. This mob-rule attitude is counterproductive, so if we can undermine it, then there could be a progress " (Townsend 2018).

"We have to vote. Very few politicians (if any) have paid any price for not only not leading, but also for holding us back from addressing this impending disaster. They have made the prospects for fighting climate change worse, and they have made sure the effects of climate change will be more destructive and happen sooner. We can 't allow that anymore. It is not enough to elect politicians who profess to understand and care about the problem; we must hold them accountable for their actions. Vote for people who offer meaningful policies, and make sure they are achieved. If they don 't, vote for someone else who will" (Schlossberg 2019).

Yes, today we vote, but free, fair and credible elections are a considerable but important challenge, logistically, financially, and politically. Poorly conducted elections, for example, those that are carried out prematurely, those without adequate inclusion, or without transparent procedures – can easily exacerbate violence.

The systems theory offers an iterative method, which is a mathematical procedure that uses an initial value to generate a sequence of improving approximate solutions for a class of problems, in which the n-th approximation is derived from the previous ones. In other words, it means the iterative voting scheme allows voters to revise their voting strategies based on the outcomes of previous iterations. An iterative voting process has a certain natural attractiveness, allowing voters to modify their stated preferences, in light of what they see about the results of an election. Assuming that all voters are equivalently entitled to make modifications, it seems an appealing way to acknowledge the strategic nature of voters, allowing them to change their votes to get results they prefer. If the process converges, we reach some stable expression of the aggregated group preference—but the process may not converge. In case of electronic voting system, this iterative method could be implemented easily.

Bowman, Hodge, and Yu (2014) using a robust computer simulation, investigated the potential of this iterative approach to solve the separability problem. The separability problem occurs when a voter's preferred outcome on one proposal depends on the outcomes of other proposals. This type of interdependence can lead to unsatisfactory or even paradoxical election outcomes, such as a winning outcome that is the last choice of every voter (Bowman et al. 2014).

Our only hope is rapid systemic change–change based on market signals that tell the ecological truth. This means restructuring the tax system: lowering income taxes and raising taxes on environmentally destructive activities, such as fossil fuel burning, to incorporate the ecological costs. Unless we can get the market to send signals that reflect reality, we will continue making faulty decisions as consumers, corporate planners, and government policymakers. Badly informed economic decisions and the economic distortions they create can lead to economic decline (Brown 2003).

Repeatedly, we witness popularity of the notion "green growth" and "decoupling of growth". The reason for that is to keep the stability of the present world economy, which strongly depends on continual growth and an expensive as well expansive financial system. Antal et al. for business transformation towards sustainability consider multi-purpose instruments including "strict environmental regulations, environmental tax reforms, the introduction of sustainability and welfare indicators, scepticism about the GDP indicator as a relevant guide for public policy, the acceptance of lower material consumption levels without severe losses of happiness, the reduction of status competition in consumption, and more cooperation at the community level" (Antal and Van den Bergh 2013).

3.13 Sustainable Development in the Developing Countries

> Humans are born free but live in chains—Jean-Jacques Rousseau, French writer and philosopher

Globalisation has led to an increase in economic and social inequalities globally, despite a decreasing trend in extreme poverty. Facing economic crises, developing countries were faced with the condition of IMF and World Bank that borrowing countries had to implement the structural adjustment programmes. "Despite their own history, the rich countries make developing countries open their borders and expose their economies to the full forces of global competition, using the conditions attached to their bilateral foreign aid and to loans from international financial institutions that they control as well as ideological influence that they exercise through intellectual dominance. In promoting policies that they did not use when they were developing countries themselves, they are saying to the developing countries, 'Do as I say, not as I did'" (Chang 2012).

The so-called "Washington Consensus", emphasising open markets and unrestricted movement of financial and physical capital, along with aligning domestic policies to the structural adjustment programmes was mandated for the developing countries and countries in transition by IMF and World Bank. The Consensus comprised liberalisation, privatisation of state infrastructure and services, fiscal austerity, and the search for competitiveness and services, as well as macroeconomic stabilisation. The expectation was that more jobs and higher rates of economic growth would be generated. This has not occurred although inequality has increased, and the incidence of social vulnerability is clearly evident. This has led to the growth of informal economy, the expansion of low-wage service industries and the increase in jobs without any legal protection" (Reardon, 2018). In reality, "these programmes required shrinking the role of government in the economy by cutting its budget, privatising SOEs (state-owned enterprises) and reducing regulations, especially on international trade. Despite making all the necessary "structural" reforms, most countries experienced a dramatic slowdown in the 1980s and at the end of the century. Only Chile did well out of neo-liberal policies, but at a considerable human cost under the Pinochet dictatorship. All the other stories of this period were economies that used state intervention extensively and liberalised only gradually. The best examples of this were Japan, economies in East Asia (South Korea, Taiwan, and Singapore) and, increasingly, China" (Chang 2015).

"My argument is somewhat different. My argument is that aid can work and that it is vital in certain circumstances. It is especially vital when people are very poor and facing life-or-death challenges, such as malaria, AIDS, safe childbirth, safe water, sanitation, or growing enough food to stay alive. Markets cannot meet the needs of very poor. The desperately poor are not consumers who will create an immediate profit. Nor can be urgent needs be met through the very meager budgets of the poor countries. And so the poor need help through other means. Then the question is whether it is possible to help effectively through international aid, without too much corruption, theft, and debilitating bureaucratic inefficiencies. My answer is yes, if done thoughtfully" (Sachs 2015).

There is no doubt that effective use of ODA (Official Development Assistance) and other public financing with a responsible organisation of a serious process of planning, road mapping, monitoring, evaluation, and strategy updates will help the poorest countries in particular, and will play a vital role in the success of the SDGs. At the heart of neoliberalism is a contradiction. While its proponents profess a belief in free unregulated markets, they favour regulations to prevent collective bodies from operating in favour of social solidarity. That is why they want control over unions, collective bargaining, professional associations and occupational guilds (Standing 2016).

"The method of making money on the basis of future expectations rather than past achievements is critical to the development of capitalism and to the boom in activity and rapid innovation. Earlier the ability of money to preserve value enabled people to accumulate wealth at a given period and use it as a call on goods and services in the future. The money created as it is in the current system automatically creates a pressure to economy to grow, both because of the need to repay interest and because the money is created in advance of being supported by products of real value. However, such products need to repay the debts created when that money is made. During the initial stage of capitalism, this provided an extraordinary boost, but today this rapid growth is causing an ecological and economical crisis. Thus, our current money system must be changed if we are to achieve sustainability. During an economic boom such investment schemes prosper, but in a crisis people seek assets of real value, for example land or productive enterprises, and those holding only paper assets in over-inflated speculative schemes are likely to lose their shirts. While traditional debt-based currencies are the "fuel" of today's capitalism, local currencies might very well become the "oil" of tomorrow's commons based economy" (Reardon et al. 2018).

"I have no quarrel with wealth per se. Many wealthy individuals are highly creative, talented, generous, and philanthropic. My quarrel is with poverty. As long as there is both widespread poverty and booming wealth at the top, and many public investments (in education, child care, training, infrastructure, and other areas) that could reduce or end the poverty, then tax cuts for the rich are immoral and counterproductive. [...] I am a firm believer in the market economy, yet my twenty-five years of work in Asia have convinced me of the value of long-term government planning of public investments for quality education, modern infrastructure, secure and low-carbon energy sources, and environmental sustainability" (Sachs 2012).

Developing countries raise concerns about the concept of green growth/sustainable development: (i) issues related to the international dimensions of green growth such as the risks of green protectionism and green conditionality for ODA; (ii) how green growth helps address poverty and other development priorities, (iii) will green growth/sustainable development efforts be impeded by high-cost barriers. An international enabling environment should send the right policy and market signals, create momentum for sustainable development by setting environmental and others regulations and standards, stimulate the demand and supply of environmental goods and services, catalyse green efforts in energy, transportation, agriculture and others sectors, make easier to access financing, technology and innovations, facilitate the international exchange of knowledge issues or the cooperation in science, technology and innovation through:

- Enhancing capacities;
- Strengthening Official Development Assistance;
- Promoting innovation and accelerating the dissemination of green technologies;
- Facilitating trade of environmental goods and services;
- Developing coherent policies (OECD 2012).

Stiglitz proposed ways to produce a more comprehensive approach to global development (Stigliz 2016):

- Increase foreign assistance from the rich countries to the poor at the value of at least 0.7% of their GDP;
- Cancellation or relief of foreign debt that should be written off;
- Genuine fair, rather than free trade, recognising the limitations of economic liberalisation and iniquities produced by global corporate monopolies and cartels;
- Protection of the global environment on which all economies ultimately depend through a sensible and workable public management of global natural resources and regulations on their usage and on actions, giving rise to externalities and costs;
- Good, democratic government, including enhanced possibilities for democratic regulation of the economics and participation in decision-making processes at all levels.

Relevant information and statistics provide the foundation for policies that promote sustainable development, and are crucial to monitoring progress and gauging results. Countries should take flexibility to focus on the indicators that reflect their own development objectives, such as building economic, environmental and social resilience and ensuring that development is inclusive.

All of us pay for sustainable development in two ways: through markets and through political institutions. "But tremendous and sometimes very bitter fights ensue over the proper balancing of financing between the market–oriented, profit–seeking investments of businesses driven by their sales of goods to consumers, and the financing that comes through the public sector" (Sachs 2015). In fact, both public and private approaches are needed and are complimentary mechanisms to finance sustainable development. One very important aspect of public finance is ODA. Yet to be successful, such aid needs to be well-targeted and well-managed. The critics argue that such type of aid, when taxpayers in one country help to provide public services in another country through the aid, is simply unnecessary and markets are always the solution. Other sceptics believe that aid is inevitably wasted. A third argument is that aid is more than wasted—it leads to a kind of dependency mentality that is demeaning and diminishes motivation. Sachs agrees with the aid sceptics that much aid is wasted and argues that this is not development aid. He supports the aid that has a purpose to achieve development objectives.

Sooner or later, economic and political crises will force a reconsideration of the principles organising the world's human economy. The rich countries propose to cap the greenhouse gas emissions of developing countries at a lower level than those of the United States and the European Union. New players in the global restructuring of capitalism today object, not unreasonably, since the West is responsible for the great bulk of the carbon dioxide in the atmosphere. Both Brazilian and Chinese leaders joked at the Copenhagen summit on global warming that the United States is like a rich man who, after gorging himself at a banquet, then invites the neighbours in for a coffee and asks them to split the bill (Hann and Hart 2011).

Indeed, why should, for example, China and Indonesia have to play the game by different rules from those followed by the developed countries in the past? It is often perceived as unfair for rich countries to lecture poor countries on resource use when the developed world is mainly responsible, by its size, history and volume of activity, for the majority of resource consumption and the problems that ensue from irresponsible development. So then, "the question becomes one of how to fairly share the burdens of achieving well-managed growth. The developed countries should help by providing technologies, finance and know-how to tackle unsustainability and climate change issues, over and above regular development assistance" (Strange and Bayley 2008).

As participants of UNEP and UNIDO projects in the developing countries, Staniškis et al. agree with the OECD scenario on how developing countries have to cope with climate change and with problems of responsible consumption and sustainable production. As mentioned in subsection 2.6, in 1999, UNEP has launched the four-year project aiming at increasing investments in resource-efficient and cleaner production innovations. The main tool for the project was a system and methods for preventive innovations generation, financing and implementation, developed by APINI scientists (see Fig. 2.7 , 2.8 and 2.9). The project goal was to demonstrate how soft investments could be stimulated by helping financial institutions to understand the essence of the RECP approach and to help local experts to develop creditworthy investment proposals. The project demonstration countries (Guatemala, Nicaragua, Vietnam, and Zimbabwe) represent a wide span of socio-economic conditions and have different industrial interests and development level. National cleaner production centres (NCPCs) have been establish by UNEP/UNIDI in all of them, and their staff was closely involved in this project. RECP innovations were generated using above-mentioned methodology in three–four key industry sectors in each country. The demonstration project in general was successful and the experience obtained and results were disseminated to other countries, for instance, Uganda, Peru, Russia, China, Ukraine and others. One of the conclusions is that donors, recipients and professionals working on project implementation realise that without better coordination, full local engagement and accountability, aid is likely to fall short of its targets (Staniškis et al. 2008).

3.14 Final Discussions and Conclusions

> Twenty-first century choice: look after our planet, and it will look after us, or don't and face the consequences... Without environmental sustainability, economic stability and social cohesion cannot be achieved.—Phil Harding, British archaeologist

In this book, three types of management/control systems are developed based on systems theory and facilitating sustainable development: (I) systems for sustainability reduction at the level of enterprises; (II) systems for transformations management/control of socio–economic systems; (III) decision-making support system for governance. We argue that for achieving sustainability, two conditions should be applied: necessary and sufficient. The methods and systems for unsustainability reduction are treated as the necessary condition, whereas the model and the system for structural transformations of socio–economic system—as the sufficient condition. The mathematical formulation of the control tasks for sustainability reduction and transformation systems clearly shows that the three types of systems are structurally related and presents the hierarchical approach for achieving sustainability on enterprise, regional or national level. Thus, the decision-making support system generates the task for transformations system leading towards sustainability and the latter gives a desired output for unsustainability reduction system. Investigations show that there are efficient methods and systems for solving the unsustainability reduction problem. Concerning the transformations of socio-economic system, many possible assumptions are described; however, profound discussions and evaluations should follow nonetheless. We hope that the models, methods and systems proposed in the book will structurise the problem more clearly and facilitate the solution.

Sometimes one can hear an opinion that "there are number of trends and proposals on evolution of private, for-profit firms in an environmentally constrained world away from profit-seeking behaviour towards more socially comprehensive goals. It is not apparent, however, that the implications of such an evolution have been defined and considered: indeed, it is arguable that progress towards sustainability can best be achieved by modifying the boundary conditions in which private corporations operate, rather than by trying to change them into organisations which reflect social goals beyond profit-seeking, i.e. that a wise policy should create incentives for corporations to behave in ways supporting the achievement of sustainable societies." From the systems theory point of view, the best "incentive"/solution for profit firms is the application of complex performance index within real boundaries and constrains (see Sect. 3.3).

As an economic system, capitalism relies on the private ownership of production means to provide incentives in order to produce goods and services that people need (or think they need). Its modus operandi is profit maximisation, assuming that a private owner can manage resources better than the community, the local government body or the national state. However, capitalism has also created serious environmental consequences, which have disproportionately affected the poor. The basic problem: the wrong kind of interdependence leads to poverty (globalisation, especially in food production) (Boyle and Simms 2009). Below are several opinions how to solve the problem.

According to philosopher Jean-Jacques Rousseau, the solution was to reject civilisation and return to the natural state. This view in its turn laid the foundation for the romantic movement of the nineteenth century to which many modern environmentalists still appeal. According to Karl Marx, the solution lays in common or public ownership.

Parson Malthus' argument was that growth in population always runs faster than growth in the resources available to feed people. Malthus was wrong, at least as far as the particulars of his claims. The global population is now over seven times what it was in Malthus' day, and this is partly because the means of subsistence expanded considerably faster than population did, i.e. the global economy is more than 80 times bigger than it was in 1800.

"We believe that we should care for others both now and in the future. First, caring for both humans and non-humans is what makes us human. Second, we need to respect those who come after us, to ensure that the planet is liveable, and perhaps in a better condition than we received it. And third, science indicates that one reason why societies have failed in the past is because human demands on Earth's natural resources outstrip the pace of supplying them, or replenishing those that can be renewed or regrown" (Diamond 2005).

Jared Diamond demonstrated a whole series of global societies that collapsed as a result of overstressing local environments and formulated a five-point framework to understand their collapse: environmental damage, climate change (man-made and non-man-made, hostile neighbours, friendly trade partners (or lack thereof), and the society's response to its environmental problems (Diamond 2005).

Garrett Hardin sounded off on the inability of technology to solve humanity's ills, and disagreed with Smith's contention that private self-interest leads ultimately and uniquely to public benefit. He concluded that "freedom in a commons brings ruin to all" (Hardin 1968).

In his book "Stakeholder Capitalism", Klaus Schwab wrote: "I hope personal experiences will help to understand my world view, which is based on the belief that the best outcomes in a society and economy result from cooperation, whether between the public and the private sector, or peoples and nations from around the world." The primary principle for the implementation of stakeholder capitalism instead of shareholder capitalism is therefore that of subsidiarity. It determines, in other words, that local stakeholders should be able to decide for themselves, except when it is not feasible. Subsidiarity supports a national or local level of decision-making for countries to determine which path will work best for them to effectively address the global goal (Schwab 2021).

And finally, capitalism's modus operandi—profit, growth and expansion—has been instrumental in causing climate change, which suggests that it would be very difficult to look within capitalism for adequate solutions to the problem. Capitalism and the environment, unless the former radically changes, are on a collision course." We believe that we can effectively design an economic system so it provisions for all, not just for today but also for tomorrow. This is the essence of sustainability;

and this is the obligation of the educated—a quid pro quo (a favour for a favour) to help make the world a better place. "The world's biggest corporations deserve credit for changing some practices in response to the escalating global environmental crisis… but trusting them to lead sustainability efforts is like trusting arsonists to be our firefighters" (Dauvergne 2018). Each year, the world's tax payers underwrite 700 billion USD of subsidies for environmentally destructive activities. There is something unbelievable about the world spending hundreds of billions of dollars annually to subsidise its own destruction (Brown 2003).

All these environmental trends are social problems as far as they affect the conditions for human life and are perceived as problems only by human beings. Shrinking biodiversity in the world's oceans, rivers and lakes, in rainforest and savannah, is not a problem for nature. Plants and animals have no awareness that their habitat is disappearing; they simply die. Human life communities register ecological problems, however, because humans are unique in being conscious not only of the past but also of the future. This alone creates a faint hope that their insight into what has gone before will lead them to think about what they will no longer be able to do. It is quite possible that many migrants in the near future will be environmental emigrants (Rifkin 2014).

The Easter Island was like a laboratory in which this or that event happens under controlled conditions. Immediate consequences for the islanders were losses of raw material (forest), losses of wild-caught food, and decreased crop yields. Lack of large timber and rope brought an end to the transport and erection of statues, and also to the construction of seagoing canoes. There is no way of compensating for such a resource collapse on an island cut off from the outside world; fishing became ever more problematic as the canoes fell into disrepair, while soil erosion on the deforested, wind-swept land made agriculture less and less of a possibility. Without food, there was no longer any fuel, and in winter, the inhabitants burned the last plants and grass. Evidently, this shrinking of the means of survival must have intensified competition in relation to food, building material, tools and symbolic representation. "If decisive proof were needed that man does not live from bread alone, the Easter islanders certainly provide it with their persistence in a cultural practice to the point of risking their own survival" (Diamond 2005).

Short history of Easter island. Nine hundred years ago, a small group of Polynesians paddled the world's greatest ocean in search of a new land. A new piece of land was discovered. They found it lush with palm trees and other endemic vegetation growing all over the island. Generations passed, and the inhabitants of what was to be known as Rapa Nui, built a civilization of art, capable of carving, raising and transporting hundreds of gigantic monolith statues, using nothing but their own hands and stone. During the 15th or sixteenth century, the civilization at this small and isolated piece of land was highly advanced. The crops were sufficiently abundant as to support a part of the population to concentrate entirely on building bigger and bigger statues. The islanders grew in numbers and much of the lush palm tree forests were cut down and burnt to clear areas for crops, for transportation of the statues and building canoes. In the seventeenth century, the resources of big trees were depleted and winds dried up the land. The islanders tried with some short success to adapt to their tree-less country but finally it collapsed.

In the beginning of the 1860's, 1,500 islanders were taken to work as slaves in guano deposits at Chincha Islands and plantations in Peru. Among the kidnapped were the ruling king as well as the wise men who knew how to read the rongo-rongo script, which today no one is left to interpret. A few of these were later released, all of which died of small-pox on the return voyage, except for two people. These two spread the disease to the rest of the Rapa Nui population. The natives had no immune system towards this foreign disease, which resulted in an aggressive decrease of the population. A few years later, only 111 people were left at the island. It should be mentioned that there have been several different attempts to explain the Easter island collapse. For instance, Hunt argues: "I believe that there is substantial evidence that it was rats, more so than humans, that led to deforestation. Our work on Anakena (white sand beach), as well as previous archaeological studies, found thousands of rat bones" (Hunt 2006).

Thus, there are two lessons to be learned: environmental devastation, which has now reached the global level, and slavery, which was the dark period of capitalism development. This was not only tragedy for those who became slaves (in most cases they did not survive) but it also depleted local societies of workers and destroyed their social system.

Just because we do not know how to create a truly sustainable society, that does not mean that we cannot do things to become less sustainable. There are many examples of environmental destruction brought about by traditional cultures. The most famous example is the Polynesian deforestation of Easter Island, precipitating the collapse of their entire society, but it is only one of many. We cannot know what went through the head of the man who felled the last tree; he probably gave a little thought and treated it as a simple necessity of life. Therefore, ours is not the first society to devastate its environment, although it is the first to do it on global scale. If you have a society in which all people are well educated and environmentally conscious, you have a better chance that the society as a whole will make choices that make the economy more prosperous and more sustainable in the long run. If you want to change how someone thinks, give up; you cannot change how another thinks. Give them a tool, the use of which will lead them to think differently. At the same time, the university, as a key institution for providing discussions, is under threat. For forty years this indispensable democratic institution has been systematically betrayed by governments and the political class, who have redirected it from its proper social and cultural functions through a relentless programme of financialisation. Docherty argues that the sector has been politicised and now works explicitly to advance a market-fundamentalist ideology that drives an ever-widening wedge between ordinary citizens and the privileged and wealthy (Docherty 2016).

So, then the question becomes one of how to fairly share the burdens of achieving well-managed growth. Developing countries have to cope with climate change and other problems they did not create, and do not have the same means as developed countries in tackling them. The developed countries should help by providing technologies, finance and know-how how to tackle these issues, over and above regular development assistance. Those technological wonders that we hope will appear in time to save us take years to develop, and there is really no break between a "now" in which we can procrastinate and a "future" when we can start to think about solutions. Even if we can avoid some of the most drastic outcomes, emergency solutions tend to be very expensive. And then we can only mitigate the negative effects of a problem, rather than erase them. Whether the crisis is starvation, pollution or flooding, those who are affected at the moment of crisis suffer the consequences of a lack of long-term planning. And in the case of species loss, there is no solution what is gone is gone forever (Strange and Bayley 2008). For instance, the estimated economic value of bees and other pollinators that are going extinct is 190 billion USD per year.

"The new economics regards the concept of "capital"—traditionally no more than money, land and labour—as far as too narrow. It is not alone in recognising that other kinds of capital are equally fundamental to human development, and often more so, whether it is social capital, intellectual capital (know-how) or human capital. Therefore, defining objectives too narrowly in terms of money threatens to undermine active communities and their social capital. The new economics also assumes that economic success is underpinned by environmental capital, the ability of the planet to support life, without which business is impossible" (Boyle and Simms 2009).

The basic vision underlying sustainable development is that of the economy as a physical subsystem of the ecosystem. A subsystem cannot grow beyond the scale of the total system of which it is a part. If the total system provides services that the subsystem cannot provide for itself, then the subsystem must avoid impinging on the parent system to an extent and in ways that would impair its ability to provide these services. The scale of the economy must remain below the capacity of the ecosystem to supply sustainably services such as photosynthesis, pollination, purification of air and water, maintenance of climate, filtering of excessive ultraviolet radiation, recycling of waste, etc. To maintain the present scale of population and per capita consumption we are consuming natural capital and counting it as income. The alternative is to stop growth in scale, and seek to overcome poverty by redistribution and qualitative increase in efficiency of resource use, rather than further quantitative increase in resource throughput. Sustainable development is development without growth in the scale of the economy beyond some point that is within biospheric carrying capacity (Daly 1997). Don't become a "man with a hammer"; there is more than one way to do economics, each with its strengths and weaknesses. ("He who has a hammer sees everything as a nail.").

The current concern about unsustainability has arisen from observation that both natural and socioeconomic systems are losing resilience, that is ability to cope with perturbations created by human activities without the appearance of fundamental, qualitative changes in the functions of these systems. Ecological concept of resilience refers to the capacity of ecosystems (or species) to cope with change. Social scientists and ecological economists applied this concept directly to human predicament. According to Lewis and Conaty, the seven principles of resilience are: diversity, modularity, social capital, innovation, overlap, tight feedback loops, and ecosystem services. Diverse ecosystems are more resilient to shocks in just the same way as diverse local economics, and people also want to invest and live in places that are real, not the bland identikit products of much of the regeneration industry. Therefore, the new economics recognises that a neighbourhood's non-monetary assets—its diversity, history, distinctiveness, neighbourliness, know-how, enthusiasm—are as important as money to its economic success. The way that money flows around the local economy, and whether it stays circulating or flows away, is as important as the total amount of money. That means we need to redefine the way we understand and measure progress, finding ways to make the invisible value of things—essential to our well-being—visible and measurable.

The university is under threat. For forty years this indispensable democratic institution has been systematically betrayed by governments and the political class, who have redirected it from its proper social and cultural functions through a relentless programme of financialisation. Docherty demonstrates that the sector has been politicised and now works explicitly to advance a market-fundamentalist ideology that drives an ever-widening wedge between ordinary citizens and the privileged and wealthy (Docherty 2016).

One large obstacle to transformation towards sustainability began with a bargain made long ago with the giant companies that promised to make our lives affluent, convenient, easy and fun. But they also made our world fragile, polluted, violent, undemocratic, insecure, and unsustainable. Moreover, they corrupted our very thoughts and words (Orr 2016).

"We have seen how difficult it is to change economic reality—whether it is low wages in poor countries, tax heavens that serve the super-rich, excessive corporate power or an overly complex financial system. Sometimes the difficulty is due to the active attempts by those who benefit from the current arrangements to defend their position through… Acknowledging the difficulties involved in changing the economic status quo should not cause to create an economy that is more dynamic, more stable, more equitable and more environmentally sustainable than what we have had for the last three decades. Yes, changes are difficult, but, in the long run, when enough people fight for them, many "impossible" things happen. Just remember: 200 years ago, many Americans thought it was totally unrealistic to argue for the for the abolition of slavery; 100 years ago, the British government put women in prison for asking for votes; 50 years ago, most of the founding fathers of today's developing nations were being hunted down by the British and the French as "terrorists"… Of course, there is "should" and there is "can". Many of us are physically too exhausted by our daily struggle for coexistence and mentally occupied with our own personal and financial affairs... Changes are difficult to make, but even big ones are possible if we try hard enough and long enough" (Chang 2015).

"Every day we invent new technologies that could make our lives and the planet's health better. Free markets, trade, and competition create so much wealth, that in theory they could make better off if there was the will to do so. But this is not the reality we are living today... For the last 30 to 50 years, the neoliberalist ideology has increasingly prevailed in large parts of the world. This approach centers on the notion that the markets know best, that the "business of business is business", and that government should refrain from setting constraining rules for the functioning of markets. Those dogmatic beliefs have proven wrong. But fortunately, we are not destined to follow them" (Schwab 2021).

"I am stating my strong belief that it is important to take action now, long before the unsustainability state of the world does indeed induce such a crisis. We cannot continue to ignore history and the obvious signs of trouble. There is still time, I hope, although the stresses on both the human and the natural worlds may be reaching a point whereby the system cannot retain its current structure, and it may jump into a new regime even less sustainable than the present". Because our house is on fire, and this should come as no surprise. Built on false promises, discounted futures, and sacrificial people, it was rigged to blow from the start. It's too late to save all our staff, but we can still save each other and a great many other species, too. Let's put out the flames and build something different in its place. Something a little less ornate, but with room for all those who need shelter and care" (Klein 2019).

We have seen how a faulty economics drives and is driven by a distorted social logic. But we have also seen that a different economics is achievable. A better and fairer social logic lies within our grasp. This is not about overthrowing society. It is not about changing human nature. It is about thinking simple steps towards the economy of tomorrow. Toward an economic fit for purpose. Towards a meaningful prosperity on a finite planet. At the heart of that economics we must place a more robust and more realistic vision of what it means to be human (Jackson 2017).

It seems that the formation of most global institutions, except UN, has never been democratic or sustainable; rather they have been constructed by economic superpowers. Perhaps we are naïve in hoping for democratic and forward-looking participation, but the pressing issue of climate change and the imperative of sustainability leave us no chance. Having said this, it is worth asking if the nation state is the best entity to achieve sustainability. There are 196 nations in the world today, each concerned with the interest of its most political and economically influential citizens. Perhaps the nation state isn't the most effective medium to achieve sustainability? Perhaps what is needed is more active cooperation on a decentralised basis?

Our corporations and industries are already having tough conversations about how they will survive this quarter. However, Lovins points out that this time of crisis gives businesses the opportunity to ask questions, not just about how they will survive the next quarter, but about how they will survive the next century. They are being given an opportunity to re-evaluate their supply chains, their employee policies, and their contribution to the health and well-being of humanity and our shared planet. The good news is that there is a tremendous business case for this re-evaluation. Back in

2014, the Carbon Disclosure Project found that "corporations that actively manage and plan for climate change secure an 18% higher return on investment (ROI) than companies that don't." Today, the business case for sustainability has only gotten stronger; companies who focused on the integrated bottom line have lower costs, lower risks, an easier time attracting and retaining talent, higher customer satisfaction, and much more (Lovins 2020).

Technologies have brought us immense progress and wealth but simultaneously are sowing the seeds for our destruction. The process of creative destruction is the essence of capitalism. Joseph Schumpeter (1942) characterised creative destruction as innovations in the manufacturing process that increase productivity, describing it as the "process of industrial mutation that incessantly revolutionizes the economic structure from within, incessantly destroying the old one, incessantly creating a new one".

"I suspect that the truly catastrophic potential of global exploitation and destruction is primarily unrelated to technology, and related instead to the expansion of population, as well as to–self interest and human nature. Technologies are just the enabling routes to self-destruction, not the cause" (Townsend 2018).

Motivating all of us—from public to politicians and industrialists—with enough urgency to make global changes will be hard, but it must be done. George Bernard Shaw said on approach to life: "Other people see things and… say Why? But I dream things that never were—and I say Why not? The problems of the world cannot possibly be solved by sceptics or cynics whose horizons are limited by the obvious reality. We need men who can dream of things that never were" (Sachs 2015).

Like any journey, there are decisions to be made. What to take, what to leave behind? When hard times come along the way, who will make the decisions and by what means and by what standards? Will we govern by triage and leave the wounded behind? "We are, in short, starting the largest, longest, most complex, and crucial transition in human history unprepared and disorganized. And it is beginning as a free for all" (Orr 2016).

3.15 Conclusions

> There must be a better way to make the things we want, a way that doesn't spoil the sky, or the rain or the land— Paul McCartney, English singer

The strategy of sustainable development is not to stop economic development, especially for the developing countries. Developing countries also need their productive capabilities to deal with consequences of climate change. Due to their climate, locations, and geography, many developing countries are going to bear the brunt of the impacts of global warming, despite having very little, if not necessarily minimal, responsibility for causing it. To better deal with poverty, inequality, and consequences of climate change, poor countries need to equip themselves with better technologies and management capabilities, which can be acquired through economic development.

At the same time, this does not mean that the rich countries should stop economic development but not in the sense of "business as usual". The structure of the economic development should be transformed according to the model and system presented in Sect. 3.3. This means that performance index for the global development should comprise all three dimensions of sustainable development, i.e. economic, environmental and social. The weight for each dimension of development performance index depends on country/region economic, environmental and social situation and should be provided by governance respectively.

The discussions on conventional economy and various suggested new modes of it, including very new "stakeholders capitalism" approach, have revealed that most of them are dealing with unsustainability reduction. As it was shown above, unsustainability reduction does not lead to sustainability. The move towards sustainable development requires to make transformations in all spheres of our life. Nonetheless, it is a fact that the state remains the most powerful organisational instrument that humankind has invented and thus big economic, environmental and social transformations are very difficult to achieve without it.

"Those who have created a lot of pollution have rarely, if ever, paid the costs. We are the ones who have paid for it, and we will continue to pay if we don 't demand that they change their behaviour. Corporations are the ones who are making money from that, and we should not support those who don 't at least tell us what they are doing: they are not held accountable for the waste they create" (Schlossberg 2019).

Subsidies often introduce economic, environmental and social distortions with unintended consequences. They are expensive for governments and may not achieve their objectives while also inducing harmful environmental and social outcomes. Socially, a reform of subsidies and incentives, especially in mining, food, fossil fuel, and energy production, will lead to a more equitable distribution of income and balanced long-run growth of local and countries' communities.

As sustainable economies do not match with the mainstream thinking, business will hardly be the driver for such a development. "This calls for better governmental leadership as well as stronger engagement of civil societies and their organisations. Strengthening social innovation and clear messages about the benefits of sustainable societies as well as risks of unsustainable ones are crucial elements on the way" (Lorek and Spangenberg 2014).

If the world is to comprehend the scale of the problem and understand the implications from the science, it will require much more effective communication from scientists. This is not a simple task, but it is also one that has not been handled well. It will involve scientists communicating with the public and decision-makers on the science and its implications with greater frequency, clarity, timeliness, transparency, and openness. To do this, scientists need to rebuild the public's trust in their competencies and repair a perceived lack of integrity, the result of repeated attacks over the last few years.

The science is inadequately incorporated in policy decision-making. For the reasons discussed above, there is a grave risk that inaction or delay will result from confusion, inadequate understanding of the science, and misunderstanding of evidence. Science funders should adapt their schemes to support broader programme structures that enable long-term, collective efforts by wider research consortiums. That will encourage sustainability science that employs interdisciplinary and transdisciplinary approaches needed to tackle complex, contested issues and trade-offs inherent to sustainable development.

Universities should fully embrace the mission of advancing sustainable societies by promoting education for sustainable development. Building the capacities and skills of the next generation of researchers and change makers is one of the biggest leverage points towards sustainability at humanity's disposal. The university of future will be inclusive of broad range of the population, actively engaged in issues that concern them, relatively open to commercial influence, and fundamentally interdisciplinary in its approach to both teaching and research. The department-based structure of the university is essentially at odds with such collaboration.

Big businesses define sustainability as the pursuit of greater technological efficiency; less waste and more recycling can reduce some of the damage from rapidly rising production and consumption. However, it will not stop the forces of planetary destruction. For that, the structural transformations insuring intergenerational equity, a respect for nature, a fair distribution of earth resources, and reasonable consumption are needed. Any chance of this happening will require governments, nonprofits, and consumers to put far less faith in corporate self-reporting and self-regulation to solve global environmental problems.

Corporate social responsibility is about how corporations have an obligation to consider the impacts on the environment, economy and society, i.e. time has came for companies to fulfil their environmental and social obligations both at home and abroad, to recognise the true costs of what they pay to employees, of what they make and buy across the entire supply chain, from sourcing to retail distribution to waste disposal.

We have to acknowledge that advertising is an important obstacle to sustainable development, because it induces people to purchase goods and services they do not need, and it contributes to culture of consumption. Changes in patterns of demand and consumption, including through regulation, promotion of responsible advertising and marketing practices, and consumer education, should be encouraged to reduce environmental impact.

Expert knowledge is absolutely necessary, but "an expert by definition knows well only a narrow field and we cannot expect them to make a sound judgement on issues that involve more than one area of life (that is most issues), balancing off different human needs, material constraints, and ethical values. The possession of expert knowledge can sometimes give you a blinkered view. This dose of scepticism about expert knowledge should be applied to all areas of life, not just economics. However, it is especially important in economics—a political argument often presented as a science" (Chang 2015).

New investments in coal, oil and gas exploration should be discouraged, as they risk leading to stranded assets. Direct and indirect fossil fuel subsidies in developed and developing countries should be phased out. The funds previously used as subsidies should be reoriented towards affordable renewable energy and energy efficiency particularly for the poor.

The transition towards a circular economy should be promoted, including waste management and planning approaches that emphasise waste prevention as opposed to end-of-pipe waste management. Export of e-waste and hazardous chemicals to countries that do not have advanced infrastructure to manage them should end.

The environmental tax reduces taxes on wages and encourages investment in such activities as wind power generation and recycling, thus simultaneously boosting employment and lessening environmental destruction. Eliminating environmentally destructive subsidies reduces both the burden on taxpayers and the destructive activities themselves.

"Sustainable development, development without growth, does not imply the end of economics – if anything, economics becomes even more important. But it is a subtle and complex economics of maintenance, qualitative improvement, sharing, frugality, and adaptation to natural limits. It is an economics of better, not bigger" (Dally 1996).

Advertisers understood that humans are not so much rational creatures as they are very good rationalisers. Accordingly, "they learned how to mine the depths of the id and to exploit our vulnerabilities – particularly those of the most susceptible. Thus, children and adolescents are targeted with the intent of making them lifelong consumers and brand loyalists. It is no accident that the first word spoken by many infants is a commercial brand name. Young children reportedly can identify more brand names than native plants and animals in their backyard" (Orr 2016).

The transition to a fair, decent, democratic, zero-carbon, and sustainable society will require transferring power to competent and thoughtful citizen-agents. It will require a public that is able and willing to think and do itself and that is thereby less dependent for its information and provisioning on sources that is little reason to trust.

There are still great opportunities in the fact that the transition to the sustainable development based on low-carbon economy coincides with the coming decade of radical structural transformation of the world. If the structural transformation is done well from the point of resource efficiency, responsible consumption and sustainable production, wastes and pollution, liveable cities, inequality and poverty, and care of forests and grasslands, it will strongly reduce the emissions.

Development finance institutions, that is, all public development banks – national, regional and multilateral – as well as business and private finance sectors, should put the burden on investors to take account of sustainability when making decisions or engaging with investees on their portfolios. Through regulatory and behaviour changes, market practices should better reflect the need to orient financial flows towards long-term sustainable development and adopt sustainability standards.

Financial system needs to be regulated because of its power and importance and current money system must be changed if we are to achieve sustainability. "During an economic boom such investment schemes prosper, but in a crisis people seek assets of real value, for example land or productive enterprises, and those holding only paper assets in over-inflated speculative schemes are likely to lose their shirts. While traditional debt-based currencies are the 'fuel' of today's capitalism, local currencies might very well become the 'oil' of tomorrow's commons-based economy" (Reardon et al. 2018).

Development cooperation is essential for sustainable development implementation: *vertical* - to initiate science-policy-society collaboration and learning at the national level with a view to co-designing context specific pathways for SD; *horizontal*—flows of goods, capital, information and people, degradation of global environmental commons connect the countries in a way the world has never been before. Multilateral collaborations, agreements among different countries and regions are thus essential and need to be strengthened. Development cooperation means doing something that is not-for-profit, or that accepts a lower profit than the market would offer. It would not happen if profit incentives alone were followed, or at least not in the same way.

"Without a change of our intent and actions, collapse of civilization is not impossible. Unfortunately, it may well be in the relatively near future. The grain of hope and positive view is that some people are now recognizing the dangers, and with immediate actions, we may have the ability to prevent a collapse of civilization and emerge with a better world order" (Townsend 2018).

"We can be sure that attempts to bring about sustainability will meet enormous resistance from many people and vested interests. However, there is not much choice about the matter. The alternative to the pursuit of sustainability is to along the present path of unsustainability, leading to disaster. But institutionalising of sustainable development, whether through national strategies or other means, will not happen if the person at the top is not determined to make it happen" (Norgaard 1994).

We hope that discussions, models, methods and systems presented in the book will more clearly structurise the problem for achieving sustainability in business and facilitate the solution, because if our developments are not sustainable, then it means we will come to suffer at some point. It may not be us, but the future generations will undoubtedly do.

We have to be open stating once again that transformations towards sustainability cannot be achieved within the traditional paradigm of economic growth and reliance on technology and by specific policies to attenuate the most unethical behaviour and nudge consumers, firms and workers in the "correct" direction, i.e. only calling upon people to behave "right" within "wrong" structures.

References

Abbosh O, Nunes P, Downes L (2019) Pivot to the future: discovering value and creating growth in a disrupted world. PublicAffairs, New York

Akenji L, Bengtsson M (2014) Making sustainable consumption and production the core of sustainable development goals. Sustainability 6:513–529. https://doi.org/10.3390/su6020513

All Party Parliamentary Group (2007) Return of the population growth factor: its impact upon the millennium development goals. House of Commons London, London

Amartya S (1999) Development as freedom. Knopf, New York

Antal M, Van den Bergh JCJM (2013) Macroeconomics, financial crisis and the environment: strategies for a sustainability transition. Environ Innov Soc Transitions 6:47–66

Applebaum B (2019) The economists' hour: how the false prophets of free markets fractured our society. Picador, London
Arillo C (2017) Consumption is the bottleneck for sustainable development. BusinessMirror
Barbier EB, Burgess JC (2017) Natural resource economics planetary boundaries and strong sustainability. Sustainability 9:1858. https://doi.org/10.3390/su9101858
Barbier EB, Burgess JC (2020) Sustainability and development after COVID-19. World Dev 135:105082. https://doi.org/10.1016/j.worlddev.2020.105082
Bennett RJ, Chorley RJ (1978) Environmental systems: philosophy, analysis and control. Princeton University Press, Princeton, NJ
Bocken NMP, Short SW, Rana P, Evans S (2014) A literature and practice review to develop sustainable business model archetypes. J Clean Prod 65:42–56. https://doi.org/10.1016/j.jclepro.2013.11.039
Bowman C, Hodge JK, Yu A (2014) The potential of iterative voting to solve the separability problem in referendum elections. Theory Decis 77:111–124. https://doi.org/10.1007/s11238-013-9383-2
Boyle D, Simms A (2009) The new economics: a bigger picture. Earthscan
Bregman R (2017) Want utopia? Start with universal basic income and a 15-hour work week. In: Wired. https://www.wired.co.uk/article/universal-basic-income-utopia. Accessed 28 May 2021
Brown LR (2003) Plan B: Rescuing a planet under stress and a civilization in trouble. WW Norton & Company
Caradonna JL (2014) Sustainability: a history. Oxford University Press, New York, NY
Castel D, Lemoine C, Durand-Delvigne A (2011) Working in cooperatives and social economy: effects on job satisfaction and the meaning of work. Perspect Interdiscip sur le Trav la santé
Chang HJ (2012) Things they don't tell you about capitalism. Bloomsbury Press
Chang HJ (2015) Economics: the user's guide. Bloomsbury Publishing USA
Coote A, Franklin J (2009) Green well fair: three economies for social justice. New Economics Foundation
Criscuolo C, Menon C (2015) Environmental policies and risk finance in the green sector: cross-country evidence. Energy Policy 83:38–56. https://doi.org/10.1016/j.enpol.2015.03.023
Crist E (2007) Beyond the climate crisis: a critique of climate change discourse
Daly HE (1997) Beyond growth: the economics of sustainable development. Beacon Press, Boston
Dauvergne P (2018) Will big business destroy our planet? Polity Press
Diamond J (2005) Collapse: how societies choose to fail or succeed. Viking Press, New York
Docherty T (2016) John higgins on academic freedom. Arts Humanit High Educ 15:193–196. https://doi.org/10.1177/1474022215613602
Eagleton T (2018) Why Marx was right. Yale University Press, New Haven & London
Ehrenfeld JR (2009) Sustainability by design: a subversive strategy for transforming our consumer culture. Yale University Press
Fioramonti L (2013) Gross domestic problem: the politics behind the world's most powerful number. Zed Books Ltd, London
Ford L (2020) Covid has magnified every existing inequality—Melinda Gates. In: Guard. https://www.theguardian.com/global-development/2020/sep/15/covid-has-magnified-every-existing-inequality-melinda-gates. Accessed 27 May 2021
Friedman TL (2006) The world is flat: The globalized world in the twenty-first century. Penguin London
Gates Bi, Gates M (2020) Goalkeepers report. COVID-19: a global perspective
Gilding P (2011) The great disruption: how the climate crisis will transform the global economy. A&C Black
Graeber D (2013) On the phenomenon of bullshit jobs: a work rant. Strike Mag 3:1–5
Greco T (2009) The end of money and the future of civilization. Chelsea Green Publishing
Hakovirta M, Denuwara N (2020) How COVID-19 redefines the concept of sustainability. Sustainability 12:3727. https://doi.org/10.3390/su12093727
Hann C, Hart K (2011) Economic anthropology. Polity Press, History, Ethnography, Critique
Hardin GR (1968) The tragedy of the commons. Science (80-) 162:1243–1248

Hawken P, Lovins AB, Lovins LH (1999) Natural capitalism: The next industrial revolution. Routledge, Boston

Hunt TL (2006) Rethinking the fall of Easter Island. Am Sci 94:412–419

Independent Group of Scientists (IGS) appointed by the Secretary-General:, Messerli P, Murniningtyas E, et al (2019) Global Sustainable Development Report 2019: The Future is Now–Science for Achieving Sustainable Development. United Nations, New York

Institute for Global Environmental Strategies (IGES) (2010) Sustainable Consumption and Production in the Asia-Pacific Region. Hayama

International Labour Organization (1998) ILO declaration on fundamental principles and rights at work and its follow-up: Adopted by the International Labour Conference at Its Eigthy-sixth Session Geneva 18 June 1998. International Labour Organization, Geneva

Jackson T (2017) Beyond consumer capitalism: foundations for a sustainable prosperity. In: Handbook on growth and sustainability. Edward Elgar Publishing

Jackson T (2016) Prosperity without growth: foundations for the economy of tomorrow. Routlege, London and New York

Jakl T, Schwager P (2008) Chemical leasing goes global. Springer Vienna, Vienna

Järvensivu P, Toivanen T, Vadén T et al (2018) Invited background document on economic transformation

Jonkutė G, Staniškis JK (2016) Realising sustainable consumption and production in companies: the SUstainable and RESponsible COMpany (SURESCOM) model. J Clean Prod 138:170–180. https://doi.org/10.1016/j.jclepro.2016.03.176

Juniper T (2016) What's Really Happening to Our Planet?: The Facts Simply Explained. Dorling Kindersley Ltd

Klein N (2015) This changes everything: Capitalism vs. the climate. Simon and Schuster

Klein N (2019) On fire: the (burning) case for a green new deal. Simon & Schuster

Kliopova I, Baranauskaitė-Fedorova I, Malinauskienė M, Staniškis JK (2016) Possibilities of increasing resource efficiency in nitrogen fertilizer production. Clean Technol Environ Policy 18:901–914. https://doi.org/10.1007/s10098-015-1068-9

Lawn PA (2000) Toward sustainable development: an ecological economics approach. CRC Press

Lewis SL, Maslin MA (2018) Human planet: How we created the Anthropocene. Yale University Press

Lorek S, Spangenberg JH (2014) Sustainable consumption within a sustainable economy–beyond green growth and green economies. J Clean Prod 63:33–44

Lovins H (2020) The Six Sustainability Lessons from COVID-19 (interview). In: Bard Coll. https://gps.bard.edu/six-sustainability-lessons-from-covid-19?utm_content=128971806&utm_medium=social&utm_source=facebook&hss_channel=fbp-112648926244. Accessed 22 May 2021

W Lutz E Striessnig A Dimitrova et al 2021 Years of good life is a well-being indicator designed to serve research on sustainability Proc Natl Acad Sci 118 https://doi.org/10.1073/pnas.1907351118

Malinauskienė M, Kliopova I, Hugi C, Staniškis JK (2018) Geostrategic supply risk and economic importance as drivers for implementation of industrial ecology measures in a nitrogen fertilizer production company. J Ind Ecol 22:422–433. https://doi.org/10.1111/jiec.12561

Mazzucato M (2018) The value of everything: making and taking in the global economy. Hachette UK, New York

Mazzucato M, Perez C (2015) Innovation as growth policy. The challenge for Europe. In: Fagerberg J, Laestadius S, Martin BR (eds) The triple challenge for Europe: Economic development, climate change, and governance. Oxford University Press, Oxford

McMichael J, Woodward A, Muir C (2019) Climate change and the health of nations. Oxford University Press

Meadows DH (2008) Thinking in systems: a primer. chelsea green publishing, White River Junction

Mont O (2002) Clarifying the concept of product–service system. J Clean Prod 10:237–245. https://doi.org/10.1016/S0959-6526(01)00039-7

Mulligan M (2014) An introduction to sustainability: environmental, social and personal perspectives. Routledge
Neßhöver C, Beckert B, Kabel C, et al (2021) Widening the European Green Deal's perspective towards a sustainable Europe. Dessau-Roßlau
New Economics Foundation (2010) The Great Transition
Norgaard RB (1994) Development betrayed. Routledge, London
OECD (2012) Green Growth and Developing Countries: A Summary for Policy Makers
Orr DW (2016) Dangerous years: Climate change, the long emergency, and the way forward. Yale University Press
Phillips C (2020) Coronavirus: live animals are stressed in wet markets, and stressed animals are more likely to carry diseases. In: Conversat. https://theconversation.com/coronavirus-live-animals-are-stressed-in-wet-markets-and-stressed-animals-are-more-likely-to-carry-diseases-135479. Accessed 5 Jul 2021
Piketty T (2014) Capital in the twenty-first century. Belknap Press of Harvard University Press, Cambridge, Massachusetts
Polanyi K, MacIver RM (1944) The great transformation. Beacon press Boston
Quammen D (2020) We Made the Coronavirus Epidemic. In: New York Times. https://www.nytimes.com/2020/01/28/opinion/coronavirus-china.html. Accessed 22 May 2021
Reardon J, Madi MAC, Cato MS (2018) Introducing a new economics. Pluto Press
Rifkin J (2014) The zero marginal cost society. Palgrave Macmillan, New York, NY
Rifkin J (2019) The green new deal: Why the fossil fuel civilization will collapse by 2028, and the bold economic plan to save life on earth. St. Martin's Press, New York
Rifkin J (2011) The third industrial revolution: how lateral power is transforming energy, the economy, and the world. Macmillan
Sachs JD (2015) The age of sustainable development. Columbia University Press
Sachs JD (2012) The Price of Civilization. Vintage Books, London
Sachs JD, Schmidt-Traub G, Mazzucato M et al (2019) Six transformations to achieve the sustainable development goals. Nat Sustain 2:805–814. https://doi.org/10.1038/s41893-019-0352-9
Schlossberg T (2019) Inconspicuous consumption: the environmental impact you don't know you have. Hachette Book Group
Schumacher EF (1973) Small is beautiful: economics as if people mattered. Blond and Briggs, London
Schumpeter JA (1942) Creative destruction. Capit Soc Democr 825:82–85
Schwab K (2021) Stakeholder capitalism. John Wiley & Sons Inc., Hoboken, NJ
Senge PM (2006) The fifth discipline. Doubleday, New York
Soddy F (1926) Virtual wealth and debt: the solution of the economic paradox
Standing G (2016) The corruption of capitalism: Why rentiers thrive and work does not pay. Biteback Publishing, London
Staniškis JK (2020) Future is now: science for achieving sustainable development – the global sustainable development report 2019. In: Zilahy G (ed) Sustainability in transforming societies: proceedings of the 26th annual conference of the international sustainable development research society (ISDRS 2020). Budapest University of Technology and Economics, Budapest, Hungary, pp 831–839
Staniškis JK, Jasch CM, Stasiškienė Ž (2005) Assessment of environmental costs for sustainable industrial development: a monograph. Technologija, Kaunas
Staniškis JK, Stasiškienė Ž, Jasch C (2008) Cleaner technologies: environmental management accounting, investment appraisal and financing. Nova Science Publishers
Stead WE, Garner Stead J (1994) Can Humankind Change the Economic Myth? Paradigm Shifts Necessary for Ecologically Sustainable Business. J Organ Chang Manag 7:15–31. https://doi.org/10.1108/09534819410061351
Stern N (2016) Why We are Waiting? The Logic, Urgency, and Promise of Tackling Climate Change. The MIT Press, Cambridge, Massachusetts, London, England
Stigliz JE (2016) Making globalization work: the next steps to global justice. Penguin, London

Strange T, Bayley A (2008) Sustainable development. OECD
Streeck W (2016) How will capitalism end?: Essays on a failing system. Verso Books
Sundararajan A (2017) The sharing economy: The end of employment and the rise of crowd-based capitalism. Mit Press, Cambridge, Massachusetts, London, England
Townsend P (2018) The dark side of technology. Oxford University Press
TWI2050–The World in 2050 (2018) Transformations to Achieve the Sustainable Development Goals. Report prepared by the World in 2050 initiative. Laxenburg
United Nations (2015) Transforming our world: the 2030 Agenda for Sustainable Development
United Nations Industrial Development Organization (UNIDO) (2020) Chemical Leasing. In: United Nations Ind. Dev. Organ. https://www.unido.org/our-focus-safeguarding-environment-resource-efficient-and-low-carbon-industrial-production/chemical-leasing. Accessed 22 May 2021
Victor P, Jackson T (2016) Towards a new, green economy
Viliani F (2021) How has COVID-19 changed the role of sustainability for organizations? In: Int. SOS. https://www.internationalsos.com/insights/how-has-covid-19-changed-the-role-of-sustainability-for-organisations. Accessed 22 May 2021
Vulkan N, Åstebro T, Sierra MF (2016) Equity crowdfunding: a new phenomena. J Bus Ventur Insights 5:37–49. https://doi.org/10.1016/j.jbvi.2016.02.001
Weaver P, Dumitru A, Lema-Blanco I, García-Mira R (2015) Transformative social innovation narrative: Timebanking
Welzer H (2017) Climate Wars. Why People will be Killed in the 21st Century. Polity Press
Yong E (2020) How the pandemic defeated America. Atl 4

Part II
Obstacles and Drivers to Transition of Organisation Towards Sustainability

Chapter 4
Theoretical Insights on Organisational Transitions Towards CSR

Eglė Staniškienė, Živilė Stankevičiūtė, Asta Daunorienė, and Joana Ramanauskaitė

To be a good economist, you must be a good political economist. It's not enough to know the diagrams of supply and demand and the mathematics of econometric regressions. You have to be able to understand social tensions and conflicts.—Paul Samuelson, Nobel Prize in economics, 1997.

4.1 Introduction

While businesses have always been dealing with sustainability issues, debates upon a wider social and environmental responsibility of business began to have serious impacts on both individual companies and policy since the 1970's. The term 'Corporate Social Responsibility' (CSR) was introduced as a general framework for business trying to address social and environmental issues related to it. In a broad sense, the literature on sustainability and business seeks to understand how creating economic value for the organisation could go hand in hand with limiting ecological impact and operating in a socially responsible way (Loorbach and Wijsman 2013). Business and its role in the society are the subject of long and controversial debates in the management literature (Mark-Herbert and Von Schantz 2007). Assuming that business and society in reality are interdependent (Porter and Kramer 2006), increasingly more organisations apply CSR as a "win–win" strategy (Singhapakdi et al. 2015) and accept the argumentation of the European Commission that CSR is in the interests of enterprises and society as a whole (European Commission 2011). The Green Paper (2001) indicated that CSR "is essentially a concept whereby companies decide voluntarily to contribute to a better society and a cleaner environment" (p.5). In this respect, CSR implies that organisations have a broader range of obligations besides economic and financial ones (Costa and Menichini 2013). Despite the earlier scepticism prevailing in business about its duty to balance the organisation's financial performance and its impact on the society and environment (Dyllick and Hockerts 2002), CSR has lately become of strategic importance for numerous organisations

J. K. Staniškis et al., *Transformation of Business Organization Towards Sustainability*, World Sustainability Series, https://doi.org/10.1007/978-3-030-93298-5_4

(Kiron et al. 2012). The fact that Fortune 500 companies currently spend more than $15 billion a year on CSR activities (Novick O'Keefe 2017) confirms the salient position of CSR in the strategic goals of organisations. The growing number and enhanced quality of CSR reports (Blasco and King 2017) also serve as evidence that being socially responsible is getting increasingly more relevant. In general, motives of businesses behind the engagement in CSR could be twofold: normative and economic ones. The normative case for CSR argues that it is a moral duty for a business to engage in CSR ("doing good"), while the business case for CSR is based on the expected returns that organisations may receive from CSR ("doing better") (Maignan and Ferrell 2004; Bhattacharya and Sen 2004). Generally speaking, the business case for CSR relies on the notion that CSR elicits organisation-favouring responses from the stakeholders. It seems that positive responses are evident in the business world, as the survey of 2,874 managers and executives from 113 countries revealed that about 31% of organisations were benefitting from sustainable business practices (Kiron et al. 2012). Favourable consumer perceptions of products, positive attitude of current and potential employees towards the organisation or business partners' willingness to do business with it are all great examples of benefits of being socially responsible (Lee et al. 2012). These examples support the idea that CSR requires the organisations' engagement with multiple stakeholders. Meanwhile, business returns from CSR practices depend on the way the stakeholders perceive the organisation's socially responsible practices (Costa and Menichini 2013). The contemporary CSR not only takes into account the inclusion of the triple bottom line of economic, social, and environmental performance (Aguinis 2011; Aguinis and Glavas 2012); creating shared value is also an integral part of the CSR (Bansal and DesJardine 2014; Carroll 2015; Ashrafi et al. 2018). To explain CSR engagement, scholars have mainly focused on stakeholder pressure for organisations to engage in socially responsible activities (Campbell 2007; Schneider et al. 2017). CSR has its roots in the stakeholder theory (Asif et al. 2013). Drawing on this theory, CSR could be defined as "a firm's voluntary consideration of stakeholder concerns both within and outside business operations" (Homburg et al. 2013). Various stakeholder groups are considered explicitly in the scientific literature: customers, suppliers, regulators, competitors, society as a whole, and business partners (Joyce and Paquin 2016; Yang et al. 2017). We will consider employees, customers in B2B context, potential employees, and consumers.

Despite various stakeholder classification schemes introduced in the literature, customers are always included among other stakeholders (Rodrigo and Arenas 2008; Parmar et al. 2010). Millennials have become the main driver for demanding that organisations should incorporate CSR into their mission and corporate values. The Nielsen Global Survey (Nielsen 2015) revealed that 67% of those surveyed preferred to work for socially responsible organisations. According to Bromley and Meyer (2017, p. 945), organisations "are transformed by new pressures to look like responsible actors", while laws and public pressure force them to "take on expanded concerns such as environmental protection, corporate social responsibility and philanthropy, employee rights and job satisfaction, workplace diversity, community engagement, and consumer safety". Environmental orientation (e.g., new environmental paradigm) and the perceived personal ability are key drivers of young adults' intentions to engage in eco-friendly behaviours (de Leeuw et al. 2015; Allen and Spialek 2018). According to Walker (2007), generation Y seeks to promote social justice; they pay particular attention to the natural environment and its preservation, poverty reduction, and solutions of ethical and moral problems. Millennials are considered to be the key consumers of today; they believe that consumption and socially responsible lifestyle habits can change the world for the better (Cone 2008). Lifestyle influences the purchasing decisions and consumption behaviour and is related to consumerism (Chan and Hu 2008). Thus, lifestyle can be viewed as an interpretive framework for the construction of social reality (Yeh and Chen 2011). Upon acquiring explicit knowledge of how specific activities affect the society (Buenstorf and Cordes 2008), consumers are likely to participate in activities such as consuming green products, recycling, considering clothing care, and adopting eco-friendly behaviour (McDonald and Oates 2006).

Employees are treated as highly important stakeholders (Rodrigo and Arenas 2008; Lee et al. 2012) as they not only determine the quality of products or services but also because the employee well-being is directly related to job performance. Scholars focus on the way the employee perceptions of CSR impacts their work attitudes and behaviour. Peterson (2004), Brammer et al. (2007), Turker (2009a), Rego et al. (2010), Stites and Michael (2011), Farooq et al. (2014) examine the direct link between CSR and employee commitment and demonstrate that the employee perception of CSR is positively related to organisational commitment.

Finding and retaining employees is a considerable task for many organisations. The evolving needs and values of workers representing different generations make employee recruitment increasingly challenging. Researchers try to answer the question, what makes an organisation attractive to a potential employee? The studies of Turban and Greening (1997), Albinger and Freeman (2000), Greening and Turban (2000), Backhaus et al. (2002), Kim and Park (2011) are based on the social identity theory and suggest that CSR creates a good reputation and sends signals to prospective job applicants about the attractiveness of the employer. Turban and Greening (1997) determined that the organisation's corporate social performance could be an important factor attracting potential employees. S.-Y. Kim and Park (2011) revealed that publicised active CSR can be shown as an indicator of attractive organisations for future and current employees.

Referring to the B2B context, customers are highly important stakeholders as suppliers need them in order to operate (Parmar et al. 2010). Supporting the notion that CSR in the B2B context really matters, we have considered the customer perception of supplier CSR expecting positive reactions of customers as a response to supplier engagement in CSR. The previous research, mainly in the B2C context, on the reactions of consumers to CSR revealed organisation-favouring responses on an array of cognitive and affective as well as behavioural outcomes (Sen and Bhattacharya 2001; Sen et al. 2006; Ng et al. 2019). For instance, Li et al. (2019) demonstrated that renewing the CSR strategy could increase the Starbucks' consumer loyalty by enhancing the customer-company identification.

The reviewed scientific literature reveals the dynamic role of different stakeholders to which business organisations must respond. The changing demands of stakeholders encourage organisations to act more responsibly; they must change and move towards sustainability. According to Markard et al. (2012) (p. 956), "sustainability transitions are long-term, multi-dimensional and fundamental transformation processes through which established socio-technological systems shift to more sustainable modes of production and consumption." Examples include the ongoing transformation of the energy sector towards renewable technologies, the emergence of electric vehicles, regional climate change initiatives and shifts in consumption practices related to food and water. Socio-technical transitions differ from technological transitions in that they include changes in user practices and institutional structures, in addition to the technological dimension (Markard et al. 2012). Sustainability transitions pose formidable challenges as an area of research and study (van den Bergh et al. 2011; Markard et al. 2012; Ferraro et al. 2015). Any sustainability transition is complex, uncertain, values-laden and political. Moreover, transitions are multi-dimensional, as they entail organisational, institutional and technological changes and implicate a variety of intertemporal obstacles and potential trade-offs across multiple stakeholders (Garud and Gehman 2012).

The relevance and timelines of research into sustainability transitions is represented not only by the number of scientific publications, but also by the Sustainability Transitions Research Network established a few years ago (http://www.transionsnetwork.org/). The analysis of ongoing studies on this topic in Lithuania reveals a trend that scientists of the management field examine the management systems of the Lithuanian organisations in the context of transformations and responsible innovation management. At the same time, researchers of the technology field suggest technological solutions to organisations in sustainability transitions. The problem is interdisciplinary and organisations in sustainability transitions should integrate the social and technological challenges. There is little empirical research on CSR drivers/motivators/obstacles/tensions/barriers in the Central and East Europe organisations; indeed, the drivers/obstacles of CSR in Europe are still relatively unknown and most of the literature is based on Anglo-American countries (US and UK) (Reverte 2009). This study aims to fill some of these gaps by explaining the socially responsible behaviour of organisations, considering CSR in relation to its drivers/barriers. This is achieved by analysing empirical evidence gathered through interviews conducted on a sample of 18 Lithuanian organisations.

More precisely, this study addresses the following research questions:

- What are the attitudes of different stakeholders towards CSR?
- Do business organisations recognise the changing demands of stakeholders and how do they do this?
- What barriers/obstacles do the organisations face in moving towards sustainability?

- What are the drivers/motivators encouraging the organisations to act more responsibly?

The aim of the second part of the book is to reveal the different stakeholders' perspectives as drivers for CSR in organisations and to address a new phase in corporate social responsibility, implying fundamental transitions towards sustainability within businesses. This part of the book addresses the obstacles the organisations face while moving from 'looking sustainable' to 'being sustainable' and the drivers of this process, in a way that sustainability is achieved through increasing broad range stakeholders' participation, so that transparency is maximised and trust can be built with lasting benefits.

The study **contributes** to the literature of CSR by exploring the perception and awareness of CSR of employees, potential employees, customers in the B2B context, and consumers. It also contributes to exploring the obstacles and drivers of organisations in sustainability transitions.

Although different definitions can be found in the scientific literature of organisational sustainability and corporate social responsibility, the terms sustainability and corporate social responsibility are used interchangeably in this part of the book.

This part of the book is organised as follows. First, in the next chapter, we provide the theoretical insights on organisational transitions towards CSR, reviewing the literature about the historical evolution and concept of CSR, stakeholder theory and its application to CSR, about the stakeholder's role in sustainability transitions, and obstacles and drivers of organisations moving towards CSR transitions. We then present the methodological and empirical part of the conducted studies: the research context, the overview of quantitative research findings and results of our interviews conducted on a sample of Lithuanian organisations. Finally, we provide a discussion and our main conclusions.

4.2 Corporate Social Responsibility: Historical Evolution of the Meaning

4.2.1 The Historical Evolution of CSR

The idea that business has a responsibility towards society beyond that of making profits for the shareholders is not new (Carroll 2009; Latapí Agudelo et al. 2019). According to Chaffee (2017), the origins of the social component in corporate behaviour can be tracked to the ancient Roman law and can be seen in the entities, which had been organised for social purposes, such as asylums, homes for the poor or for the elderly. Going further, in the centuries that followed, some social reforms were carried out; welfare schemes aimed at protecting and retaining the employees were implemented and by the 1920s, the debates among businesspeople started regarding their responsibility to balance the maximisation of profits with creating and sustaining an equilibrium with the demands that their labour force, clients or community made (Latapí Agudelo et al. 2019). However, despite some advancement, CSR is mostly a product of the post-World War II period, especially from the early 1950s up to the present time (Carroll 2015). Consistently, given the changes in social consciousness inspired by the social movements of the 1960s in the US regarding civil rights, women's rights, consumers' rights and environmental movement, CSR has grown in importance and significance (Carroll and Shabana 2010). Today, CSR is a global concept "that has progressed from the interplay of thought and practice" (Carroll 2015) (p. 87). Moreover, as social expectations towards corporate behaviour have been changing constantly (Latapí Agudelo et al. 2019), the concept of CSR is dynamic and acquires new shapes and facets, which reflect the maturity of the society.

Considering that the history of CSR is vast and long and having in mind the scope of this study, it is necessary to reveal the evaluation of the concept of CSR. As such, the review disclosing the evaluation of the concept of CSR during the last 70 years is provided below, based on the most relevant academic publications, historical events, and some practical examples from business following each decade. The summary of the evaluations encompasses only some of the fundamental aspects through a chronological timeline.

1950s. This was the starting point of the modern era of CSR (Carroll 2015). Actually, Howard R. Bowen's publication of his landmark book "Social Responsibilities of the Businessman" (1953) best marks the beginnings of the modern period of CSR. Bowen (1953) believed that the large organisations of that time concentrated great power, were vital decision-makers and accordingly their actions had a tangible impact on the society, affected the lives of citizens in numerous ways (Carroll 2009; Latapí Agudelo et al. 2019). Consistent with these beliefs, Bowen (1953) raised the question: "What responsibilities to society may businessmen reasonably be expected to assume?" (p. xi). Although grammatically the sentence was not confusing or complicated, is seems that the answer was not so straightforward. Even nowadays, politicians, governments, societies, and business people struggle with the precise answer to the mentioned question and due to this, sometimes the consensus regarding the nature of CSR is missing. Bowen (1953) himself defined the social responsibilities of business executives as follows: "the obligations of businessmen to pursue those policies, to make those decisions, or to follow those lines of action which are desirable in terms of the objectives and values of our society" (Bowen 1953) (p. 6). Referring to the definition, the explicit focus is on the responsibility of businesspeople to the society to work for its good, while linking CSR with benefits for businesses themselves is not much expressed (Carroll and Shabana 2010). Carroll (2009) invited to call Bowen "the Father of Corporate Social Responsibility" as he published the first academic work focused specifically on the doctrine of CSR. Moreover, Bowen (1953) was ahead of his time with his new approach to management, which aimed at improving the business response to its social impact.

During the 1950s, other authors were concerned with corporate behaviour and its responses to social issues too (Carroll 2009). According to Frederick (2006), there were three main ideas in the 1950s: the idea of balancing competing claims to corporate resources, the idea of corporate managers as public trustees, and the acceptance of philanthropy as a manifestation of business support of good causes (Frederick 2006). However, as stated by Carroll (2009), the period between the 50 s and 60 s was one of more "talk" than "action" in respect of CSR as there were very few actions of the organisations, beyond philanthropy, and the business executives were just learning to be comfortable with the CSR talk.

1960s. The interest in CSR during the 1960s was largely influenced by social movements of that period (especially in the USA) and, accordingly, by the growing awareness in the society (Latapí Agudelo et al. 2019). People started to become aware of the possible threatening consequences posed by the pollution, rapid population growth, and resource depletion to the environment and human health (Du Pisani 2006). Scientific information about the impact the big companies were having on the natural environment and the limits of economic growth was published in books such as Rachel Carson's "The Silent Spring" (1962), Paul Ehrlich's "The Population Bomb" (1968), Edward Goldsmith's "A Blueprint for Survival" (1974), and Fritz Schumacher's "Small is Beautiful" (1973). These books served as an influential source of information inviting people to various protest actions.

Turning to academic publications, the decade of 1960s was characterised by the growth in attempts to state more precisely what CSR meant (Carroll 2009). One of most prominent authors in this period was Keith Davis (1960), who claimed that CSR should be seen in the managerial context as the "businessmen's decisions and actions taken for reasons at least partially beyond the firm's direct economic or technical interest" (p. 70). According to Davis (1960), social responsibility has two rather different faces. On the one hand, businesspeople have a broad obligation to the community with regard to economic developments affecting the public welfare. On the other hand, the businesspeople's obligation is to nurture and develop human values (Davis 1960). The significance of Davis' ideas is that he claimed that the "social responsibilities of businessmen need to be commensurate with their social power" (p. 71) and that the avoidance of such would lead to a decrease of the organisation's social power (Davis 1960).

Another prominent author in the 1960s was Frederick (1960) who proposed a new theory of business responsibility based on five requirements. The first one was about having a criterion of value (in this case, for economic production and distribution). The second requirement argued that business responsibility should be based on the latest concepts of management and administration (for instance, at that time, the "Great Man" theory of management was being replaced with the concept of manager as coordinator and planner). The third requirement required to acknowledge the historical and cultural traditions behind the current social context. The fourth requirement invited to recognise that the behaviour of an individual businessperson was a function of their role within the society and its social context. Finally, the last requirement urged to admit that responsible business behaviour did not happen automatically, but, on the contrary, was the result of deliberate and conscious efforts (Frederick 1960).

The 1970s were also well known for scepticism with respect to CSR. The most famous sceptic was Milton Friedman, a Nobel laureate in economics, who argued that the corporations in a free capitalist system should limit themselves to the pursuit of purely economic benefits (Friedman 1962). This notwithstanding, by the end of the decade, a strong pressure on corporations to behave according to the social expectations became obvious (Waterhouse 2017).

1970s. Generally speaking, the period of 1970s started with a low level of confidence in business to fulfil the needs and wants of the society (Waterhouse 2017). During the first Earth Day celebration in 1970, 20 million people across the USA demanded to fight against the pollution caused mainly by the oil industry corporations or power plants (Latapí Agudelo et al. 2019). In response to this, the federal government of the USA established several institutions (such as Occupational Safety and Health Administration), which not only addressed, but also to some extent formalised the responsibilities of business regarding the social concerns (Latapí Agudelo et al. 2019).

It is noteworthy that two publications of the Committee of Economic Development of the USA made a huge impact on the development of the concept of CSR. The first one, called 'A New Rationale for Corporate Social Policy', explored the extent to which it was justified for the corporations to get involved in social problems (Baumol 1970). The second publication, called 'The Social Responsibilities of Business Corporation', dealt with the new expectations the society started placing on the business sector (Committee for Economic Development 1971): "Today it is clear that the terms of the contract between society and business are, in fact, changing in substantial and important ways. Business is being asked to assume broader responsibilities to society than ever before and to serve a wider range of human values. Business enterprises, in effect, are being asked to contribute more to the quality of American life than just supplying quantities of goods and services. Inasmuch as business exists to serve society, its future will depend on the quality of management's response to the changing expectations of the public" (Committee for Economic Development 1971) (p. 16).

As regards business level, it is important to mention that in this period, some of the today's most renowned organisations with respect to CSR were established, such as The Body Shop or Ben & Jerry's (Latapí Agudelo et al. 2019). Gradually, the formalisation and integration of the policies and practices that addressed the social issues in the daily activities of business have started to grow. According to Carroll (2015), the organisations entered an era that might be called "managing corporate social responsibility". However, the popularity of the term CSR created some issues. The most challenging of them was related to the precise meaning of CSR. The unrestricted use of the term elicited an uncertainty regarding the definition (Latapí Agudelo et al. 2019). The situation was partly solved by Carroll (1979) who provided one of the most influential definitions stating that: "The social responsibility of business encompasses the economic, legal, ethical, and discretionary expectations that society has of organisations at a given point in time" (Carroll 1979) (p. 500). This definition made at least two significant contributions. First, it provided a clear and concise conceptualisation that could be applicable under any context. Second, it did not view the economic and social objectives as incompatible trade-offs, but rather as an integral part of the business framework of total social responsibility (Lee 2008; Latapí Agudelo et al. 2019).

Summing up, the social movements, new legislation and academic writings in the period of 1970s provided the background for the CSR operationalisation aspects, which became the main focus in the 1980s.

1980s. The 1980s were characterised by the reduced pressure on the corporations from the side of governments in the USA and the United Kingdom, as for Reagan and Thatcher administrations, the growth and accordingly the strength of the economies of their countries depended on their ability to maintain a free-market environment while having as little as possible intervention from the state (Pillay 2015). However, the organisations still needed to react to different stakeholder groups, which expected the business to fulfil the social expectations of the time (Lee 2008; Latapí Agudelo et al. 2019). As the state regulatory framework eased off, the focus turned very much to business ethics and the operationalisation of CSR with the creation of new frameworks and models (Lee 2008; Latapí Agudelo et al. 2019). Moreover, following the work of R. Edward Freeman (1984) on stakeholder theory, the term stakeholder started gaining its relevance and popularity acknowledging the need to address the preferences of shareholders, employees, consumers, and other stakeholders. Hence, the alternative or complementary concepts to CSR such as stakeholder management or business ethics emerged during the 1980s sparking a broader discussion around the organisation's behaviour with regard to society (Lee 2008; Latapí Agudelo et al. 2019).

Incidentally, the context of the 1980s was full of social concerns and events, which also reflected the view and approach of the society to corporate behaviour. Among the most influential ones there were the creation of the European Commission's Environment Directorate-General (1981), establishment of the World Commission on Environment and Development (1983), and publication of the report "Our Common Future", which provided a definition of sustainable development (1987). Although these events did not relate directly to CSR, they reflected a growing sense of awareness of the community worldwide with regard to environmental protection and sustainable development, and indirectly to corporate behaviour (Lee 2008; Latapí Agudelo et al. 2019).

1990s. In the beginning of the 1990s, three main trends emerged in CSR (actually, they are relevant these days too): globalisation, institutionalisation, and strategic reconciliation (Carroll 2015).

Globalisation can be defined as "a process of intensification of cross-border social interactions due to declining costs of connecting distant locations through communication and the transfer of capital, goods, and people" (Scherer and Palazzo 2011) (p. 901). While expanding business globally, organisations from the US or West European countries generally moved to cheap labour countries with an absence of legal infrastructure. In doing this, the companies wanted to achieve legitimacy in the developing countries and additionally strived to not lose their global market position. In respect to the first point, being a socially responsible organisation was the surest way to legitimacy. Concerning the second aspect, global competitiveness and global visibility as well vulnerability of corporations' international image were inextricably linked. Due to such linkage, companies were forced to plan carefully their responsibility and ethics initiatives wherever they were doing business (Carroll 2015). It should be frankly admitted that business attention to CSR has not been entirely voluntary as most of the organisations took such action after being surprised by public response to issues that they had not previously thought were in the area of their responsibilities (the example of Nike company) (Porter and Kramer 2006). Still, for the organisations, being socially responsible was the surest path to survive in the long run and to balance the challenges and opportunities of the globalisation process.

As regards the institutionalisation of CSR, during the 1990s, it became stronger as CSR practices became commonplace, more formalised, and more deeply integrated into business practices (Carroll 2015). The process of institutionalising CSR into business thinking and practice was definitely a multi-decade transformation (Carroll 2015); however, during this period, it was accelerated as a result of taking a more active role from the investor side. The key question of CSR became framed as follows: what organisations were doing for themselves (investors) versus what they were doing for "outside" stakeholders in terms of consumers, women or minorities (Carroll 2015). As a good illustration of the institutionalisation of CSR serves the foundation in 1992 of the association Business for Social Responsibility (Latapí Agudelo et al. 2019). The same association is now a team of sustainable business experts working with its global network of more than 250 member companies to build a just and sustainable world (https://www.bsr.org/en/about). The core of the mission is to "work with business to create a just and sustainable world. We envision a world in which everyone can lead a prosperous and dignified life within the boundaries of the Earth's natural resources" (https://www.bsr.org/en/about).

Turning to strategic reconciliation, four academic contributions to CSR could be highlighted.

First, Wood (1991) introduced the model of Corporate Social Performance, which intended "to address (a) motivating principles, (b) behavioural processes, and (c) observable outcomes of corporate and managerial actions relating to the firm's relationships with its external environment" (p. 693). Following Wood (1991), the basic idea of CSR is that business and society are interwoven rather than distinct entities and accordingly the society has some expectations for business behaviour and outcomes. Three aspects and sources of expectations could be distinguished: expectations placed on all businesses because of their roles as economic institutions; expectations placed on particular organisations because of what they are and what they do; and expectations placed on managers as moral actors within the organisation (Wood 1991). Thus, the principles of Corporate Social Performance include legitimacy (institutional level), public responsibility (organisational level), and managerial discretion (individual level). As regards behavioural processes, Wood (1991) introduced three processes of a responsible organisation, namely environmental assessment (organisation monitors and assesses environmental conditions), stakeholder management (organisation attends to the numerous stakeholder demands placed on it), and issue management (organisation designs plans and policies to respond to changing conditions).

Turning to the last element of the model of Corporate Social Performance, three types of outcomes were proposed: social impacts of corporate behaviour (regardless of the motivation for such behaviour or process by which the behaviour occurs); programmes the organisations use to implement responsibility and/or responsiveness; and policies developed by organisations to handle stakeholder interests and accordingly social issues (Wood 1991).

Second, Carroll (1991) presented the Pyramid of Corporate Social Responsibility arguing that four categories or components of CSR, namely economic, legal, ethical, and philanthropic, might be depicted as a pyramid. Economic responsibility refers to the foundation for the other levels of the pyramid, seeing that historically, business organisations were created as economic entities designed to provide goods and services to the society and to be committed to being as profitable as possible. Legal responsibility refers to the importance of the business performing in a manner consistent with the government expectations and law and to complying with various regulations. Ethical responsibilities refer to the organisation's behaviour beyond the law-abiding duties. Philanthropic responsibility includes actively engaging in practices, which promote human welfare and improve the quality of life of the society. Thus, the organisations respond to the society's expectation that businesses should be a good corporate citizen (Carroll 1991).

Third, Burke and Logsdon (1996) aimed at revealing the benefits of strategic implementation of CSR while providing evidence that CSR 'pays off' for the organisation as well as for the organisation's stakeholders and society. Five strategy dimensions were identified, which help assess the value created for the organisation by the CSR programmes: centrality (how close or fitting CSR is to the organisation's mission and objectives), specificity (the ability to gain specific benefits for the organisation), proactivity (being able to create policies in anticipation of social trends), voluntarism (the discretionary decision-making process that is not influenced by external compliance requirements), and visibility (the relevance of the observable and recognisable CSR for all stakeholders) (Burke and Logsdon 1996).

Fourth, "the triple bottom line" concept introduced by Elkington (1994) indicated that organisations needed to display socially and environmentally responsible behaviour that could be positively balanced with their economic goals. In such case, new types of economic, social, and environmental partnership were needed (Elkington 1998). The debates around CSR provided some ideas how the partnership between the three dimensions could be implemented.

2000s. During this period, two waves in response to CSR debates were evident: the recognition and expansion of CSR and its implementation; and strategic approach to CSR (Latapí Agudelo et al. 2019).

Speaking of the recognition and expansion of CSR, the push on political and institutional levels was strongly expressed and visible (Carroll 2015). Among the variety of initiatives, several of them are extremely worth mentioning here due to their magnitude and succession.

One of the most notable and influential initiatives (outcomes of actions on political level) refers to the United Nations Global Compact (UNGC) launched in July 2000. Currently, the UNGC includes more than 12,000 organisations in more than 160 countries representing nearly every sector and size (https://www.unglobalcompact.org/). The UNGC introduced ten principles in the areas of human rights, labour, environment, and anti-corruption, which attracted attention towards CSR worldwide.

Another initiative approved on the political level refers to the Millennium Declaration (2000) with its eight Millennium Development Goals. Although the Millennium Development Goals were not directly linked to CSR, the global recognition of CSR nonetheless became stronger (Latapí Agudelo et al. 2019). Turning to the initiatives on the EU level, in 2001, the EC presented a Green Paper called "Promoting a European Framework for Corporate Social Responsibility" declaring that the EU was therefore concerned with "corporate social responsibility, a voluntary business approach which is expected to contribute to the strategic goal decided in Lisbon: "to become the most competitive and dynamic knowledge-based economy in the world, capable of sustainable economic growth with more and better jobs and greater social cohesion" (p. 3). The Green Paper claimed that CSR provided the foundations of an integrated approach that combined economic, environmental and social interests to their mutual benefit (Commission of the European Communities 2001). Moreover, CSR was introduced as a way "of managing change and of reconciling social development with improved competitiveness" (Commission of the European Communities 2001) (p. 7).

With respect to the second wave, during the 2000s, a weighty contribution to the concept of CSR is seen throughout the academic literature. Here, just few of the works will be mentioned trying to provide some novel or generalising perspectives towards CSR.

In the early years of the twenty-first century, Smith (2001) argued that corporate practices have changed in response to public interest or threat of consumer boycotts and these changes have often had a positive social impact. According to Lantos (2001), CSR entails the obligation stemming from "the implicit "social contract" between business and society for firms to be responsive to society's long-run needs and wants, optimizing the positive effects and minimizing the negative effects of its actions on society" (p. 600). Both aspects, in terms of minimising harm and promoting benefits, were underlined while arranging and implementing the CSR activities.

Furthermore, van Marrewijk (2003) explored the timeless question to whom businesses owed their responsibilities and concluded that different scholars have referred to a sequence of three approaches (shareholder, stakeholder and societal), each including and transcending one other, showing past responses to the mentioned question. The shareholder approach indicates that the social responsibility of business is to increase its profits (Friedman 1962). The stakeholder approach points that organisations should balance a multiplicity of stakeholder interests and shareholders represented only one group of stakeholders (Freeman 1984). Based on the societal approach, organisations were responsible to society as a whole, of which they were an integral part (van Marrewijk 2003). Acknowledging the diversity of the challenges the organisations need to respond to, van Marrewijk (2003) argued that organisations might adopt and commit to five levels or interpretations of CSR. The first level is *Compliance-driven CSR* (Blue). CSR at this level refers to providing welfare to society, within the limits of regulations from the authorities. Meanwhile, the motivation for CSR is driven by the fact that CSR is perceived as a duty and obligation. The second level is called *Profit-driven CSR* (Orange). At this level, CSR means the integration of social, ecological and ethical aspects into business operations while contributing to the financial bottom line. The core driver for being socially responsible is profit. The third level is labelled *Caring CSR* (Green) and implies the balancing of economic, social and ecological concerns, which are all three important in themselves. The organisations choose this level of CSR when they follow the approach that social responsibility, human potential, and care for the planet are important as such. The fourth level deals with *Synergistic CSR* (Yellow). On this level, CSR "consists of a search for well-balanced, functional solutions creating value in the economic, social and ecological realms of corporate performance, in a synergistic, win-together approach with all relevant stakeholders" (van Marrewijk 2003) (p. 102). The CSR is driven by recognition that sustainability is important in itself. The fifth level refers to *Holistic CSR* (Turquoise). Holistic CSR represents full integration of CSR in every aspect of the organisation with the intention to contribute "to the quality and continuation of life of every being and entity, now and in the future" (van Marrewijk 2003) (p. 103). In this case, the driver for CSR is the perception and the acceptance that sustainability is the only alternative since all beings and phenomena are mutually interdependent (van Marrewijk 2003). Summing up, although each organisation has the right to choose its interpretation and level of commitment to CSR, the holistic interpretation enables more successfully to cope with the most current challenges inside and outside the organisations.

Finally, analysing the period of 2000s, it is important to mention the work of Porter and Kramer (2006), as it provided a new insight on business and society interrelations stating that business and society needing each other was not a cliché, but the basic truth. The starting point of Porter and Kramer (2006) was that "the prevailing approaches to CSR are so fragmented and so disconnected from business and strategy as to obscure many of the greatest opportunities for companies to benefit society" (p. 2). In order to solve this issue and to make the situation less complicated, the duality of relationships was underlined arguing that the interdependence between a company and society took two forms: inside-out linkages and outside-in linkages. While looking inside-out, organisations map the social impact of their value chain and identify the negative and the positive and effect of these activities on society and only then focus on the ones with the greatest strategic value. Meanwhile, the organisations should look outside-in to understand the impact of their social context on their productivity and ability to carry out the strategy, especially in the long run (Porter and Kramer 2006). Ultimately, strategic CSR was proposed as it involved both inside-out and outside-in dimensions working in tandem. As strategy is about making choices, strategic CSR is about choosing which social issues to focus on. Organisations were called not to address hundreds of social issues, but only a few to make a real difference for the society (Porter and Kramer 2006).

2010's. This period is characterised by several features such as: political agreements and documents, for instance Paris Agreement and Agenda 2030; further evolvement of the concept of CSR and the relevance of creating a shared value (Porter and Kramer 2006; Freudenreich et al. 2020) and the growing scepticism about the appropriateness of corporate actions (Schrempf-Stirling et al. 2016). Further analysis of the mentioned features starts with the explanation of the growing scepticism, seeing that it serves as a destroying force for business and society relationship, and for business legitimacy (Schrempf-Stirling et al. 2016).

In 2019, Porter and Kramer wrote: "In recent years business increasingly has been viewed as a major cause of social, environmental, and economic problems. Companies are widely perceived to be prospering at the expense of the broader community" (p. 323). For decades, the discussions about the responsibilities of business have been inspired by an increasing number of environmental and social scandals and by the growing awareness of unfair or even discriminatory business behaviour (Scherer and Palazzo 2007). Moreover, as declared by the European Commission (2011), the economic crisis and its social consequences have to some extent contributed to lower consumer confidence and trust levels in business. In turn, among society, the scepticism about business intentions and capabilities to behave in a socially responsible way has been increasing. Moreover, the stimulus for the growth of such scepticism lies in the eroding power balance between organisations and governments as the organisations organise their activities globally and governmental regulations of the appropriate countries remain compulsory (Schrempf-Stirling et al. 2016). Under such circumstances, organisations have some space for irresponsible behaviour, especially in countries, where governments are either not willing or not able to regulate the business activities. Nevertheless, the currently growing sensitivity of the society to unsustainable business models (Scherer et al. 2013) is constantly motivating the organisations to address the bundle of issues in the fields of environment protection, social fairness and equity, and economic prosperity thus diminishing the scepticism level.

The second important feature of the 2010's is related to the concept of shared value creation while committing to CSR (Porter and Kramer 2019; Freudenreich et al. 2020). Shared value can be defined as "policies and operating practices that enhance the competitiveness of a company while simultaneously advancing the economic and social conditions in the communities in which it operates. Shared value creation focuses on identifying and expanding the connections between social and economic progress" (Porter and Kramer 2019) (p. 327). Value creation requires an active contribution of all stakeholders (Dentoni et al. 2016); moreover, business would lose its resources and partners if value creation was not mutually beneficial for all actors (Freudenreich et al. 2020). Following the shared value idea, organisations can create economic value by creating social value. Porter and Kramer (2019) proposed three distinct ways for creating shared value: by reconceiving products and markets, by redefining productivity in the value chain, and by creating supportive industry clusters where the company operates. The idea of shared value directs the organisations towards the right kind of profit, namely the profit which creates social benefits rather than diminishes them; it reconnects the company success and community success; it resets the boundaries of capitalism and opens up many ways to serve new needs, gain efficiency, create differentiation, and expand markets (Porter and Kramer 2019).

The third important aspect covers several agreements, which have established high ambitions for the journey towards sustainable development and business responsibility. The Paris Agreement, a legally binding international treaty on climate change requiring economic and social transformation, is one of them (United Nations Framework Convention on Climate Change 2015). Another agreement is the 2030 Agenda for Sustainable Development (UN General Assembly 2015), which proposed 17 sustainable development goals (SDGs). Even though the SDGs did not represent any commitments for the private sector, the countries committed to creating specific regulations and policies that would encourage the private organisations to implement new business practices or to improve their current ones in order to move towards the SDGs implementation (Latapí Agudelo et al. 2019). Turning to the EU level, it is important to mention that the Directive 2014/95/EU requires large companies of public interest to disclose non-financial and diversity information. Summing up, all these legal documents are aimed at further and explicit business engagement in CSR activities.

Table 4.1 provides the summary of the main features of each of the 10 years' timeline.

Summing up, there is a long and varied history of the evolution of CSR, influenced by political events, legal documents, scientific debates and, most important, by changes in society's consciousness and expectations towards organisations and their role in respect to society. Further, definitional treatment of CSR is presented.

4.2.2 Definitional Variety and Contents of CSR

Generally speaking, CSR refers to an explicit framework for better understanding of the business and society relationship (Carroll 2015; Han and Lee 2016). CSR is based on the presumption that an organisation needs to behave in a socially responsible way (Asif et al. 2013).

As it was mentioned before, the role of business in society has been a matter of discussion for several centuries. During this period, the power of business has increased as well as the awareness of the general public requiring to behave in a socially responsible way (Porter and Kramer 2019). Seeing that business and its surrounding community are intertwined, there is no doubt that the success of business depends largely on the society (Boesso and Michelon 2010; Lee et al. 2012). As such, CSR may provide a general framework to structure the responsible use of corporate power and social involvement (Turker 2009b). Furthermore, CSR can help the organisations "to find a way to realise real sustainable business in view of their central role in the global economic and financial stability" (Costa and Menichini 2013) (p. 150).

This notwithstanding, CSR means many things to many people: what is understood as CSR has developed over time; it varies with region, country and culture; it is different for different types of organisations; and is entirely in the eye of the paradigm beholder (Voegtlin and Greenwood 2016). This makes it difficult to find a

Table 4.1 The main aspects of the CSR history

Period	The most relevant and influential ideas, aspects of particular period as regards CSR
1950s	Bowen's (1953) definition of CSR: "the obligations of businessmen to pursue those policies, to make those decisions, or to follow those lines of action which are desirable in terms of the objectives and values of our society" (Bowen 1953) (p. 6)
1960s	Businesspeople have a relevant obligation towards society in terms of economic and human values, and to a certain extent social responsibility could be linked to economic returns for the organisation (Davis, 1960) The growing power of businesspeople was acknowledged (Frederick, 1960) A new understanding of CSR acknowledged the relevance of the relationship between organisations and society (Davis 1960; Frederick 1960) Scepticism towards CSR (Friedman 1962)
1970s	The beginning of this period is characterised by a low level of confidence in business to fulfil the needs of society (Latapí Agudelo et al., 2019) The development of CSR concept was strongly influenced by the Committee of Economic Development of the USA (Latapí Agudelo et al., 2019) The beginning of era of "managing corporate social responsibility" (Carroll, 2015) Carroll's definition of CSR: "The social responsibility of business encompasses the economic, legal, ethical, and discretionary expectations that society has of organisations at a given point in time" (Carroll 1979) (p. 500)
1980's	Stakeholder theory (Freeman 1984) Report "Our Common Future" (WCED 1987)
1990's	Three main trends in CSR emerged: globalisation, institutionalisation, and strategic reconciliation (Carroll 2015) "Corporate Social Performance" (Wood 1991) "Pyramid of Corporate Social Responsibility" (Carroll 1991) CSR "pays off" for the organisation as well for its stakeholders (Burke and Logsdon 1996) "The triple bottom line" concept introduced by Elkington (1994)
2000's	The United Nations Global Compact (2000) EU "Green Paper" (2001) Holistic interpretation of CSR provided by van Marrewijk (2003) Insights of Porter and Kramer (2006) on business and society interrelations
2010's	Growing scepticism about the appropriateness of corporate actions (Schrempf-Stirling et al. 2016) Agenda 2030 with 17 sustainable development goals (UN General Assembly 2015) Shared value creation while committing to CSR (Porter and Kramer 2019)

commonly accepted definition of CSR. As such, based on literature review, Dahlsrud (2008) found at least 37 different definitions of CSR revealing potentially a significant problem: "if competing definitions have diverging biases, people will talk about CSR differently and thus prevent productive engagements" (Dahlsrud 2008) (p. 1).

Originally, CSR was used as a general term arguing that managers ought to seriously consider their impacts on the society. The definition of Bowen (1953) was perhaps the first academic definition of CSR (see chapter "The historical evolution of CSR") outlining a specific set of the principles for organisations to fulfil their social responsibilities. Later in the literature, the focus turned to the actions taken by organisations and their leaders in respect of society along with business's own interests (Carroll 2015). Two core aspects of CSR were emphasised, namely to protect and improve the welfare of society. Protection refers to the business need to avoid its negative impacts such as employee discrimination, air pollution or unsafe products. The improvement implies that organisations need to create positive benefits for the society. Simultaneously, the early debates on CSR captured a one more notion arguing that organisations not only had economic and legal obligations but also certain responsibilities that extended beyond those obligations, though these were not enounced explicitly (Carroll 2015).

Turning to the definitional diversity of CSR, some of more contemporary and recent definitions of CSR are provided in Table 4.2. Further, some elaboration on these definitions is presented.

According to the Green Paper (Commission of the European Communities 2001) (Table 4.2), at least three aspect are highly important while defining CSR: integration of social, environmental and economic aspects; interaction with stakeholders; and voluntary nature. According to van Marrewijk (2003), CSR relates to such phenomena as transparency, stakeholder dialogue and sustainability reporting.

In 2011, the European Commission acknowledged that many organisations in the EU have not yet fully integrated environmental and social concerns into their core strategy and daily activities, and in turn proposed a modern understanding of CSR (Table 4.2). According to it: "to fully meet their corporate social responsibility, enterprises should have in place a process to integrate social, environmental, ethical, human rights and consumer concerns into their business operations and core strategy in close collaboration with their stakeholders" (p. 6) with two aims: firstly, maximising the creation of shared value not only for their shareholders, but also for their other stakeholders and society at large; secondly, identifying, preventing and mitigating their possible adverse impacts. As such, CSR is treated as a complex phenomenon, the commitment to which largely depends the size of the organisation and the nature of its operations. Generally speaking, the EU encourages the organisations to adopt strategic approach to CSR and "to explore the opportunities for developing innovative products, services and business models that contribute to societal wellbeing and lead to higher quality and more productive jobs" (p. 6).

More recently, Voegtlin and Greenwood (2016) argued that while defining CSR the changing institutional conditions of a globalising business environment and their implications for an organisation's CSR should be emphasised.

Acknowledging the definitional diversity, a feature common to all definitions provided in Table 4.2. is that all of them are closely interrelated with the concept of "stakeholder". One of the chapters of this book is dedicated to presenting the stakeholder theory and describing some of the stakeholders, namely employees, potential employees, B2B context and consumers, in more detail.

Besides the stakeholder relevance, looking more precisely into the meaning of CSR, it should be admitted that there are many different ways to think about what CSR includes. In this section, the focus is limited to categorisations suggested by Carroll (1991), Garriga and Melé (2004), Dahlsrud (2008) and Bhardwaj et al. (2018).

Table 4.2 Definitions of CSR

Author, year	Definition of CSR
Smith (2001)	"CSR refers to the obligations of the firm to its stakeholders, people affected by corporate policies and practices. These obligations go beyond legal requirements and the firm's duties to its shareholders. Fulfilment of these obligations is intended to minimize any harms and maximize the long-run beneficial impact of the firm on society" (p.2)
Green Paper (Commission of the European Communities 2001)	"A concept whereby companies integrate social and environmental concerns in their business operations and in their interaction with their stakeholders on a voluntary basis"(p. 7)
van Marrewijk (2003)	"Company activities – voluntary by definition – demonstrating the inclusion of social and environmental concerns in business operations and in interactions with stakeholders" (p. 102)
European Commission (2011)	"The responsibility of enterprises for their impacts on society"
Aguinis (2011)	"Context-specific organisational actions and policies that take into account stakeholders' expectations and the triple bottom line of economic, social, and environmental performance" (p. 858)
Aguinis and Glavas (2012)	"Context-specific organisational actions and policies that take into account stakeholders' expectations and the triple bottom line of economic, social, and environmental performance" (p. 933)
Costa and Menichini (2013)	"A business approach that considers how firm activities impact on society and that companies have a broader range of obligations besides economic and financial ones, taking into account also legal, ethical and philanthropic responsibilities" (p. 150)
Voegtlin and Greenwood (2016)	"A shifting political contest between business, government and civil society actors over governance of the corporation" (p. 180)

Carroll (1991) introduced four dimensions of CSR: economic, legal, ethical, and philanthropic. The economic responsibility of business is "to produce goods and services that society desires and to sell them at a profit" (Carroll 1979) (p. 500). Economic dimension refers to creating jobs and promoting the creation of innovative products and services (Jeon et al. 2020). By doing so, businesses fulfil their primary responsibility as economic units in the society (Carroll and Shabana 2010). Legal responsibility implies the positive and negative obligations put on companies by the laws and regulations of the society where they run the business (Carroll and Shabana 2010). Ethical dimension refers to the organisations' responsibility to be fair and just while making implementing decisions, beyond their legal obligation (Carroll 1991). The avoidance of social harm is also a part of ethical dimension (Jeon et al. 2020). Philanthropy implies the organisations' actions that are carried out in response to the society's expectation that businesses should be good corporate citizens and actively engage in acts or programmes to promote human welfare or goodwill (Carroll 1991), and the quality of life in the community (Jeon et al. 2020). All these dimensions were proposed to be in hierarchical order with economic dimension at the lowest and philanthropic dimension at the highest level (Carroll 1991). The Pyramid of CSR comprises distinct dimensions; however, these dimensions are not mutually exclusive; on the contrary, taken together they constitute a whole. Still, a consideration of the separate dimensions might help the leaders in the organisations see and realise that "the different types of obligations are in a constant but dynamic tension with one another" (Carroll 1991) (p. 42). Thus, the tensions between dimensions as organisational reality were already acknowledged by earlier writings on CSR.

While analysing the definitions of CSR (a total of 37), Dahlsrud (2008) identified five dimensions, which were named to reflect the content of CSR, namely the stakeholder dimension, the social dimension, the economic dimension, the voluntariness dimension, and the environmental dimension. For evaluating the relative use of each dimension, a dimension ratio was calculated by dividing the dimension score by the sum of frequency counts for all the definitions. The results revealed that all dimensions, except for environmental, had comparable dimension ratios above 80%. In the case of the environmental dimension, the ratio was significantly lower, at 59% (Dahlsrud 2008). This could be explained by the fact that the environmental dimension was not included in the early definitions of CSR, and this might have resulted in the current definitions not including it either (Dahlsrud 2008). Nonetheless, as emphasised by Dahlsrud (2008), all the dimensions are necessary in order to understand how CSR is defined. Moreover, the analysis of CSR definitions enabled to draw several conclusions, which are extremely relevant from scientific and practical points of view. First, although all definitions apply different phrases, they are predominantly congruent, making the lack of one universally accepted definition less problematic than it might seem at a first glance. Second, CSR definitions are describing a phenomenon, but do not provide any guidance on how to manage the challenges within this phenomenon. As the result of this, "the challenge for business is not so much to define CSR, as it is to understand how CSR is socially constructed in a specific context and how to take this into account when business strategies are developed" (Dahlsrud 2008) (p. 6).

Following Bhardwaj et al. (2018), there are two types of CSR: company ability-relevant CSR (CSR-CA) and company ability-irrelevant CSR (CSR-NCA). When the organisation invests in CSR-CA, doing so helps it to improve its product development and/or manufacturing capabilities, while CSR-NCA does not have such impact (Sen and Bhattacharya 2001). In order to better understand the distinction between these two types of CSR, the examples mentioned by Bhardwaj et al. (2018) are provided below. Thus, a perfect illustration of CSR-CA would be the Ben & Jerry's implementation of fair-trade norms in their production and creating a dairy farm sustainability programme. These actions might eventually enhance the company's performance and bring in better quality products. Meanwhile, a company like TOMS, which donates a pair of shoes to a child every time a customer purchases its product, applies CSR-NCA, as such policy would not improve the company's ability per se (Bhardwaj et al. 2018).

Garriga and Melé (2004) proposed to categorise the CSR theories into four groups: instrumental theories, political theories, integrative theories, and ethical theories.

Instrumental theories assume that the organisation is an instrument for wealth creation and that this is its sole social responsibility (Garriga and Melé 2004). Thus, CSR is viewed only as a way to create wealth through the organisational image enhancement (Costa and Menichini 2013). In the political theories' group, the social power of an organisation is emphasised, specifically in its relationship with the society; this leads the organisation to accept social rights and duties (Garriga and Melé 2004). Integrative theories underline the integration of the society and organisation's demands, based on the consideration that business success depends on society's welfare and vice-versa (Garriga and Melé 2004). Finally, the ethical theories propose that the relationship between organisations and society is embedded with ethical values (Garriga and Melé 2004).

Summing up, currently, CSR has become a significant part of the business paradigm (Supanti and Butcher 2019), addressing different ways of thinking about what CSR includes. At this point, a question arises about the future of CSR. Carroll (2015) proposed three scenarios, namely the Gloomy Scenario, the Hopeful Scenario, and the Probable Scenario, concerning the further progress of CSR. Irrespective of the scenario, stakeholders play a predominant role. Accordingly, the next chapter focuses on the stakeholder theory and describes some of the stakeholders.

4.3 Stakeholder Theory: The Essence of Theory and Its Application to CSR

4.3.1 General Review of Stakeholder Theory

Generally speaking, stakeholder theory is a theory concerned with the relationship between an organisation and its stakeholders (Fernando and Lawrence 2014). There is no agreement when the term was mentioned for the first time; however, it gained popularity with the works of Freeman (1984) who defined a stakeholder as "any group or individual who can affect or is affected by the achievement of the firm's objectives" (p. 49). Recently, a more precise definition was suggested by Post et al. (2002) arguing that stakeholders were "individuals and constituencies that contribute, either voluntarily or involuntarily, to its wealth-creating capacity and activities, and who are therefore its potential beneficiaries and/or risk bearers" (p. 8). Customers, suppliers, employees, financiers (stockholders, bondholders, banks, etc.), and communities can all serve as obvious examples of stakeholders.

According to Parmar et al. (2010), the stakeholder theory makes it possible to deal effectively with three problems. First, from a stakeholder perspective, business can be understood as a set of relationships among groups that have a stake in the activities that make up the business (Walsh 2005). To understand a business is not only to know how these relationships work, but also to know how these relationships change over time (Parmar et al. 2010). It is the executive's job to manage these relationships while creating as much value as possible for stakeholders and, finally, to manage the distribution of that value (Freeman 1984). If trade-offs have to be made, the executives must figure out how to make them for all stakeholders (Freeman et al. 2007). Second, although effective management of stakeholder relationships helps businesses survive and thrive, it is also a moral endeavour because it concerns questions of values and benefits for individuals or their groups (Phillips 2003). Third, the kind of management that "focuses attention on the creation, maintenance, and alignment of stakeholder relationships better equips practitioners to create value and avoid moral failures" (Post et al. 2002) (p. 406).

Some scholars have suggested that the perspective of Freeman (1984) was too broad, because at the end all social players are directly or indirectly affected by the actions of organisations (Rodrigo and Arenas 2008). As a result, a debate began to address how and why stakeholders were relevant to an organisation and this, in turn, led to various stakeholder classification schemes and different kinds of interpretations. Further, the various interpretations are provided and later the focus is shifted onto the stakeholder classification schemes.

As regards different interpretation of stakeholders, Fernando and Lawrence (2014) highlight several assumptions which indicate the scope and provide overall insights for the stakeholder theory. These assumptions can be summarised as follows (Fernando and Lawrence 2014):

1. Stakeholders are identified from the vantage point of one focal organisation.
2. In order to achieve the goals, an organisation needs to manage its stakeholders effectively.
3. Stakeholders constitute different categories and often such categories have conflicting interests.
4. An organisation must be able to balance conflicting interests of those stakeholders. Such duty captures both environments of the particular organisation, namely external environment and internal environment.
5. When expending something, stakeholders exert pressure on an organisation.
6. The ability of stakeholders to pressurise an organisation depends on the organisational attributes of stakeholders.
7. An organisation has financial, environmental and social responsibilities to its stakeholders.

As a result of these assumptions, two major branches of stakeholder theory are introduced in the literature, namely the ethical (moral or normative) branch, and the managerial (positive) branch (Fernando and Lawrence 2014).

Following the ethical branch of the stakeholder theory, all the stakeholders have the same right to be treated fairly by an organisation irrespectively of their power (Deegan 2009). Rather than considering only powerful stakeholders (Deegan and Unerman 2006), the ethical perspective urges to take all stakeholders of a particular organisation into consideration. Based on the ethical branch, the organisation is "not viewed as a mechanism which drives the maximisation of shareholders' wealth, but, rather, as one which meets the expectations of all stakeholders" (Fernando and Lawrence 2014) (p. 159).

Meanwhile, the managerial perspective of the stakeholder theory proposes that the organisation should attempt to meet the expectations only of those stakeholders who control the critical resources required by the particular organisation. As such, an organisation is expected to be accountable to its economically powerful stakeholders, rather than all stakeholders as in the ethical perspective (Fernando and Lawrence 2014). The main challenge the organisations face is to have a precise answer to two questions: how organisations should decide to whom they are responsible, and to what extent that responsibility extends (O'Riordan and Fairbrass 2008).

The two mentioned perspectives are closely related to stakeholder classifications, trying to answer the question concerning the role the stakeholders play in the life of a particular organisation. As regards the classification, a huge variety was proposed in the literature: strategic and moral stakeholders (Goodpaster 1991); external and internal stakeholders (Pearce II 1982; Carroll 1989); latent, expectant, and definitive stakeholders (Mitchell et al. 1997); subgroups of stakeholders such as shareholders, employees, and customers (Preston and Sapienza 1990); single issue, and multiple issues stakeholders (Wood 1994); supportive, marginal, non-supportive, mixed-blessing stakeholders (Savage et al. 1991); voluntary and involuntary stakeholders (Clarkson 1995); and primary and secondary stakeholders (Clarkson 1995). The main aspect of these categorisations is to emphasise that there are various stakeholder groups with different and sometimes conflicting expectations (Fernando and Lawrence 2014).

One of the predominant methods of classification in literature is the so called "stakeholder salience" suggested by Mitchell et al. (1997), who used the Freeman's stakeholder concept and provided an approach that helps to identify "who or what really counts" and to assess the degree to which managers pay attention to their stakeholders. This classification distinguishes between three attributes of stakeholders as explained below (Mitchell et al. 1997; Jongbloed et al. 2008).

The stakeholder's power to influence the organisation—here power defines a relationship among social actors in which a social actor A can get another social actor B to do something that B would not have otherwise done.

The legitimacy of the stakeholder's relationship with the organisation—legitimacy is defined as a generalised perception or assumption that the actions of an entity are desirable, proper, or appropriate within some socially constructed system of norms, values, beliefs and definitions.

The urgency of the stakeholder's claim on the organisation—urgency represents the degree to which the stakeholder claims call for immediate action.

These three attributes may help identify the crucial stakeholders to deal with. The presence or absence of the power, legitimacy and urgency attributes translates into a stakeholder typology. It is also relevant to underline that power, legitimacy and urgency can change as they are not static, but dynamic. This implies that particular stakeholders can move from one class to another by gaining or losing particular attributes.

Summing up, stakeholder theory might be employed and used to understand the relationship between an organisation and its environment (internal and external) while analysing various business issues, policies, activities, practices or initiatives, including CSR. The further chapter is dedicated to revealing the stakeholder theory for CSR.

4.3.2 Stakeholder Theory for CSR

In general, CSR has its roots in the stakeholder theory (Carroll 1991; Lee et al. 2013a). The stakeholder theory is organised under two core questions: what is the purpose of the organisation; and what responsibility does management have to stakeholders? (Freeman 2000; Theodoulidis et al. 2017). Although the stakeholder theory indicates the generation of value as a core driver of the organisation, it nonetheless also admits that this value is to be shared by a group of stakeholders (Theodoulidis et al. 2017). Thus, although the stakeholder theory is about value creation (Freeman 2000), profit is still one of the possible outcomes of the value creation process (Theodoulidis et al. 2017).

Considering stakeholder theory, the success of CSR initiatives depends on the organisations' capacity to build strong relationships with their stakeholders (De Roeck et al. 2014). Meeting the demands of stakeholders is necessary for at least two reasons: sustaining a continued supply of resources and for legitimation (Asif et al. 2013). The stakeholders are capable "of inflicting unacceptable damage on the viability of the organisation" (Garvare and Johansson 2010) (p. 737), in case their interests are not met. Moreover, while addressing stakeholder expectations, their engagement becomes highly relevant (Aguinis 2011). Dealing with internal and external stakeholders' expectations means that organisations are encouraged and forced to behave more proactively instead of carrying out certain initiatives reactively (Aguinis 2011).

Generally speaking, stakeholders are essential to the successful functioning of the organisation because they provide resources, form the industry structure and make up the socio-political arena (Post et al. 2002). As it was discussed when presenting the diversity of CSR definitions, the stakeholder element is underlined in the majority of the definitions. The focus on stakeholders is strongly addressed in the European level documents. For instance, the Green Paper (Commission of the European Communities 2001) indicated that "corporate social responsibility is a process by which companies manage their relationships with a variety of influential stakeholders who can have a real influence on their licence to operate, the business case becomes apparent" (p. 4). Based on the Green Paper (Commission of the European Communities 2001), CSR is about attracting and retaining talented staff, about building relationships with customers and accordingly about managing risk and assuring reputation. The later definition of CSR provided by the EC in 2011 also fits nicely with the stakeholder approach that views CSR as an extension of corporate governance (Theodoulidis et al. 2017).

Campbell (2007) suggests that organisations are acting in a socially responsible manner when they undertake two key activities, calling them minimum behavioural standard in respect of the stakeholders. First, they do not knowingly do anything that could harm their key stakeholders, i.e. investors, employees, customers, suppliers, or the local community. Second, if organisations do cause harm to their stakeholders, they must then rectify it whenever the harm is discovered and brought to their attention (Campbell 2007).

As this book deals with four stakeholder groups, namely employees, potential employees, customers in B2B context, and consumers, these are briefly described below.

Employees

Recently, the number of studies where the focus is placed on employees in relation to organisations' CSR has been growing rapidly (Supanti and Butcher 2019; Bouraoui et al. 2019, 2020; Kim et al. 2020; Wang et al. 2020).

Essentially, the literature describes employees as primary stakeholders (Mitchell et al. 1997). Employees are relatively highly salient stakeholders to whom the organisation owes a perfect duty (Greenwood 2007). Employees represent an essential stakeholder group who are relevant resources contributing to overall business performance and determining long-term survival of the particular organisation (De Roeck et al. 2014).

Employees have a continuing investment in the organisation in terms of investment of experience and specialised skills (Greenwood 2007). Furthermore, by taking a job, employees might need to invest by changing the living place, suspending previous relationships or by acquiring new competencies through learning. They may become financially dependent on the organisation, while the organisation is likely to form the basis of their economic livelihood through their income or share ownership (Greenwood 2007).

Analysing employees in respect to CSR, it is worth highlighting the dual nature of such relationship. First, employees play a vital role in reaching the organisation's goals in terms of CSR. Second, the employee perceptions of CSR initiatives implemented by the organisation can shape the employee attitudes and behaviours (Bouraoui et al. 2019).

Concerning the first aspect, the employees "constitute" the organisation as they are a "resource of the corporation, they represent the company towards other stakeholders, and they act in the name of the corporation" (Crane and Matten 2004) (p. 224). Employees are among the important stakeholders of the organisation because they determine the quality of the product and/or service the customers receive (Lee et al. 2012).

Turning to the second aspect, there is a wide consensus among scholars that employees perceive the CSR initiatives implemented in the same company differently (Aguinis and Glavas 2013). As Ong et al. (2018) (p. 45) state, "individuals vary in the extent to which they value such activities, and therefore individuals vary in the extent to which they are likely to be influenced by their organisations' CSR activities".

According to Sirgy et al. (2001), employees have needs, such as health, safety, self-actualisation etc.; they seek fulfilment through work. This means that employees derive satisfaction from their jobs to the extent that their jobs meet these needs (Sirgy et al. 2001). As such, CSR initiatives can serve as a fundamental source for the satisfaction of basic, growth or other needs of employees. For instance, fair treatment of workers and compensation policies, promotion of work and life balance can directly signal to the employees that the organisation is safe to invest their effort and time, giving them a sense of well-being (Kim et al. 2020). Moreover, employees can evaluate the motives behind CSR better than consumers and other external stakeholders because of their tacit knowledge of the organisation and this may influence the way the employees relate to the organisation (Fryzel and Seppala 2016).

In general, CSR reputation and practices of a firm can be used to attract, retain and motivate the employees, i.e. "win the war for talent" (Bhattacharya et al. 2008). As stated by Lee et al. (2013), employees who have a favourable view of the organisation's CSR tend to have positive views about the organisation in other areas, including leadership or the organisation's competitiveness. Thus, CSR may serve as a good reason to stay and work hard (Donia et al. 2019).

Potential employees

Finding and retaining the employees is a considerable task for many organisations. The evolving needs and values of employees representing different generations make employee recruitment increasingly challenging. Researchers try to answer the question what makes an organisation attractive to a potential employee. Millennials are becoming the main driver for demanding that organisations should incorporate CSR into their mission and corporate values. The Nielsen Global Survey (Nielsen 2015) revealed that 67% preferred to work for socially responsible organisations. According to Bromley and Meyer (2017) (p. 945), organisations "are transformed by new pressures to look like responsible actors" and laws and public pressure force them to "take on expanded concerns such as environmental protection, corporate social responsibility and philanthropy, employee rights and job satisfaction, workplace diversity, community engagement, and consumer safety".

Report by Deloitte (Deloitte Touche Tohmatsu 2016) focusing on the work values of the millennials demonstrated that while attention to the social and environment responsibility was a highly relevant concern to them, 95% of respondents described organisations as appearing to have "no ambition beyond making money" (p. 9). Two years later, a report by Deloitte (Deloitte Touche Tohmatsu 2018) once again revealed that young workers were eager "for business leaders to be proactive about making a positive impact in society—and to be responsive to employees' needs" (p. 2).

As stated by Cable and Judge (1996), job seekers prefer organisations that have the same "personality" as they do. Several studies have investigated the influence of CSR on organisational attractiveness for potential employees (Zhang et al. 2020). Turban and Greening (1997) demonstrated that the organisation's corporate social performance could be an important attracting factor for potential employees. According to Birth et al. (2008), potential employees are more attracted by employers with ethical integrity and socially responsible behaviour. An organisation's social initiatives may attract potential employees by serving as a signal of working conditions in a particular organisation (Turban and Greening 1997).

Customers in the B2B context

Actually, the characteristics of B2B (business-to-business) market are very different from those of the B2C (business-to-consumer) market (Han and Lee 2021). One or several consumers in B2C context could hardly inflict great damage to the organisation; while in the B2B context, where the organisation's operations depend on the supplier's reliability, unfavourable consequences can result from choosing the "wrong" business partner (Homburg et al. 2013). Thus, customers in the B2B context are highly important stakeholders as organisations need them in order to operate (Parmar et al. 2010). Traditionally, on-time delivery, price, and technical quality have been recognised as the most important factors in the B2B market (Han and Lee 2021). However, recently the situation has been changing as not only economic factors but also some social factors are being recognised as important factors in the B2B context (Han and Lee 2021). Partnership between the organisation and its customers, selection and analysis of customers, and engagement in fair trading transactions with customers may serve as CSR actions towards customers in the B2B context (Jamali 2008). Summing up, increasingly more organisations make sustainability practices a requirement for their customers in the B2B context (Guo et al. 2019).

Consumers

Consumers are also among the most crucial stakeholders of an organisation (Jeon et al. 2020). As the success of a particular organisation depends to a great extent on consumers, good relations with them are desirable (Turker 2009b). CSR can be used to influence the consumer feelings, thoughts, and consequently buying behaviours (Turker 2009b). When making purchasing decisions, consumers take the organisations' CSR initiatives into consideration (Bhardwaj et al. 2018). In a global survey conducted by Nielsen, 50% of 29,000 respondents had an intention of paying a higher price for the products and services developed by companies that invest in CSR (The Nielsen Global Survey 2013).

It is worth mentioning and highly relevant to understand that consumers and employees are interrelated stakeholders (Turker 2009b). Employees seem to be the agents of their employers, at least from a consumer point of view. On the one hand, if an organisation produces unsafe products or misleads its consumers, employees may also feel the shame for this behaviour. And vice versa, if an organisation pays attention to its consumers by providing high quality products or fair information, employees may also be proud of being members of that particular organisation (Turker 2009b).

4.4 Obstacles and Drivers of CSR Transitions

4.4.1 CSR Transitions

Seeing that the previous section disclosed the stakeholder importance when organisations are moving towards CSR, in this section we will analyse the CSR transitions, seeking to answer the question how the CSR transitions are understood in the scientific literature.

Researchers have formulated the meaning of "transition" in various ways. Pisano et al. (2014) argue that "transition" implies smaller elements of the transformation phase; Davies (2013) considers that transition involves a larger picture, including social, political, economic, and cultural change. Transition researchers define the term holistically in a variety of technological, material, organisational, institutional, political, economic, and socio-cultural ways (Geels and Schot 2007). According to Grin et al. (2010), transitions are major, non-linear changes in societal cultures, structures and practices that arise from the coevolution between economy, society, and ecology. Loorbach and Wijsman (2013) propose that transitions can be viewed as a shift from one dynamic equilibrium (e.g. a fossil-based centralised energy system) to another (e.g. a renewable energy-based, decentralised system). They are the result of interacting developments at different levels of scale that, under specific conditions, might over time fundamentally alter the dominant practices, paradigms, and structures. Usually, transitions take a very long predevelopment phase in which there is a gradual build-up of pressure on a dominant regime, which may be understood as the dominant structure, culture and practices in a societal system. This pressure stems from an internal dysfunction of the regime, increasing competition of alternatives, or a changing external context. When these pressures start to reinforce each other, a relatively rapid systemic change might occur. In research, transitions are thus visualised as processes of multi-level (Geels 2002), multi-phase (Rotmans et al. 2001) changes. Skellern et al. (2017) reveal the arguments of human geographers such as Hicks (2014) who see the transition as an evolutionary, adaptation process in which an organism becomes better able to live in its habitat.

In the context of sustainability, transitions seem inevitable from the perspective of limited resources, ecological thresholds (Rockström et al. 2009) and changing economic and demographic landscapes. They do however also provide possibilities for innovation, green economic growth and new business. Taking sustainability issues into consideration can lead to difficulties. Companies often propose very biased and self-serving external accounts and struggle to change their internal plans, processes and instruments (Passetti et al. 2018). The shifting from the current way of work to a different mode requires addressing a whole list of regulations, current market state etc. Sustainability transitions are necessarily about interactions between technology, policy/power/politics, economics/business/markets, and culture/discourse/public opinion (Geels 2011). According to Keijzers (2002), sustainability efforts change the internal governance of the firm as new environmental preservation demands come forward, and the external governance of businesses as new initiatives need to be employed to ensure adequate and sustainable social, technological, and infrastructural conditions for production (Loorbach and Wijsman 2013). One of the reasons for transitions to occur is caused by the need of organisations to apply sustainability decisions and policies. The thought that sustainability is increasingly turning into a key challenge for organisations (Merad et al. 2014) should be considered not only by the sustainability and environmental field researchers. Leaders of organisations must face the challenges of the today's world, accept them, and integrate sustainability-oriented models to their strategies and development goals. Addressing environmental issues is not just a good advertisement for the organisation in the market. Environmental strategies reflect the organisation's stance on the natural environment and also represent a good substitution of a firm's management capabilities (Delmas et al. 2011). Organisations applying sustainable actions do make a difference in their field locally. Spreading of knowledge and best practices can influence others to take part in the sustainability movement.

Sustainability transitions refer to purposeful, long-term, multi-dimensional, fundamental transformations of socio-technical systems towards more sustainable modes of production and consumption, requiring participation of different types of actors (Markard et al. 2012; Lyytimäki et al. 2019; Schlaile and Urmetzer 2019). Kern and Markard (2016) distinguished numerous features describing sustainability transitions: value-laden and contested, e.g. trade-offs such as low-carbon vs. nuclear risks; conflicting views; key role for public policies; purposive transitions associated with sustainability targets; power & politics central; vested interests; winners & losers; coalitions & alliances; complex, uncertain, long-term; context-dependent: different pathways; multi-dimensional, systemic interaction, e.g. interaction of multiple technologies.(Kern and Markard 2016). Characteristics of sustainability transitions described by Köhler et al. (2019) included the following: multi-dimensional and co-evolution; multi-actor process; stability and change; long-term process; open-endedness and uncertainty; values, contestation, and disagreement; normative directionality. These characteristics indicate the transdisciplinary nature of the research on sustainability transitions.

The theoretical model of sustainability transitions was proposed by Geels (2002) and is still valid and broadly used today. Drawing on the literature on transition pathways (Geels and Schot 2007; De Haan 2010) that identifies different ideal–typical patterns of fundamental systemic change and following the typology of Boons (Boons 2009), there are at least three basic pathways through which the existing regimes could change: optimisation (stability), reconfiguration (dynamic) and system innovation (transformative). The question is to what extent frontrunner practices in terms of effective and successful transformative strategies might be transferred to other businesses and what can be done to influence the speed and direction of these processes (Loorbach and Wijsman 2013).

Sustainability transitions integrate the environment value stream in the socio-technical system. They are aimed at reconfiguring the organisation within a nexus of governance, technology and practice. Socio-technical transitions do not just change the structure of the existing system; they also affect societal domains such as planning and policy-making and the dominant way in which and what an organisation produces or provides services (Geels 2011; Skellern et al. 2017). Transitions are understood as non-linear processes; they entail multiple, interdependent developments (Köhler et al. 2017).

Summing up, sustainability transitions are viewed in this study as long-term changes that are multi-dimensional and fundamental transformations of organisational systems towards more sustainable modes of productions and consumption.

In the next section, we will analyse the drivers for sustainability.

4.4.2 Sustainability Drivers

Based on the previous sections, we can argue that different stakeholders could be seen as important drivers towards sustainability, but there are more of those. It is considered that motivation, commitment, identification, or pro-social behaviours manifest differently when sustainability issues are confronted by an organisation (Howard-Grenville et al. 2014). Matten and Moon (2008) suggest that national institutional frameworks consisting of specific combinations of political, financial, educational, and cultural systems affect CSR behaviours. From this perspective, the legal requirements and demands of the various stakeholder groups and society acquire importance as drivers for CSR. Campbell (2007) considers that economic conditions affect the degree to which corporations act in socially responsible ways, but this relationship is mediated by a variety of institutional factors that constrain and enable these behaviours. Agudo-Valiente et al. (2017) notice that corporate social responsibility differs across sectors (Cramer 2005) and one possible explanation of that is the presence of contextual factors (Zhao et al. 2012). Qian Zhang et al. (2019) have analysed CSR drivers in the construction sector and found that the identified drivers could be categorised into three sub-themes, namely: policy pressure; market pressure; and innovation and technology development. These drivers are mainly external, while the internal drivers were named as motivations. The motivations for CSR implementation reflect the innate willingness and comprehensive concerns of organisations. These include "financial benefits, branding, reputation, and image, human resource benefits, supplier-induced benefits, persuasion and inspiration, relationship building, policy benefits, organisational culture and awareness, strategic business direction, and availability of resource and capability" (Zhang et al. 2019) (p. 580).

Looking through the lens of instrumental stakeholder theory, various scholars mention the possibility of obtaining long-term benefits through CSR (Walsh et al. 2003) or cost savings (Miras-Rodríguez et al. 2015). CSR could be used to improve the relationships with stakeholders (Maignan and Ralston 2002; Campbell 2006, 2007) or company reputation (Selvi et al. 2010; Govindasamy and Suresh 2017), obtain greater community solidarity (Hemingway and Maclagan 2004), prevent incident-induced reputation risks (Margolis et al. 2009) and promote customer loyalty (Pérez and Rodríguez del Bosque, 2015; cf. Agudo-Valiente et al., 2017).

Larger firms have more resources for the development of CSR, making the firm size a key factor and a predictor of CSR (Muller and Kolk 2010). Gamerschlag et al. (2011) point out size-related CSR behaviour differences. When speaking of big corporations, leadership and the business case are the most important drivers internally, whilst the most important external drivers include reputation, customer needs, and regulation and legislation (Lozano 2015). Nevertheless, some actors may be self-motivated by a genuine concern for sustainability, or by new opportunities arising from sustainability challenges (Hajer et al. 2015).

Agudo-Valiente et al. (2017) performed a literature review and categorised the drivers promoting CSR into subjective and objective. Objective drivers for CSR included stakeholder pressure, institutional framework, reputation management, leading corporations, and sectorial trends. The subjective CSR drivers identified based on the literature review are integrating ethics, sustainable development, and organisational commitment to transparency.

Based on the research of Engert et al. (2016), "the drivers are: legal compliance; competitive advantage; cost reduction; economic performance; innovation; social and environmental responsibility; risk management; corporate reputation; and quality management". From these drivers it is visible that sustainability transitions and the decisions made regarding them must be incorporated in the strategic level of the organisations.

Summing up, there is no consensus on a uniform classification of drivers, and in this book, drivers of corporate sustainability transitions are seen from the stakeholders' perspective. For example, leadership is very important for the employees to perform their best; regulation and legislation are considered when choosing business partners, whereas reputation is crucial to all stakeholders.

In the next section, we will analyse the obstacles and tensions that occur when organisations are moving towards CSR.

4.4.3 Tensions in Sustainability Transitions

Tensions exist at multiple levels, occur simultaneously, and influence emotions whenever they result in anxiety, discomfort, uncertainty or even paralysis of the actors confronted by them (Putnam et al. 2016).

Sustainability depends on internal and external characteristics of the organisation that should or must be preserved and on the definition of what is a suitable state of the system (Merad et al. 2014). The definition of corporate sustainability holds tension as an inherency. Tensions in sustainability may exist between present and future temporal contexts or between competing elements of economy, society, and environment (Slawinski and Bansal 2012). In line with Merad et al. (2014), companies' environmental practices are shaped by different external and internal pressures or critiques (Howard-Grenville 2006; Delmas and Toffel 2008; Wright and Nyberg 2017). The integration of environmental concerns within organisational structures, decision-making and organisational values creates tension between the business and sustainability (Passetti et al. 2018). Emerging challenges of the environment cause organisations to change internally, adapting to the sustainability issues, therefore some characteristics of organisations cannot be preserved. This is one of the reasons the tensions arise from—the change that corporate sustainability calls for requires the firms to fundamentally alter their current patterns of activity (Hahn et al. 2015).

Given the contradictions between economic and societal demands (Margolis and Walsh 2003), companies attempting to integrate CSR into their business strategy and operations also expose themselves to internal CSR tensions that require systematic managerial attention (Smith et al. 2013; Hahn et al. 2015; Ozanne et al. 2016; cf. Siltaloppi et al. 2020).

Most people and organisations alike react to change unfavourably. Fear of change is easily explained—people feel insecure when their stable environment starts to change. It becomes unstable and requires more effort to perform routine tasks during the transitional period. The outcome of change is also unknown. Therefore, most organisations, when asked about change, describe it as "difficult", "confronting", "hard-hitting", "stressful", "tense", "painful", "uncomfortable", stating that it involves "strain", "stress", "unpleasantness" etc. (Kabanoff et al. 1995). Only organisations expressing collegial values associated change with positive rather than negative terms and approached them enthusiastically (Kabanoff et al. 1995).

Despite the negative emotional impact on the actors, tensions do in fact give rise to both negative and positive outcomes. The ability to perceive tensions and make them salient through communication enables humans to re-assess the existing conditions (van Bommel 2018) and develop "outside-of-the-box" solutions (Harari 2011; Wannags and Gold 2020).

Engaging with grand challenges such as climate change is particularly problematic for businesses, given the long-term, complex nature of these problems and the underlying tension between economic growth and its material consequences (Wright and Nyberg 2017). One of the main reasons for organisations addressing sustainability-related changes timidly is the possible negative impact on the firm's financial performance (McWilliams and Siegel 2000; Damania 2001; Delmas et al. 2011), although some studies find a positive link between environmental approaches and economic results (Russo and Fouts 1997; Delmas et al. 2011). However, since the environmental issues are real, it is time to pay more attention to breakthrough, diffusion, tipping points, and thresholds, because environmental problems require accelerated transitions (Köhler et al. 2017).

Agudo-Valiente et al. (2017) revealed the subjective and objective CSR barriers based on the literature review. Subjective barriers include philanthropy, public relations exercise, sceptical view only for large corporations, and utopic value. Objective barriers for CSR are availability of financial resources, availability of temporary resources, availability of human resources, difficulties involved in interpreting CSR, and low institutional interest.

Wannags and Gold (2020) reviewed the existing approaches to the classification of tensions and trade-offs. They found that Smith and Lewis (2011) have defined categories of paradoxes experienced in organisations in general, whereas Hahn et al. (2018) and Haffar and Searcy (2017) focus on contradictions within the field of corporate sustainability. Similarly to Putnam et al. (2016), Hahn et al. (2015) view tensions through the lens of institutional theory, which accentuates the workings of institutional logics (cf. Thornton and Ocasio, 2008) at each pole in a paradox. Smith and Lewis (2011) and Haffar and Searcy (2017) prefer to focus on dynamic capabilities, highlighting the notion of dynamics that enable tensions to manifest in trade-offs, or alternatively allow for the resolution of these tensions" (Wannags and Gold 2020) (p. 3).

Smith and Lewis (2011) identified four categories of paradoxical tensions—belonging, performing, organising, and learning. Belonging tensions arise because organisational actors strive for both self-expression and group affiliation (Lewis 2000). These issues of identifying promote tensions in the areas of organisational culture, values, roles, and membership. Performing tensions arise from the plurality of stakeholders in the organisation's operations and result in competing strategies and goals. Organising tensions arise as complex organisational systems create competing designs, structures, processes, and practices to achieve the desired outcomes. Learning tensions manifest when organisational beliefs and assumptions fail to keep pace with the contextual change (Ozanne et al. 2016).

Haffar and Searcy (2017) propose that all corporate sustainability trade-offs result from one type of tension, namely the tension between private and shared values, which resembles the paradoxical dynamic between acting in the interests of shareholders only at the organisational level, and in the interests of society and the planet as a whole at the systemic level. They classify trade-offs according to their performance dimensions (economic, environmental and societal change), time horizons and conflicting stakeholder demands, and by whether they refer to conflicting performance targets or implementation approaches (scope depth or measurement management).

Hahn et al. (2015) propose four categories of tensions, including the underlying logics and potential management responses. These four categories are: (1) personal versus organisational sustainability agenda, (2) corporate short-term versus corporate long-term orientation, (3) isomorphism versus structural and technological change, and (4) efficiency versus resilience. In addition, they propose a variety of tension dimensions, i.e. space, time, and changes in economic, environmental or societal conditions, and they distinct levels as well: tension categories (1) and (2) typically occur at individual and/or organisational level, while tension categories (3) and (4) typically occur at organisational and/or systemic level (cf. Wannags and Gold, 2020). Wannags and Gold (2020) carried out a systematic, content analysis-based literature review and added two more categories of tensions. These new categories are tensions between the desire for sustainable consumption and actual unsustainable consumption behaviour and tensions between legitimacy in context A and context B.

The field of tensions between the consumers' desire for sustainability and actual unsustainable consumption behaviour category provides "intra-organisational" insights into consumers as an actor group. Antonetti and Maklan (2014) discovered that consumers with an interest in sustainability could find themselves in a moral dilemma between self-interest and support for sustainability causes if there was no sustainable consumption option that fulfilled their needs. Deciding in favour of self-interest entails guilt for the consumer, whereas choosing in favour of sustainability may fill the consumer with pride but fails to fulfil their needs. In the data analysed by Wannags and Gold (2020), the tension between legitimacy in context A and legitimacy in context B is found in an inter-organisational context, although tensions between the head offices and subsidiaries of a multinational enterprise are also conceivable. For the most part, the data highlights tensions between a focal company and a supplier.

Summarising the part of theoretical insights, we can see the dynamics of the concept of CSR in different historical periods and the changing perception of CSR by stakeholders. The latter also poses new challenges for business organisations, encouraging them to behave more responsibly, to move from less unsustainable towards sustainable business practices. There are different drivers and motivators for organisations to move towards sustainability, and because of this, we have paid special attention to the role of different stakeholders in this book. The sustainability transitions of organisations undoubtedly face certain tensions and barriers too. Looking at the classifications of drivers and tensions in the scientific literature, we see that there is no consensus among scientists, as drivers and tensions are strongly context-dependent. In the next section, we will present the results of research conducted by the authors of the book, which will reveal the drivers and barriers encountered by organisations operating in the context of Eastern Europe.

References

Agudo-Valiente J, Garcés-Ayerbe C, Salvador-Figueras M (2017) Corporate social responsibility drivers and barriers according to managers' perception evidence from Spanish firms. Sustainability 9:1821. https://doi.org/10.3390/su9101821

Aguinis H (2011) Organizational responsibility: Doing good and doing well. APA handbook of industrial and organizational psychology, Vol 3: Maintaining, expanding, and contracting the organization. American Psychological Association, Washington, pp 855–879

Aguinis H, Glavas A (2012) What we know and don't know about corporate social responsibility. J Manage 38:932–968. https://doi.org/10.1177/0149206311436079

Aguinis H, Glavas A (2013) Embedded versus peripheral corporate social responsibility: psychological foundations. Ind Organ Psychol 6:314–332. https://doi.org/10.1111/iops.12059

Albinger HS, Freeman SJ (2000) Corporate social performance and attractiveness as an employer to different job seeking populations. J Bus Ethics 28:243–253. https://doi.org/10.1023/A:1006289817941

Allen MW, Spialek ML (2018) Young millennials, environmental orientation, food company sustainability, and green word-of-mouth recommendations. J Food Prod Mark 24:803–829. https://doi.org/10.1080/10454446.2017.1415827

Antonetti P, Maklan S (2014) Exploring postconsumption guilt and pride in the context of sustainability. Psychol Mark 31:717–735. https://doi.org/10.1002/mar.20730
Ashrafi M, Adams M, Walker TR, Magnan G (2018) How corporate social responsibility can be integrated into corporate sustainability: a theoretical review of their relationships. Int J Sustain Dev World Ecol 25:672–682. https://doi.org/10.1080/13504509.2018.1471628
Asif M, Searcy C, Zutshi A, Fisscher OAM (2013) An integrated management systems approach to corporate social responsibility. J Clean Prod 56:7–17
Backhaus KB, Stone BA, Heiner K (2002) Exploringthe relationship between corporate social performance and employer attractiveness. Bus Soc 41:292–318. https://doi.org/10.1177/0007650302041003003
Bansal P, DesJardine MR (2014) Business sustainability: it is about time. Strateg Organ 12:70–78. https://doi.org/10.1177/1476127013520265
Baumol WJ (1970) A new rationale for corporate social policy. Heath Lexington Books
Bhardwaj P, Chatterjee P, Demir KD, Turut O (2018) When and how is corporate social responsibility profitable? J Bus Res 84:206–219. https://doi.org/10.1016/j.jbusres.2017.11.026
Bhattacharya CB, Sen S, Korschun D (2008) Using corporate social responsibility to win the war for talent. MIT Sloan Manag Rev 49
Bhattacharya CB, Sen S (2004) Doing better at doing good: when, why, and how consumers respond to corporate social initiatives. Calif Manage Rev 47:9–24. https://doi.org/10.2307/41166284
Birth G, Illia L, Lurati F, Zamparini A (2008) Communicating CSR: practices among Switzerland's top 300 companies. Corp Commun Int J 13:182–196. https://doi.org/10.1108/13563280810869604
Blasco JL, King A (2017) The road ahead. The KPMG survey of corporate responsibility reporting 2017
Boesso G, Michelon G (2010) The effects of stakeholder prioritization on corporate financial performance: an empirical investigation. Int J Manag 27:470
Boons F (2009) Creating ecological value: an evolutionary approach to business strategies and the natural environment. Edward Elgar Publishing, Cheltenham, UK, Northampton, MA, USA
Bouraoui K, Bensemmane S, Ohana M (2020) Corporate social responsibility and employees' affective commitment: A moderated mediation study. Sustainability 12(14):5833. https://doi.org/10.3390/su12145833
Bouraoui K, Bensemmane S, Ohana M, Russo M (2019) Corporate social responsibility and employees' affective commitment. Manag Decis 57:152–167. https://doi.org/10.1108/MD-10-2017-1015
Bowen HR (1953) Social responsibilities of the businessman. Harper & Row, New York, New York, USA
Brammer S, Millington A, Rayton B (2007) The contribution of corporate social responsibility to organizational commitment. Int J Hum Resour Manag 18:1701–1719. https://doi.org/10.1080/09585190701570866
Bromley P, Meyer JW (2017) They are all organizations: the cultural roots of blurring between the nonprofit, business, and government sectors. Adm Soc 49:939–966. https://doi.org/10.1177/0095399714548268
Buenstorf G, Cordes C (2008) Can sustainable consumption be learned? a model of cultural evolution. Ecol Econ 67:646–657. https://doi.org/10.1016/j.ecolecon.2008.01.028
Burke L, Logsdon JM (1996) How corporate social responsibility pays off. Long Range Plann 29:495–502. https://doi.org/10.1016/0024-6301(96)00041-6
Cable DM, Judge TA (1996) Person-organization fit, job choice decisions, and organizational entry. Organ Behav Hum Decis Process 67:294–311. https://doi.org/10.1006/obhd.1996.0081
Campbell JL (2006) Institutional analysis and the paradox of corporate social responsibility. Am Behav Sci 49:925–938. https://doi.org/10.1177/0002764205285172
Campbell JL (2007) Why would corporations behave in socially responsible ways? an institutional theory of corporate social responsibility. Acad Manag Rev 32:946–967. https://doi.org/10.5465/amr.2007.25275684

Carroll AB (1979) A three-dimensional conceptual model of corporate performance. Acad Manag Rev 4:497–505. https://doi.org/10.5465/amr.1979.4498296
Carroll AB (1989) Business and society ethics and stakeholder management. South Western Publishing, Cincinnati, OH
Carroll AB (1991) The pyramid of corporate social responsibility: toward the moral management of organizational stakeholders. Bus Horiz 34:39–48
Carroll AB (2015) Corporate social responsibility: the centerpiece of competing and complementary frameworks. Organ Dyn 44:87–96. https://doi.org/10.1016/j.orgdyn.2015.02.002
Carroll AB, Shabana KM (2010) The business case for corporate social responsibility: a review of concepts, research and practice. Int J Manag Rev 12:85–105. https://doi.org/10.1111/j.1468-2370.2009.00275.x
Carroll AB (2009) A history of corporate social responsibility: concepts and practices. Oxford University Press
Carson R (1962) Silent spring. Houghton Mifflin Harcourt
Chaffee EC (2017) The origins of corporate social responsibility. Univ Cincinnati Law Rev 85:353
Chan K, Hu F (2008) Attitudes toward material possessions among Chinese children. Young Consum 9:49–59. https://doi.org/10.1108/17473610810857318
Clarkson ME (1995) A stakeholder framework for analyzing and evaluating corporate social performance. Acad Manag Rev 20:92–117. https://doi.org/10.5465/amr.1995.9503271994
Commission of the European Communities (2001) GREEN PAPER: Promoting a European framework for Corporate Social Responsibility. Eur Comm
Committee for Economic Development (1971) Social responsibilities of business corporations. Committee for Economic Development
Cone (2008) The 2006 Cone millennial cause study
Costa R, Menichini T (2013) A multidimensional approach for CSR assessment: the importance of the stakeholder perception. Expert Syst Appl 40:150–161. https://doi.org/10.1016/j.eswa.2012.07.028
Cramer J (2005) Experiences with structuring corporate social responsibility in Dutch industry. J Clean Prod 13:583–592. https://doi.org/10.1016/j.jclepro.2003.12.008
Crane A, Matten D (2004) Business ethics: a European perspective: managing corporate citizenship and sustainability in the age of globalization. Oxford University Press, Oxford
Dahlsrud A (2008) How corporate social responsibility is defined: an analysis of 37 definitions. Corp Soc Responsib Environ Manag 15:1–13. https://doi.org/10.1002/csr.132
Damania R (2001) When the weak win: the role of investment in environmental lobbying. J Environ Econ Manage 42:1–22. https://doi.org/10.1006/jeem.2000.1147
Davies AR (2013) Cleantech clusters: transformational assemblages for a just, green economy or just business as usual? Glob Environ Chang 23:1285–1295. https://doi.org/10.1016/j.gloenvcha.2013.07.010
Davis K (1960) Can business afford to ignore social responsibilities? Calif Manage Rev 2:70–76. https://doi.org/10.2307/41166246
de Leeuw A, Valois P, Ajzen I, Schmidt P (2015) Using the theory of planned behavior to identify key beliefs underlying pro-environmental behavior in high-school students: Implications for educational interventions. J Environ Psychol 42:128–138. https://doi.org/10.1016/j.jenvp.2015.03.005
De Roeck K, Marique G, Stinglhamber F, Swaen V (2014) Understanding employees' responses to corporate social responsibility: mediating roles of overall justice and organisational identification. Int J Hum Resour Manag 25:91–112. https://doi.org/10.1080/09585192.2013.781528
Deegan C (2009) Financial accounting theory. McGraw Hill, North Ryde, NSW, Australia
Deegan C, Unerman J (2006) Financial accounting theory
Delmas MA, Toffel MW (2008) Organizational responses to environmental demands: opening the black box. Strateg Manag J 29:1027–1055. https://doi.org/10.1002/smj.701

Delmas MA, Hoffmann VH, Kuss M (2011) Under the tip of the iceberg: absorptive capacity, environmental strategy, and competitive advantage. Bus Soc 50:116–154. https://doi.org/10.1177/0007650310394400

Deloitte Touche Tohmatsu (2016) The 2016 Deloitte Millennial Survey: winning over the next generation of leaders

Deloitte Touche Tohmatsu (2018) Deloitte Millennial Survey: millennials disappointed in business, unprepared for Industry 4.0. Deloitte Touche Tohmatsu, Place of publication not identified

Dentoni D, Bitzer V, Pascucci S (2016) Cross-sector partnerships and the co-creation of dynamic capabilities for stakeholder orientation. J Bus Ethics 135:35–53. https://doi.org/10.1007/s10551-015-2728-8

Donia MBL, Ronen S, Tetrault Sirsly C-A, Bonaccio S (2019) CSR by any other name? the differential impact of substantive and symbolic CSR attributions on employee outcomes. J Bus Ethics 157:503-523. https://doi.org/10.1007/s10551-017-3673-5

Du Pisani JA (2006) Sustainable development–historical roots of the concept. Environ Sci 3:83–96. https://doi.org/10.1080/15693430600688831

Dyllick T, Hockerts K (2002) Beyond the business case for corporate sustainability. Bus Strateg Environ 11:130–141. https://doi.org/10.1002/bse.323

Ehrlich PR (1968) The population bomb

Elkington J (1994) Towards the sustainable corporation: win-win-win business strategies for sustainable development. Calif Manage Rev 36:90–100. https://doi.org/10.2307/41165746

Elkington J (1998) Partnerships fromcannibals with forks: The triple bottom line of 21st-century business. Environ Qual Manag 8:37–51. https://doi.org/10.1002/tqem.3310080106

Engert S, Rauter R, Baumgartner RJ (2016) Exploring the integration of corporate sustainability into strategic management: A literature review. J Clean Prod 112:2833–2850. https://doi.org/10.1016/j.jclepro.2015.08.031

European Commission (2011) Communication from the commission to the European parliament, The Council, The European economic and social committee and the committee of the regions, 15

Farooq O, Payaud M, Merunka D, Valette-Florence P (2014) The impact of corporate social responsibility on organizational commitment: exploring multiple mediation mechanisms. J Bus Ethics 125:563-580. https://doi.org/10.1007/s10551-013-1928-3

Fernando S, Lawrence S (2014) A theoretical framework for CSR practices: integrating legitimacy theory, stakeholder theory and institutional theory. J Theor Account Res 10:149–178

Ferraro F, Etzion D, Gehman J (2015) Tackling grand challenges pragmatically: robust action revisited. Organ Stud 36:363–390. https://doi.org/10.1177/0170840614563742

Frederick WC (1960) The growing concern over business responsibility. Calif Manage Rev 2:54–61. https://doi.org/10.2307/41165405

Frederick WC (2006) Corporation, be good!: the story of corporate social responsibility. Dog Ear Publishing, Indianapolis

Freeman RE (1984) Strategic management: a stakeholder approach. Pitman, Boston

Freeman RE (2000) Business ethics at the millennium. Bus Ethics Q 10:169–180. https://doi.org/10.2307/3857703

Freeman RE, Harrison J, Wicks AC (2007) Managing for stakeholders: Business in the 21st century

Freudenreich B, Lüdeke-Freund F, Schaltegger S (2020) A stakeholder theory perspective on business models: value creation for sustainability. J Bus Ethics 166:3-18. https://doi.org/10.1007/s10551-019-04112-z

Friedman M (1962) Capitalism and freedom. University of Chicago Press, Chicago

Fryzel B, Seppala N (2016) The effect of CSR evaluations on affective attachment to CSR in different identity orientation firms. Bus Ethics A Eur Rev 25:310–326. https://doi.org/10.1111/beer.12116

Gamerschlag R, Möller K, Verbeeten F (2011) Determinants of voluntary CSR disclosure: empirical evidence from Germany. Rev Manag Sci 5:233–262. https://doi.org/10.1007/s11846-010-0052-3

Garriga E, Melé D (2004) Corporate social responsibility theories: mapping the territory. J Bus Ethics 53:51–71

Garud R, Gehman J (2012) Metatheoretical perspectives on sustainability journeys: evolutionary, relational and durational. Res Policy 41:980–995. https://doi.org/10.1016/j.respol.2011.07.009
Garvare R, Johansson P (2010) Management for sustainability–a stakeholder theory. Total Qual Manag Bus Excell 21:737–744. https://doi.org/10.1080/14783363.2010.483095
Geels FW (2002) Technological transitions as evolutionary reconfiguration processes: a multi-level perspective and a case-study. Res Policy 31:1257–1274. https://doi.org/10.1016/S0048-7333(02)00062-8
Geels FW (2011) The multi-level perspective on sustainability transitions: responses to seven criticisms. Environ Innov Soc Transitions 1:24–40. https://doi.org/10.1016/j.eist.2011.02.002
Geels FW, Schot J (2007) Typology of sociotechnical transition pathways. Res Policy 36:399–417. https://doi.org/10.1016/j.respol.2007.01.003
Goodpaster KE (1991) Business ethics and stakeholder analysis. Bus Ethics Q 1:53–73. https://doi.org/10.2307/3857592
Govindasamy V, Suresh K (2017) Exploring approaches to drivers and barriers of corporate social responsibility implementation in academic literature. SHS Web Conf 33:00021. https://doi.org/10.1051/shsconf/20173300021
Greening DW, Turban DB (2000) Corporate social performance as a competitive advantage in attracting a quality workforce. Bus Soc 39:254–280. https://doi.org/10.1177/000765030003900302
Greenwood M (2007) Stakeholder engagement: beyond the myth of corporate responsibility. J Bus Ethics 74:315–327. https://doi.org/10.1007/s10551-007-9509-y
Grin J, Rotmans J, Schot J (2010) Transitions to sustainable development: new directions in the study of long term transformative change. Routledge
Guo S, Hwang S, Wang C (2019) Effect of B2B advertising on firm's market value: CSR as a strategic complement. J Bus Ind Mark 35:895–908. https://doi.org/10.1108/JBIM-01-2019-0013
De Haan H (2010) Towards transition theory. Erasmus University Rotterdam
Haffar M, Searcy C (2017) Classification of trade-offs encountered in the practice of corporate sustainability. J Bus Ethics 140:495–522. https://doi.org/10.1007/s10551-015-2678-1
Hahn T, Pinkse J, Preuss L, Figge F (2015) Tensions in corporate sustainability: towards an integrative framework. J Bus Ethics 127:297–316. https://doi.org/10.1007/s10551-014-2047-5
Hahn T, Figge F, Pinkse J, Preuss L (2018) A paradox perspective on corporate sustainability: descriptive, instrumental, and normative aspects. J Bus Ethics 148:235–248. https://doi.org/10.1007/s10551-017-3587-2
Hajer M, Nilsson M, Raworth K et al (2015) Beyond cockpit-ism: Four insights to enhance the transformative potential of the sustainable development goals. Sustain 7:1651–1660. https://doi.org/10.3390/su7021651
Han S-L, Lee JW (2021) Does corporate social responsibility matter even in the B2B market?: Effect of B2B CSR on customer trust. Ind Mark Manag 93:115–123. https://doi.org/10.1016/j.indmarman.2020.12.008
Harari YN (2011) Sapiens: a brief history of humankind. Vintage, London, United Kingdom
Hemingway CA, Maclagan PW (2004) Managers' personal values as drivers of corporate social responsibility. J Bus Ethics 50:33–44. https://doi.org/10.1023/B:BUSI.0000020964.80208.c9
Hicks D (2014) A geography of hope. Geography 99:5–12. https://doi.org/10.1080/00167487.2014.12094385
Homburg C, Stierl M, Bornemann T (2013) Corporate social responsibility in business-to-business markets: How organizational customers account for supplier corporate social responsibility engagement. J Mark 77:54–72. https://doi.org/10.1509/jm.12.0089
Howard-Grenville JA, Buckle SJ, Hoskins BJ, George G (2014) Climate change and management. Acad Manag J 57:615–623. https://doi.org/10.5465/amj.2014.4003
Jamali D (2008) A stakeholder approach to corporate social responsibility: A fresh perspective into theory and practice. J Business Ethics 82(1):213–231. https://doi.org/10.1007/s10551-007-9572-4

Jeon MM, Lee S, Jeong M (2020) Perceived corporate social responsibility and customers' behaviors in the ridesharing service industry. Int J Hosp Manag 84:102341. https://doi.org/10.1016/j.ijhm.2019.102341

Jongbloed B, Enders J, Salerno C (2008) Higher education and its communities: interconnections, interdependencies and a research agenda. High Educ 56:303–324. https://doi.org/10.1007/s10734-008-9128-2

Joyce A, Paquin RL (2016) The triple layered business model canvas: a tool to design more sustainable business models. J Clean Prod 135:1474–1486. https://doi.org/10.1016/j.jclepro.2016.06.067

Kabanoff B, Waldersee R, Cohen M (1995) Espoused values and organizational change themes. Acad Manag J 38:1075–1104. https://doi.org/10.2307/256621

Keijzers G (2002) The transition to the sustainable enterprise. J Clean Prod 10:349–359. https://doi.org/10.1016/S0959-6526(01)00051-8

Kern F, Markard J (2016) Analysing energy transitions: combining insights from transition studies and international political economy. In: The Palgrave handbook of the international political economy of energy. Springer, pp 291–318

Kim S-Y, Park H (2011) Corporate social responsibility as an organizational attractiveness for prospective public relations practitioners. J Bus Ethics 103:639–653. https://doi.org/10.1007/s10551-011-0886-x

Kim H, Rhou Y, Topcuoglu E, Kim YG (2020) Why hotel employees care about corporate social responsibility (CSR): using need satisfaction theory. Int J Hosp Manag 87:102505. https://doi.org/10.1016/j.ijhm.2020.102505

Kiron D, Kruschwitz N, Haanaes K, Velken IVS (2012) Sustainability nears a tipping point. MIT Sloan Manag Rev 53:69

Köhler J, Geels FW, Kern F et al (2019) An agenda for sustainability transitions research: state of the art and future directions. Environ Innov Soc Transitions 31:1–32. https://doi.org/10.1016/j.eist.2019.01.004

Köhler J, Geels F, Kern F, et al (2017) A research agenda for the Sustainability Transitions Research Network

Lantos GP (2001) The boundaries of strategic corporate social responsibility. J Consum Mark 18:595–632. https://doi.org/10.1108/07363760110410281

Latapí Agudelo MA, Jóhannsdóttir L, Davídsdóttir B (2019) A literature review of the history and evolution of corporate social responsibility. Int J Corp Soc Responsib 4:1. https://doi.org/10.1186/s40991-018-0039-y

Lee MP (2008) A review of the theories of corporate social responsibility: its evolutionary path and the road ahead. Int J Manag Rev 10:53–73. https://doi.org/10.1111/j.1468-2370.2007.00226.x

Lee YK, Sally KY, Lee KH, Li D (2012) The impact of CSR on relationship quality and relationship outcomes: a perspective of service employees. Int J Hosp Manag 31:745-756. https://doi.org/10.1016/j.ijhm.2011.09.011

Lee EM, Park S-Y, Lee HJ (2013) Employee perception of CSR activities: Its antecedents and consequences. J Bus Res 66:1716–1724. https://doi.org/10.1016/j.jbusres.2012.11.008

Lewis MW (2000) Exploring paradox: toward a more comprehensive guide. Acad Manag Rev 25:760–776. https://doi.org/10.5465/amr.2000.3707712

Li Y, Liu B, Huan TC (2019) Renewal or not? Consumer response to a renewed corporate social responsibility strategy: Evidence from the coffee shop industry. Tour Manag 72:170–179. https://doi.org/10.1016/j.tourman.2018.10.031

Loorbach D, Wijsman K (2013) Business transition management: exploring a new role for business in sustainability transitions. J Clean Prod 45:20–28. https://doi.org/10.1016/j.jclepro.2012.11.002

Lozano R (2015) A holistic perspective on corporate sustainability drivers. Corp Soc Responsib Environ Manag 22:32–44. https://doi.org/10.1002/csr.1325

Lyytimäki J, Vikström S, Furman E (2019) Voluntary participation for sustainability transition: experiences from the 'commitment to sustainable development 2050.' Int J Sustain Dev World Ecol 26:25–36. https://doi.org/10.1080/13504509.2018.1460632

Maignan I, Ferrell OC (2004) Corporate social responsibility and marketing: an integrative framework. J Acad Mark Sci 32:3–19. https://doi.org/10.1177/0092070303258971

Maignan I, Ralston DA (2002) Corporate social responsibility in Europe and the U.S.: insights from businesses' self-presentations. J Int Bus Stud 33:497–514. https://doi.org/10.1057/palgrave.jibs.8491028

Margolis JD, Elfenbein HA, Walsh JP (2009) Does it pay to be good...and does it matter? A Meta-analysis of the relationship between corporate social and financial performance. SSRN Electron J. https://doi.org/10.2139/ssrn.1866371

Margolis JD, Walsh JP (2003) Misery loves companies: rethinking social initiatives by business. Adm Sci Q 48:268. https://doi.org/10.2307/3556659

Markard J, Raven R, Truffer B (2012) Sustainability transitions: an emerging field of research and its prospects. Res Policy 41:955–967. https://doi.org/10.1016/j.respol.2012.02.013

Mark-Herbert C, Von Schantz C (2007) Communicating corporate social responsibility–brand management. EJBO-Electronic J Bus Ethics Organ Stud 12:4–11

van Marrewijk M (2003) Concepts and definitions of CSR and corporate sustainability: Between agency and communion. J Bus ethics 44:95–105. https://doi.org/10.1023/A:1023331212247

Matten D, Moon J (2008) "Implicit" and "Explicit" CSR: a conceptual framework for a comparative understanding of corporate social responsibility. Acad Manag Rev 33:404–424. https://doi.org/10.5465/amr.2008.31193458

McDonald S, Oates CJ (2006) Sustainability: consumer perceptions and marketing strategies. Bus Strateg Environ 15:157–170. https://doi.org/10.1002/bse.524

McWilliams A, Siegel D (2000) Corporate social responsibility and financial performance: correlation or misspecification? Strateg Manag J 21:603–609. https://doi.org/10.1002/(SICI)1097-0266(200005)21:5%3c603::AID-SMJ101%3e3.0.CO;2-3

Merad M, Dechy N, Marcel F (2014) A pragmatic way of achieving highly sustainable organisation: governance and organisational learning in action in the public French sector. Saf Sci 69:18–28. https://doi.org/10.1016/j.ssci.2014.01.002

Miras-Rodríguez MDM, Carrasco-Gallego A, Escobar-Pérez B (2015) Has the CSR engagement of electrical companies had an effect on their performance? A closer look at the environment. Bus Strateg Environ 24:819-835. https://doi.org/10.1002/bse.1848

Mitchell RK, Agle BR, Wood DJ (1997) Toward a theory of stakeholder identification and salience: defining the principle of who and what really counts. Acad Manag Rev 22:853–886. https://doi.org/10.5465/amr.1997.9711022105

Muller A, Kolk A (2010) Extrinsic and intrinsic drivers of corporate social performance: evidence from foreign and domestic firms in Mexico. J Manag Stud 47:1–26. https://doi.org/10.1111/j.1467-6486.2009.00855.x

Ng TWH, Yam KC, Aguinis H (2019) Employee perceptions of corporate social responsibility: effects on pride, embeddedness, and turnover. Pers Psychol 72:107–137. https://doi.org/10.1111/peps.12294

Nielsen NV (2015) The Nielsen Global survey of corporate social responsibility and sustainability

Novick O'Keefe L (2017) CSR grows in 2016 as companies embrace employees' values. In: Huffingt. Post. https://www.huffpost.com/entry/csr-grows-in-2016-as-comp_b_13657368. Accessed 25 Oct 2019

O'Riordan L, Fairbrass J (2008) Corporate social responsibility (CSR): models and theories in stakeholder dialogue. J Bus Ethics 83:745–758. https://doi.org/10.1007/s10551-008-9662-y

Ong M, Mayer DM, Tost LP, Wellman N (2018) When corporate social responsibility motivates employee citizenship behavior: The sensitizing role of task significance. Organ Behav Hum Decis Process 144:44–59. https://doi.org/10.1016/j.obhdp.2017.09.006

Ozanne LK, Phipps M, Weaver T et al (2016) Managing the tensions at the intersection of the triple bottom line: a paradox theory approach to sustainability management. J Public Policy Mark 35:249–261. https://doi.org/10.1509/jppm.15.143

Parmar BL, Freeman RE, Harrison JS et al (2010) Stakeholder theory: the state of the art. Acad Manag Ann 4:403–445. https://doi.org/10.5465/19416520.2010.495581

Passetti E, Cinquini L, Tenucci A (2018) Implementing internal environmental management and voluntary environmental disclosure: Does organisational change happen. Account Audit Account J 31:1145–1173. https://doi.org/10.1108/AAAJ-02-2016-2406
Pearce JA II (1982) The company mission as a strategic tool. Sloan Manag Rev 23:15
Pérez A, Rodríguez del Bosque I (2015) An integrative framework to understand how CSR affects customer loyalty through identification, emotions and satisfaction. J Bus Ethics 129:571–584. https://doi.org/10.1007/s10551-014-2177-9
Peterson DK (2004) The relationship between perceptions of corporate citizenship and organizational commitment. Bus Soc 43:296–319. https://doi.org/10.1177/0007650304268065
Phillips R (2003) Stakeholder theory and organizational ethics. Berrett-Koehler Publishers, San Francisco, CA
Pillay R (2015) The changing nature of corporate social responsibility: CSR and development–the case of Mauritius. Routledge, London andNew York
Pisano U, Lepuschitz K, Berger G (2014) Sustainability transitions at the international, European and national level: Approaches, objectives and tools for sustainable development governance. Austria, Vienna
Porter ME, Kramer MR (2006) The link between competitive advantage and corporate social responsibility. Harv Bus Rev 84:78–92
Porter ME, Kramer MR (2019) Creating shared value. In: Lenssen GG, Smith NC (eds) Managing Sustainable Business. Springer, Netherlands, Dordrecht, pp 323–346
Post JE, Preston LE, Sachs S (2002) Managing the extended enterprise: the new stakeholder view. Calif Manage Rev 45:6–28. https://doi.org/10.2307/41166151
Preston LE, Sapienza HJ (1990) Stakeholder management and corporate performance. J Behav Econ 19:361–375. https://doi.org/10.1016/0090-5720(90)90023-Z
Putnam LL, Fairhurst GT, Banghart S (2016) Contradictions, dialectics, and paradoxes in organizations: a constitutive approach. Acad Manag Ann 10:65–171. https://doi.org/10.5465/19416520.2016.1162421
Rego A, Leal S, Cunha MP et al (2010) How the perceptions of five dimensions of corporate citizenship and their inter-inconsistencies predict affective commitment. J Bus Ethics 94:107–127. https://doi.org/10.1007/s10551-009-0252-4
Reverte C (2009) Determinants of corporate social responsibility disclosure ratings by Spanish listed firms. J Bus Ethics 88:351–366. https://doi.org/10.1007/s10551-008-9968-9
Rockström J, Steffen W, Noone K et al (2009) A safe operating space for humanity. Nature 461:472–475. https://doi.org/10.1038/461472a
Rodrigo P, Arenas D (2008) Do employees care about CSR programs? a typology of employees according to their attitudes. J Bus Ethics 83:265–283. https://doi.org/10.1007/s10551-007-9618-7
Rotmans J, Kemp R, van Asselt M (2001) More evolution than revolution: transition management in public policy. Foresight 3:15–31. https://doi.org/10.1108/14636680110803003
Russo MV, Fouts PA (1997) A resource-based perspective on corporate environmental performance and profitability. Acad Manag J 40:534–559
Savage GT, Nix TW, Whitehead CJ, Blair JD (1991) Strategies for assessing and managing organizational stakeholders. Acad Manag Perspect 5:61–75. https://doi.org/10.5465/ame.1991.4274682
Scherer AG, Palazzo G (2007) Toward a political conception of corporate responsibility: business and society seen from a habermasian perspective. Acad Manag Rev 32:1096–1120. https://doi.org/10.5465/amr.2007.26585837
Scherer AG, Palazzo G (2011) The new political role of business in a globalized world: a review of a new perspective on CSR and its implications for the firm, governance, and democracy. J Manag Stud 48:899–931. https://doi.org/10.1111/j.1467-6486.2010.00950.x
Scherer AG, Palazzo G, Seidl D (2013) Managing legitimacy in complex and heterogeneous environments: sustainable development in a globalized world. J Manag Stud 50:259–284

Schlaile MP, Urmetzer S (2019) Transitions to sustainable development. In: Leal Filho W, Azul AM, Brandli L et al (Eds) Decent work and economic growth. Springer International Publishing, Cham, pp 1–16

Schneider A, Wickert C, Marti E (2017) Reducing complexity by creating complexity: a systems theory perspective on how organizations respond to their environments. J Manag Stud 54:182–208. https://doi.org/10.1111/joms.12206

Schrempf-Stirling J, Palazzo G, Phillips RA (2016) Historic corporate social responsibility. Acad Manag Rev 41:700–719. https://doi.org/10.5465/amr.2014.0137

Schumacher EF (1973) Small is beautiful: economics as if people mattered. Blond and Briggs, London

Selvi Y, Wagner E, Türel A (2010) Corporate social responsibility in the time of financial crisis: Evidence from Turkey. Ann Univ Apulensis-Series Oeconomica 12:

Sen S, Bhattacharya CB (2001) Does doing good always lead to doing better? consumer reactions to corporate social responsibility. J Mark Res 38:225–243. https://doi.org/10.1509/jmkr.38.2.225.18838

Sen S, Bhattacharya CB, Korschun D (2006) The role of corporate social responsibility in strengthening multiple stakeholder relationships: a field experiment. J Acad Mark Sci 34:158–166. https://doi.org/10.1177/0092070305284978

Siltaloppi J, Rajala R, Hietala H (2020) Integrating CSR with business strategy: a tension management perspective. J Bus Ethics 1–21. https://doi.org/10.1007/s10551-020-04569-3

Singhapakdi A, Lee D-J, Sirgy MJ, Senasu K (2015) The impact of incongruity between an organization's CSR orientation and its employees' CSR orientation on employees' quality of work life. J Bus Res 68:60–66. https://doi.org/10.1016/j.jbusres.2014.05.007

Sirgy MJ, Efraty D, Siegel P, Lee D-J (2001) A new measure of quality of work life (QWL) based on need satisfaction and spillover theories. Soc Indic Res 55:241–302. https://doi.org/10.1023/A:1010986923468

Skellern K, Markey R, Thornthwaite L (2017) Identifying attributes of sustainable transitions for traditional regional manufacturing industry sectors–A conceptual framework. J Clean Prod 140:1782–1793. https://doi.org/10.1016/j.jclepro.2016.07.183

Slawinski N, Bansal P (2012) A matter of time: The temporal perspectives of organizational responses to climate change. Organ Stud 33:1537–1563

Smith NC (2001) Changes in corporate practices in response to public interest advocacy and actions. In: Bloom PN, Gundlach GT (eds) Handbook of marketing and society. Sage Publications Thousand Oaks, CA, London and New Delhi, pp 140–161

Smith WK, Lewis MW (2011) Toward a theory of paradox: a dynamic equilibrium model of organizing. Acad Manag Rev 36:381–403

Smith WK, Gonin M, Besharov ML (2013) Managing social-business tensions: a review and research agenda for social enterprise. Bus Ethics Q 23:407–442

Stites JP, Michael JH (2011) Organizational commitment in manufacturing employees: relationships with corporate social performance. Bus Soc 50:50–70. https://doi.org/10.1177/0007650310394311

Supanti D, Butcher K (2019) Is corporate social responsibility (CSR) participation the pathway to foster meaningful work and helping behavior for millennials? Int J Hosp Manag 77:8–18. https://doi.org/10.1016/j.ijhm.2018.06.001

Theodoulidis B, Diaz D, Crotto F, Rancati E (2017) Exploring corporate social responsibility and financial performance through stakeholder theory in the tourism industries. Tour Manag 62:173–188. https://doi.org/10.1016/j.tourman.2017.03.018

Thornton PH, Ocasio W (2008) Institutional logics. In: Greenwood R, Oliver C, Sahlin K, Suddaby R (eds) The SAGE Handbook of Organizational Institutionalism. Sage Publications, London, pp 99–128

Turban DB, Greening DW (1997) Corporate social performance and organizational attractiveness to prospective employees. Acad Manag J 40:658–672. https://doi.org/10.5465/257057

Turker D (2009a) Measuring corporate social responsibility: a scale development study. J Bus Ethics 85:411–427. https://doi.org/10.1007/s10551-008-9780-6

Turker D (2009b) How corporate social responsibility influences organizational commitment. J Bus Ethics 89:189–204. https://doi.org/10.1007/s10551-008-9993-8

UN General Assembly (2015) Transforming our world: the 2030 agenda for Sustainable Development. 2015. United Nations 1–35

United Nations Framework Convention on Climate Change (2015) Paris agreement. 1–27

United Nations (2000) UN Global Compact. https://www.unglobalcompact.org/. Accessed 15 Mar 2021

van Bommel K (2018) Managing tensions in sustainable business models: Exploring instrumental and integrative strategies. J Clean Prod 196:829–841. https://doi.org/10.1016/j.jclepro.2018.06.063

van den Bergh JCJM, Truffer B, Kallis G (2011) Environmental innovation and societal transitions: Introduction and overview. Environ Innov Soc Transitions 1:1–23. https://doi.org/10.1016/j.eist.2011.04.010

Voegtlin C, Greenwood M (2016) Corporate social responsibility and human resource management: a systematic review and conceptual analysis. Hum Resour Manag Rev 26:181–197. https://doi.org/10.1016/j.hrmr.2015.12.003

Walker K (2007) How socially responsible organizations can attract Millennials and obtain a competitive advantage. In: Administrative Sciences Association of Canada Conference. Ottawa

Walsh JP (2005) Book review essay: taking stock of stakeholder management. Acad Manag Rev 30:426–438. https://doi.org/10.5465/amr.2005.16387898

Walsh JP, Weber K, Margolis JD (2003) Social issues and management: our lost cause found. J Manage 29:859–881. https://doi.org/10.1016/S0149-2063_03_00082-5

Wang Y, Xu S, Wang Y (2020) The consequences of employees' perceived corporate social responsibility: a meta-analysis. Bus Ethics A Eur Rev 29:471–496. https://doi.org/10.1111/beer.12273

Wannags LL, Gold S (2020) Assessing tensions in corporate sustainability transition: From a review of the literature towards an actor-oriented management approach. J Clean Prod 264:121662. https://doi.org/10.1016/j.jclepro.2020.121662

Waterhouse BC (2017) The personal, the political and, the profitable: business and protest culture, 1960s-1980s. Financ Hist 14

Wood DJ (1991) Corporate Social Performance Revisited. Acad Manag Rev 16:691–718. https://doi.org/10.5465/amr.1991.4279616

Wood DJ (1994) Business and society. Harper Collins, New York, USA

Wright C, Nyberg D (2017) An inconvenient truth: how organizations translate climate change into business as usual. Acad Manag J 60:1633–1661. https://doi.org/10.5465/amj.2015.0718

Yang M, Evans S, Vladimirova D, Rana P (2017) Value uncaptured perspective for sustainable business model innovation. J Clean Prod 140:1794–1804. https://doi.org/10.1016/j.jclepro.2016.07.102

Yeh N-C, Chen Y-J (2011) On the everyday life information behavior of LOHAS consumers: a perspective of lifestyle. J Educ Medi Libr Sci 48:489–510

Zhang Q, Oo BL, Lim BTH (2019) Drivers, motivations, and barriers to the implementation of corporate social responsibility practices by construction enterprises: a review. J Clean Prod 210:563–584. https://doi.org/10.1016/j.jclepro.2018.11.050

Zhang Q, Cao M, Zhang F et al (2020) Effects of corporate social responsibility on customer satisfaction and organizational attractiveness: A signaling perspective. Bus Ethics A Eur Rev 29:20–34. https://doi.org/10.1111/beer.12243

Zhao Z-Y, Zhao X-J, Davidson K, Zuo J (2012) A corporate social responsibility indicator system for construction enterprises. J Clean Prod 29–30:277–289. https://doi.org/10.1016/j.jclepro.2011.12.036

Chapter 5
Empirical Insights on Obstacles and Drivers of CSR-Committed Organisations to Sustainability Transitions

Eglė Staniškienė, Živilė Stankevičiūtė, Asta Daunorienė, and Joana Ramanauskaitė

5.1 Research Context—Economy in Transition

Lithuania is a country in transition, having had to shift from a planned economy system to market economy system, and this implied radical reforms aimed at turning the society upside down (Pilia 2017). Lithuania survived a long period of occupation, experienced privatisation and goods' famine. Moreover, thirty years ago, this country restored its independence, and fifteen years ago, it joined the European Union (EU). Based on the Eurostat database, before the country joined the EU, Lithuania's GDP per capita stood at just 42% of the EU average, while in 2011 it increased to 62% and to 78% in 2017. The growth of income level was similar to that of GDP. In 2017, the income level comprised 79% of the EU average. However, the inequality of income distribution was one of the highest in the EU (Liobikienė et al. 2020).

Lithuania is in a geopolitically active area with transit roads and most northern ice-free port of the Baltic Sea, which places the country in a position to successfully develop its economy. After regaining independence, Lithuania was unwilling to open up for foreign capital, especially for the enterprises considered to be of strategic importance. Mygind finds a reason in the fact that the opposition parties particularly feared Russian investors and were against the implementation of liberal rules, which would allow foreigner investors to buy land (Mygind 1994). Soon after opening-up, Lithuania started increasing the exchange of goods with the rest of the world (Pilia 2017). Currently, Lithuania belongs to the EU and NATO, providing both incentives and safety for successful independent development (Cameron 2009; Fisher 2010).

Lithuania has adopted its National Strategy for Sustainable Development in 2003, with later revisions in 2009 and 2011. Furthermore, the vision outlined in the main long-term planning strategies—'Lithuania 2030' and National Development Programme (NDP) 2020—accentuates the importance of sustainable development. Finally, national policy-makers have now an ambition to actively integrate the SDGs into planning processes by setting the sustainable development principles as the base for the upcoming new NDP 2030 as well as the new General Plan of the Territory

J. K. Staniškis et al., *Transformation of Business Organization Towards Sustainability*, World Sustainability Series, https://doi.org/10.1007/978-3-030-93298-5_5

of the Republic of Lithuania (Punytė and Simonaitytė 2018a). In its journey towards sustainable development, Lithuania has shown manifold positive results within the priority areas for development, established considering its national interests and local contexts. The Lithuanian government has initiated several executive reforms regarding the pension system, increasing support for young families and creating new opportunities for employment. For example, Lithuania is becoming a hub for young entrepreneurs to pursue their innovative (and often socially conscious) start-up ideas. Additionally, the integrated healthcare services are being developed with a greater attention towards improving accessibility and quality. High quality education for children, youth and adults is another national priority. Here, Lithuania is leading in the EU in regard to the proportion of the population obtaining secondary and tertiary education. Reforms are being planned to ensure further improvement in the quality and accessibility of education for all. In 2016, renewable energy contributed 26% of all energy consumed in Lithuania and is thus an important component in improving the country's energy independence. Moreover, Lithuania has been performing well on international environmental indices, yet with the globally growing threats of climate change, pollution, habitat and biodiversity loss, it is important that the protection and restoration of nature, whose health is key to human well-being, receive even more attention. Finally, Lithuania is increasing its role in international development cooperation, thus contributing to achieving the SDGs globally, too (Punytė and Simonaitytė 2018a).

Sustainable development of organisations was encouraged by creating the Lithuanian Responsible Business Association (LAVA—Lietuvos Atsakingo Verslo Asociacija) in 2005, and extending the work of the National Network of Responsible Business Enterprises (NAVĮT—Nacionalinis Atsakingo Verslo Įmonių Tinklas). The direction taken by the companies represents interests of consumers, shareholders, and other stakeholders, such as governments, policymakers, general society, etc. There are studies that investigate transitions of companies from planned to free market in the post-soviet bloc (Fisher 2010; Tõnurist 2015), but not so much concerning the sustainable development of the enterprises or countries. Search on the Web of Science Core Collection did not provide any results for keywords post-soviet, sustainability transitions, organisation/business, and their synonyms. This might be due to the difference between sustainability and market transition, where sustainability is regarded as socio-technical, and market as a socio-economic system (Fisher 2010), though these systems are tightly interlinked and co-dependent.

The Soviet Union did leave a mark that might be felt in the contemporary society of the countries that were affected. The cultural legacy of this regime had multiple indirect negative impacts (Rodrigo et al. 2015; Dawson 2019) in a broad variety of areas. Dawson (2019) named several attributes linked with the mentality of post-soviet countries: "passivity, circumspection, distrust, and a widespread indifference to environmental issues, pervading society, including governance systems, at multiple levels" (p. 56). Therefore, instead of gradual transition from one regime to another, these countries experienced shock-therapy (Fisher 2010; Brown et al. 2012) and some authors in geography field refer to this transition more of as 'a form of transformation' (Smith 1996; Lynn 1999; Brown et al. 2012). However, experiencing these transitions, or transformations, possibly prepared the countries for sustainable transitions. Rodrigo et al. (2015) conducted a study on transition dynamics at a spatial level in 99 countries and distinguished four groups of countries by their approach to sustainability issues: crossroaders, compliers, athletes, and laggards. The cluster of compliers contains fifteen ex-eastern bloc nations (except for Uzbekistan) that do particularly well in quality of governance, but not in creating wealth cleanly, though they try to follow more sustainable paths, regardless of not being exposed so much to the sustainable development concept. Growing economically strong, the compliers' cluster does not possess effective and efficient energy management; however, they are addressing their effect on socio-environmental conditions, trying to improve it and comply with the standards raised by alliances such as the EU. Complier countries already belonging to the EU or in the process of becoming a member must meet higher standards, although they are not yet addressing their energy and CO_2 emission appropriately. Lithuania is presented as one of the exemplar cases complying to sustainability issues that uses some governmental pressures, adopts improved industrial practices, takes moderate steps to reduce poverty and CO_2 emissions, but uses energy quite inefficiently, however, trying to develop more sustainably (Rodrigo et al. 2015).

Some challenges particular to Lithuania include limited reach to public transport; 10.6% of the population do not have access to indoor sanitation; the income of 20% of the richest people in the country was 7.1 times higher than 20% of the poorest people (Punytė and Simonaitytė 2018b); 20th place in the SDG Index of 2018 of 27 EU countries. Sustainable development principles in Lithuania are established through the main strategic planning documents of the country: Lithuania's Progress Strategy 'Lithuania 2030'; 2014–2020 National Progress Programme; National Strategy for Sustainable Development adopted in 2003; and the White Paper on Lithuanian Regional Policy prepared in 2017. The necessity of sustainable development is also mentioned in the Law on Territorial Planning of the Republic of Lithuania. However, it is noteworthy that the topic of sustainable development lacks coherence and specificity in these documents, while the key issue in this regard is strategy—the National Strategy for Sustainable Development—is more of a recommendatory nature (Punytė and Simonaitytė 2018b). Correspondingly, the Lithuanian National Strategy for Sustainable Development has not been updated since 2011, and its implementation reports have not been submitted since 2014.

Nevertheless, participation in the EU had a major impact on Lithuania's development patterns. Post-soviet countries that are members of the EU established market economy faster and performed wider-ranged reforms (Cameron 2009). Integration in the EU can be seen as a stimulus for improving governance, having examples set by older members, absorbing sustainable development goals as part of the strategy (Leal Filho et al. 2016). Following the lead of more prominent EU countries, having transitional experiences and competencies, should provide these transitioning economies with a boost for integrating sustainable development principles in their routines. The similarities between socio-technical and socio-economical system transformations should help transfer the multilevel concept of sustainable development to the market transition (Fisher 2010). However, as the above analysis suggests, it is not to be expected from an incumbent regime to actively engage and invest into radical innovations as top-down initiatives do not intentionally generate niches and evolutionary bottom-up processes (Geels and Schot 2007). Nonetheless, transition management is based on experiments, whereas eastern bloc of post-soviet countries did not experiment and rely on niche management for the transition to the market economy (Fisher 2010). Therefore, it is important to address initiative grounds of sustainable transitions, which could lay in incumbent actors in the context.

5.2 Result Overview of Previous Quantitative Research of CSR Perception from Different Stakeholder Perspectives

To understand different perspectives of CSR for organisations based on the notion of stakeholders' role, the book authors have conducted several studies to indicate the stakeholders' perception of CSR. As these studies have already been published, in this section we will provide only an overview of the key findings, which reveals the importance of the role of different stakeholders in moving towards sustainability.

The first study (Staniškienė et al. 2018b) aimed at envisioning **employees** as inner drivers for sustainability transitions. It sought to identify the linkage between the CSR importance to the employee and employee perception of CSR and organisational trust. The second study (Staniškienė et al. 2018a) analysed **potential employees** as drivers for CSR in organisations. It aimed at exploring the relationships between the attractiveness of CSR performance, organisational commitment, and environmental attitudes among the potential job seekers. The third study (Staniškienė et al. 2019b) was related to corporate social responsibility and B2B context**.** Its goal was to reveal the linkage between the customer perception of supplier CSR and interorganisational trust in the **B2B context**. The fourth study (Staniškienė et al. 2019a) examined the sustainable lifestyle practices of **millennials** and their motivation to promote green products to their friends. The millennials were treated as a sustainability agents who make an impact on the future attitude of organisations to CSR.

The studies were constructed on validated research variables in line with the stakeholder theory. In the table below, we provide the basic information about the conducted quantitative studies (see Table 5.1).

Study of employees

Respondents' profile. In the respondents' group, there were more females than males. Forty-three per cent of employees were at the age between 46 and 55; the other 34% of employees were at the age group from 26 and 45. Only 28% were employed in managerial positions, while the majority of respondents (72%) were employed in other positions.

The research related to employees as inner drivers for sustainability transitions intended to test three key hypotheses:

Table 5.1 Basic information about the conducted quantitative studies

Target stakeholders	Research period	Sample size	Research objective	Main theoretical constructs
Employees	January–April 2018	N = 398	To reveal the linkage between CSR and organisational trust taking the employee perspective into consideration	CSR (Park and Levy 2014) Organisational trust (Lee et al. 2012) CSR importance to the employee (Korschun et al. 2014)
Potential employees	December 2017–February 2018	N = 298	To investigate the relationships between environmental attitudes, attractiveness of CSR performances, and organisational commitment among potential employees	Perception of CSR (Turker 2009b) Organisational commitment (Mowday et al. 1982; Turker 2009a) Environmental attitudes (NEP scale) (Dunlap et al. 2000; Lundmark 2007)
Customers in B2B context	May–July 2018	N = 384	To reveal the linkage between the customer perception of supplier CSR and interorganisational trust in the B2B context	Interorganisational trust (McKnight et al. 2002) Customer perceptions of organisational CSR (Park and Levy 2014)
Consumers	September 2018–February 2019	N = 453	To investigate the relationships between the millennials' sustainable lifestyle practices and motivation to promote green products to their friends	Green purchase intention (Lee et al. 2015) Clothing care Recycling Eco-citizenship Fair trade Pro-environmental attitude Motivation of the millennials to promote green products to their friends (Smith 2010)

H1. Employee perception of CSR is positively related to organisational trust.

H2. Employee perception of CSR is positively related to CSR importance to the employee.

H3. CSR importance to the employee is positively related to organisational trust.

The results of the regression analysis indicated a very strong linkage between CSR, CSR importance to employees, and organisational trust items. The research data allowed to support the hypothesis H1. H2 and H3 were rejected. The research findings supported most of the previous studies in that the perception of CSR was multidimensional and each dimension had a different effect in determining the employee attitudes. For example, a positive effect between CSR and organisational trust was found and this finding confirmed the contention of Park et al. (2014) that generally, CSR perception could impact organisational trust. Moreover, the study suggests that the more positively employees perceive their organisation's CSR activities concerning customers, the more they trust in the organisation. The same applies in case of another CSR dimension, namely CSR-employees. It seems that employees view the care of them as a mandatory aspect in order to put their trust in an organisation. The expectations of employees that the organisation will make decisions and perform considering their interests and welfare increase if employees perceive the organisation's actions concerning them more positively. Thus, treating employees fairly, encouraging them to develop their skills or providing the conditions for work-life balance will help the employees to trust the organisation more.

Particular attention was paid to the CSR importance to employees; however, the research results did not allow for proving the hypothesis. This finding contradicts the studies of Reed (2004) and Korschun et al. (2014) on the relationship between the CSR importance to the employees and CSR perception. From employee perspective, factors such as helping local communities where the organisation operates and the perception of better world did not lead to understanding the CSR role.

From the theoretical perspective, this study supported the idea that all dimensions of CSR should be considered when measuring the perceptions related to CSR (Lee and Jackson 2010). Understanding the unique role of each dimension is important seeking to gain the employee trust. According to Chin, organisations need to build sustainable relationships between managers and employees in order to achieve long-term results. According to the author, relationship building is one of the most challenging aspects of the CSR programme (Chin, 2019).

From the managerial point of view, organisations should acknowledge the importance of CSR. A proved hypothesis provides a comprehensive tool for executives to test the employee perception of CSR and its relation to organisational trust. Supporting previous research (Lee and Jackson 2010) the findings demonstrate that organisations create value for themselves by increasing organisational trust and investing resources in CSR activities. Moreover, an empirical research of Vilanova Pichot (2015) of eight organisations implementing responsible strategic initiatives showed that CSR was one of the key principles why employees wanted to work in the analysed organisations. The study revealed that business leaders recognised that CSR gave trust to organisations and that organisations were committed to sustainability by translating it into new initiatives. Even more, the research results were in line with the findings of research by Lemoine et al. (2015), who emphasised the employee's awareness of the positive effects of their actions on others and the impact of positive experiences on the desire to change the lives of others.

Study of potential employees

Respondent profile. In this study, students were surveyed, with the majority of them being female. The main age groups were 18–22 (38%) and above 30 years (37%). Students were distributed quite evenly by their degree of studies – 45% of respondents were undergraduate students and 55% were graduate students.

Analysing potential employees as drivers for CSR in organisations study, three research hypotheses were tested.

H1. Potential employee perception of CSR is positively related to organisational commitment.

H2. Potential employee perception of CSR is positively related their environmental attitude (NEP scale).

H3. Environmental attitude (NEP scale) of the potential employees is positively related to their organisational commitment.

The research results indicated a strong linkage between the potential employee perception of CSR and their intention to commit to a socially responsible organisation. The research data allowed for supporting the hypothesis H1.

This study continued the exploration of the role of CSR perception in organisational attractiveness for potential employees that had been started by Turban and Greening (1997), Greening and Turban (2000), and Backhaus et al. (2002). Reporting the results of previous studies, the authors confirmed that potential job seekers considered CSR important to the overall assessment of an organisation and that they would like to commit to a socially responsible organisation. The authors found that women reported a significantly greater concern for CSR than men. The research findings revealed that the issues that were directly related to the daily lives of potential employees were more important than broader issues. This could not be confirmed by the refuted hypotheses H2 and H3. The results showed no relationship between the potential employee perception of CSR and organisational commitment. Similar results were presented by Hanson-Rasmussen and Lauver (2017). Their results showed no relationship with the job attractiveness for the respondents from the USA and India, and a negative relationship for respondents from China. The research results complement the analysis by Vilanova Pichot (2015) which revealed that some studies suggested that a good CSR reputation could help an organisation attract talented employees.

Moreover, a study of Jones et al. (2016) shows that information about the organisation's CSR practices informs about its values and in particular affects the perception of of the potential employees of the value and employer's social orientation, which in turn informs the potential employees about the treatment to be expected.

Customers in B2B context

Respondent profile. In the respondent group, there were more females than males. Thirty per cent of the customers were at the age between 18 and 25; 29% of customers were aged 33–41. Almost half of the respondents (45%) had been working in a particular organisation for 3–5 years.

To test the customer perception of supplier CSR and interorganisational trust in the **B2B** context, three hypotheses were raised.

H1. Customer perception of CSR towards environment and community will be positively related to their trust in supplier, including competence trust (H1a), benevolence trust (H1b), and integrity trust (H1c).

H2. Customer perception of CSR towards employees will be positively related to their trust in supplier, including competence trust (H2a), benevolence trust (H2b), and integrity trust (H2c).

H3. Customer perception of CSR towards customers will be positively related to their trust in supplier, including competence trust (H3a), benevolence trust (H3b), and integrity trust (H3c).

All hypotheses raised were confirmed. Concerning the hypothesis H1, customer perception of CSR towards environment and community was found to have a statistically significant positive effect on interorganisational trust. Turning to different components of interorganisational trust, CSR towards environment and community had the most significant positive effect on competence trust and a less significant effect on benevolence trust.

Concerning H2, it was confirmed in the same manner as in the case of H1. A positive relationship was found between the customer perception of CSR towards employees and interorganisational trust. The most significant positive effect was detected on competence trust and a less significant effect on benevolence trust.

H3 was confirmed as well. However, the most significant positive effect was established on the competence trust and a less significant effect on integrity trust.

A positive relationship was found between the customer perceptions of CSR directed at environment and community and the trust in suppliers. However, the received results contradicted the findings of Homburg et al. (2013), where the supplier's CSR engagement targeted at secondary stakeholders (community) had no significant effect on the customer trust in suppliers. Turning to the B2C context, the current findings support the study of Park et al. (2014), where the consumer perception of ethical responsibility generates consumer integrity trust and the perception of philanthropic responsibility generates the consumer benevolence trust. Referring to employees as important stakeholders, Lee et al. (2012) found that employee perception on CSR would have a positive effect on organisational trust. Thus, the results from the B2C contexts and the current results send a reliable message that the extent to which the suppliers show their concern for community and environment can serve as a valuable signal of the suppliers' trustworthiness (Flammer 2015).

Consistent with the expectations, customer perceptions of the supplier's CSR directed at employees demonstrated the existence of a positive relationship with trust. Thus, it pays off to take care of the employees, as customers draw a parallel between the supplier behaviour towards employees and towards them. Based on the social exchange theory, if the suppliers treat their employees fairly and respectfully, the customers expect to receive the same treatment (Homburg et al. 2013). The current findings are in line with the study of Homburg et al. (2013) where the organisation's CSR engagement targeted at primary stakeholders (employees and customers) had a positive effect on trust. Generally, the current findings strongly support the overall notion that employees are highly important stakeholders of organisations and tackling the employee concerns seriously is a key target for each organisation willing to survive in the long-term (Guest 2017).

As predicted, the findings revealed that the perceived CSR towards customers results in a higher level of trust. Since the organisational success in the B2B context depends to a great extent on customers, suppliers try to build and maintain good relationships with them (Turker 2009a). As in the previous case, the current findings support the study of Homburg et al. (2013) where the organisation's CSR engagement targeted at customers had a positive effect on trust. Thus, customer decisions concerning the trustworthiness of potential suppliers when making the purchasing decisions and long-term relational commitment can be influenced by the customers' perception of the supplier's engagement in CSR.

Study of consumers

Respondent profile. This study targeted the millennials. Th majority of respondents were female (70.4%) aged between 17 and 22 (50.1%) living in the cities (86.5%). Almost all of the respondents had an active social status; they either worked (44.4%), or studied (54.7%), had up to 3 years of working experience (44.8%) and adequate income (61.6%). When asked about the corporate social responsibility of the organisation that the respondents were employed in, the majority (54.5%) of respondents had no information about it or were unemployed. Thirty-two percent of the respondents worked in socially responsible organisations.

To analyse the **millennials** and their motivation to promote green products to their friends, six hypotheses were raised.

H1-H6: Millennials' sustainable lifestyle practices (H1: green purchase intention; H2: clothing care; H3: recycling; H4: eco-citizenship; H5: fair trade; and H6: pro-environmental attitude) positively affect green word of mouth (WOM) behaviours.

Research results showed that all sustainable lifestyle practices had a significant positive relationship with the motivation to promote green products to friends. Moreover, the results confirmed a strong linkage between sustainable lifestyle practices and motivation to promote green products to friends: green purchase intention (supporting hypothesis H1), recycling (H3) and pro-environmental attitude (H6). Millennials who were more prone to commit their time and money to sustainable actions, had a strong tendency to talk about it with their friends, in other words to promote sustainable—green—products.

Respondents in the age group of 31–39-year-olds generally practiced sustainable lifestyle more than their younger counterparts. This group also had the highest green purchase intention out of all age groups.

The major difference between men and women was observed in recycling, where women were more likely to recycle than men. Women were also more likely to promote green products to their friends than men, as also shown in the findings of the research by Smith (2010). People with more working experience have a higher tendency to lead sustainable lifestyle practices, while respondents who experienced a constant lack of money were more likely to promote green products to their friends.

Research results revealed that respondents were not inclined to participate in activities related to eco-citizenship, but they upheld the pro-environmental attitude. Consequently, these results propose that people are not willing to actively participate in green/eco activities, but support the eco-movement mentally. Thus, the respondents acknowledged the issues of environment and necessity to support the pro-environmental ideas. Contrary to expectations, this study supported the research conducted by Anderson et al. (2018) which confirmed that the millennials would pay a price premium for positive CSR behaviour of the organisation. The findings of this study were also in line with the research results presented by S. Kim and Austin (2019) showing that organisations engaged in socially responsible business practices could have a direct impact on meeting and exceeding the expectations of their millennial consumers.

The research results show that "businesses will have to urgently address this disconnect between the younger generation's perceptions about businesses' priorities and what business objectives should be. Companies will need to craft policies which realign and better balance the profit motive drive with millennial's twenty-first-century concerns and outcomes" (Ahmad 2019) (p. 5).

5.3 Obstacles and Drivers of CSR-Committed Organisations to Sustainability Transitions

5.3.1 Data Sample and Collection

More than 100 organisations, which were members of the Lithuanian Association of Responsible Business or participated in the UN Global Compact initiative or declared that they had implemented ISO 14001 standards, were contacted by email and invited to take part in the research. The core idea was that these organisations were familiar with the sustainability-related thinking. The emails contained the general information about researchers and the aim of the research. The examples of research questions were provided as well as the ethical aspects and possibility to suspend the participation in the research even after it has been started. The emails were sent to official email box of such organisations or to particular employees responsible for sustainability or communication issues. In case it was not possible to find out from the organisation's webpage who was in charge of sustainability matters, the email was also sent to the top-level managers arguing that sustainability should be on their agenda. A total of 18 organisations representing private and public sectors agreed to join the research.

In order to reveal the obstacles and drivers of CSR-committed organisations to sustainability transitions, a decision was made to conduct a qualitative research. The reasons behind the use of the qualitative research lies in several aspects. Qualitative research is useful for exploring new topics or understanding complex issues to explain the behaviour and beliefs of people (Hennink et al. 2020). Next, qualitative research is suitable to understand the actual behaviour of human interactions, meanings and procedures that establish actual organisational surroundings (Alam 2020). Hence, "qualitative research can deliver full and comprehensive images of definite actions in actual settings that recover and preserve the real senses that performers ascribe to these activities and environments" (Alam 2020) (p. 4). Generally speaking, qualitative research provides detailed information regarding the social processes, management issues, concepts of organisation, behaviour of the performers, and human communications and their meanings (Gephart 2004).

A semi-structured interview was used as the data-gathering technique for the research. In the period between November 2019 and August 2020, 18 interviews were conducted with the employees who dealt with sustainability issues in a particular organisation. The employees were asked to view the sustainability through the lens of their organisation and not as individuals. The majority of the interviews were conducted via Zoom, except for one interview, which was conducted on-site. It is worth mentioning that in the previous analyses of some researchers, it was revealed that compared to the traditional face-to-face interview, during online video conferencing interviews the participants were more expressive and open and a good relationship between and among participants was built quicker (Deakin and Wakefield 2014; Torrentira 2020). An average interview lasted 55 min. Interviews were conducted in the Lithuanian language. Each interview was recorded using Zoom features or Dictaphone.

As semi-structured interview was applied, an interview guide, including both closed-ended and open-ended questions, was prepared. Naturally, the interviewer had a certain amount of room to adjust the sequence of the questions to be asked and to add questions based on the context of the participants' responses (Zhang and Wildemuth 2009).

The interview guide was divided into three sections: the first section comprised the questions regarding the organisation's understanding of CSR, the second section consisted of the questions concerning the obstacles that emerged during the transition towards sustainability and the third section was about the drivers that pushed the organisations forward. Interview questions ranged from broad ("How is corporate social responsibility understood in your organisation?" to more specific ("Have you considered other alternatives?"). Samples of interview questions are provided below:

- How is corporate social responsibility understood in your organisation? How do you understand sustainable development? How close or different are these concepts? Does and how does this understanding manifest itself in your organisation's strategy, values, vision, and mission?
- Please provide some examples of social/environmental/economic initiatives or solutions that your organisation is implementing.
- Why were such decisions made? Have you considered other alternatives? What doubts arose as to what hindered/gave you the impetus to make these decisions? Do you remember the initiatives that you considered but failed to implement?
- What could further encourage your organisation to pay more attention to sustainability?

As it was mentioned before, 18 Lithuanian organisations agreed to take part in the research. Later, data analysis allowed for concluding that the criterion of data saturation was fulfilled. Data saturation is a significant and broadly argued aspect of sampling in qualitative research (Carollo and Guerci 2017; Alam 2020). Data saturation, according to G. Guest et al. (2006) refers to "the point in data collection and analysis when new information produces little or no change to the codebook" (p. 65). As at the end of 18 interviews, it was not possible to generate any new information, codes, or themes from the data source; hence, this leads to the conclusion that data saturation was not an issue in this research.

The characteristics of organisations surveyed are provided in Table 5.2.

Table 5.2 reveals whether the organisation is participating in the UN Global Compact initiative. Members of LAVA are identified, as well as organisations that publicly identify their compliance with the ISO 14001 standard. The date of the latest sustainability or sustainability-related report is provided. Some organisations link their sustainability initiatives and/or related documentation to their parent companies. The position held in the company by the interviewee is provided. There were eight large, five medium, three small, and two very small organisations while the majority of them (7) operated in the support services sector (consultation firms, etc.). Size of the company was determined according to the employee number in line with the guidelines laid down in article 4 of the Law on Financial Reporting by Undertakings of the Republic of Lithuania (2001):

- very small < 10;
- small 10–50;
- medium 50–250;
- large > 250.

5.3.2 Data Analysis

All transcripts were recorded using literal transcription strategy. An average transcript takes 9 pages of single-spaced A4 format in pt. 11.

Table 5.2 Characteristics of sample organisations

Company	Sector	Size	UN Global compact	LAVA	ISO14001	Sustainability report	Respondent
OrgA	General Industrials	Medium	+	+	+	2019	Communication coordinator
OrgB	Support services	Small	+	+	+	2019	Director
OrgC	Support services	Very small	+	–	–	2020	Director
OrgD	Academic	Large	+	–	–	2018–2019	Sustainability coordinator
OrgE	Gas, Water & Multiutilities	Large	+	+	–	2019	Communication manager for sustainable development
OrgF	Academic	Large	+	–	–	2019	Deputy director
OrgG	General Industrials	Large	–	–	+	–	Communication
OrgH	General Industrials	Medium	–	–	+	–	Project manager
OrgI	Support services	Very small	–	–	–	–	Consultant
OrgJ	Support services	Medium	–	–	–	–	Project manager
OrgK	Support services	Small	–	–	–	–	Advisor for climate, environment, and digitalisation
OrgL	Public institution	Medium	–	–	+	2019	Specialist
OrgM	Support services	Small	–	–	–	–	Environment and sustainable development policy specialist
OrgN	Bank	Large	–	+	–	2014	Project manager
OrgO	Personal goods	Medium	–	+	–	–	Director
OrgP	Support services	Large	–	–	–	–	Integrated Communications
OrgR	Bank	Large	+	+	–	2019	Sustainability manager
OrgS	Industrial transportation	Large	–	–	–	2019	HR manager

The current research applied an inductive approach to data analysis and followed the approach developed by Gioia et al. (2013). They recommended the following procedure: (1) the researchers should read all the interview transcripts to acquire an overall feeling; (2) all the transcripts should be read multiple times and extract significant concepts (1st order analysis) along the way; (3) the 1st order concepts are to be organised into subthemes called attributes (the 2nd order analysis) (4) the researchers should go through the formulated attributes multiple times and organise them into themes (3rd order level); (5) after organising the themes, they are to be integrated into an exhaustive description of obstacles and drivers; (6) the researchers return to some of the participants and validate the description of structure with them.

Thus, the present data analysis began with reading the transcripts several times separately by each researcher. The search for the obstacles that emerged during the transition towards sustainability and for the drivers that pushed the organisations forward was the main focus. Each researcher separately performed the 1st and 2nd order analysis producing significant concepts and subthemes. To agree to this initial coding, all researchers compared their interpretations and discussed the similarities and differences. After finalising the subthemes, the analysis was continued with the focus on how these subthemes interacted with and related to one another. Thus, subthemes were organised into themes, while all researchers worked together.

After the themes were organised, an exhaustive description of obstacles and drivers were prepared which is further presented in the next section.

5.3.3 Barriers and Drivers of the Interviewed Organisations

This section is aimed at exploring the results of interviews, categorising barriers and drivers into themes that are defined by subthemes—the attributes. Each sub-section dedicated to exploring the barriers and drivers of corporate sustainability transitions is constructed as follows: the theme of barrier/driver is identified; an illustration of a mind map of attributes defining that barrier/driver is presented; finally, the barrier/driver and its attributes are described and explained, providing examples and citations of the respondents of this research. The first part of the results describes the barriers, the second part—the drivers of corporate sustainability transitions.

Barriers

Discussing what barriers and tensions rise in the organisations might become cumbersome and awkward, since it might reveal the flaws of the organisations or even expose what they would like to hide and make them reluctant to reveal their issues. This section explores the paths of organisations towards sustainability, and the obstacles that kept them from reaching their goals and are still making them struggle. Nonetheless, there are stimulating factors—drivers, too—pushing the organisations to make sustainable or, at least, less unsustainable choices. Since the topic of barriers and drivers of organisational transition toward sustainability is a sensitive one, the identities of the organisations and the respondents will be kept depersonalised. However,

Fig. 5.1 Identified themes of the CSR barriers

all the selected organisations have a certain interest in sustainability, they are already taking some measures to improve their everyday activities, not only responding to the emerging trend, but also making the effort to drive that change, to be the leaders towards sustainability, having various incentives for this behaviour in mind. This notwithstanding, they do encounter barriers and struggles in their endeavours, which are going to be explored in further paragraphs (Fig. 5.1).

Determination

One of the key intrinsic barriers reflects on the sheer willpower of the organisation to make the change and participate in sustainable transitions. The attributes of the barrier of determination are presented in the Fig. 5.2. Five organisations revealed that they struggled with motivation to act in a more sustainable manner, their attitudes did not fit with the values that sustainability demanded, there was no willpower, no one took the initiative to move in that direction and contribute to being more sustainable (OrgD, OrgM, OrgL, OrgK, OrgN). This might be caused by different engagement of the people, since not everyone wanted to take part or "not all of them wanted to be involved equally" (OrgD), whereas others took part in certain initiatives only that they felt were pleasant and fun for the group, and brought value for that exclusive

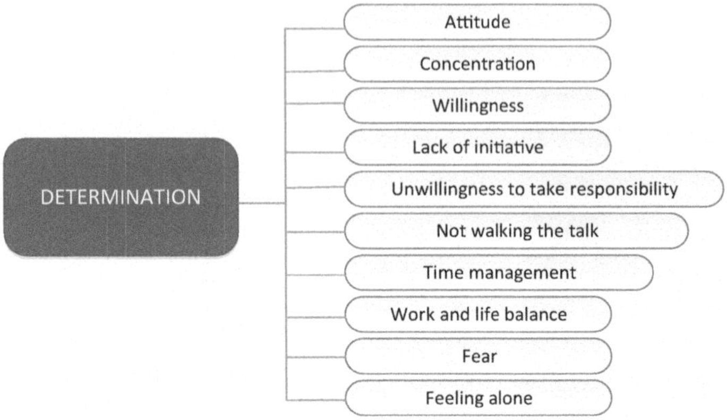

Fig. 5.2 Attributes of the barrier of determination

set of people (OrgB). Nonetheless, this can be traced back to the responsibility each individual takes (OrgL) and the mindset they bring to the environment, because in some cases, sustainability "is no longer important after the office door closes" (OrgL). Sustainability-related issues are numerous and to tackle them or even take them into consideration might seem too complicated or indeed terrible (OrgE), unworthy of getting started with, because it is more valuable to spend the time on other initiatives (OrgM). Being able to materialise the company's vision into action—walk the talk (OrgM, OrgK)—determines the success of sustainable transition that relies heavily on the set of core values that can either propel the transition or hinder it.

Organisational Culture

Organisational culture as a barrier is defined by such attributes as values of historical heritage, lack of transparency, show-off culture etc. that are depicted in Fig. 5.3. Organisational culture is shaped by the values of stakeholders and certain values are

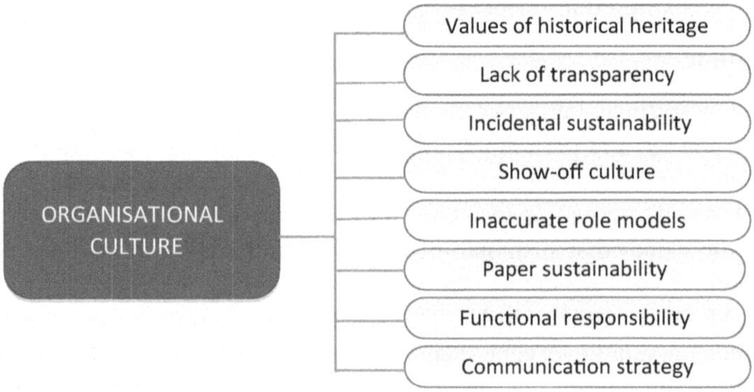

Fig. 5.3 Attributes of the barrier of organisational culture

believed to be more pro-sustainability than others. Even though it is not discussed very widely, some organisations reveal that they still possess some remnants of the soviet era and can see the traces of it (OrgG, OrgH). Essentially, this has to do with transparency, going completely open to the public with their actions and the impact the organisation has on the environment, which is not the case with the majority of Lithuanian organisations. It is not a common practice to share reports including statistical data on the company with the public; few organisations choose to participate in networks such as Global Reporting Initiative, UN Global Compact, etc. It could be because this topic is not yet visible in the country and it would not increase prominence or have any other valuable outcome, because "if no one notices, or says that it does not matter, why do it at all" (OrgO). Therefore, one of the challenges that organisations are dealing with and will have to address in the future, if they wish to be sustainable, is the transparency of their activities.

Often, organisations are encouraged to participate in sustainability issues by their partners, parent companies, or shareholders, majority of which give a nudge from the Nordic part of Europe. Scandinavian stakeholders are those who set the example how to run the business, be profitable and sustainable at the same time. However, Lithuanian organisations are often given examples of enterprises such as IKEA to follow, and even if it is a great example, their set of rules and organisational culture might not fit here, "because what all those promoters of social initiatives in business are doing now, they still rely very often on the wrong practices, and not on the examples that would be understood by this audience of ours, which is more cruel, probably, more rude, much more down to earth" (OrgB). If, due to the pressure from Scandinavian stakeholders, organisations decide that it is reasonable to address sustainability and take on certain initiatives, which do not correlate with their values, it might be established "only on a paper levels of organisations < ... > . [Whereas,] it has to touch the attitude. An organisational approach must emerge" (OrgF). Sustainability does not work as a formality only.

As it was mentioned previously, to have an impact, organisations must walk the talk, they should "do nothing outside without tidying up [their] garden and talking to some internal audiences on some topics" (OrgR). Additionally, organisations decide to set boundaries on their actions towards sustainability, not going in all directions at once (OrgP, OrgR, OrgN and others), thus only concentrating and limiting their approach to certain issues, however, making it easier for the organisation to keep track of managing sustainability and preserving their values that allow to lower the resistance from the employees when entering the change towards sustainability. However, when the organisations decide to approach sustainability, communication is of the essence, starting from inner stakeholders, to reaching larger society and stakeholders in other environments.

Reasons why the organisations choose to communicate their sustainable activities might differ. Some of them want to boast about their activities (OrgJ), promoting themselves and possibly gaining competitive advantages or other benefits, while others consider thoroughly what and when to communicate, trying not to reveal too much, which might be perceived as self-promotion or advertisement (OrgO). One of the interviewed organisations tries "to construct a message in a way so that it does not sound like corporate boasting"; they expect that such message regarding sustainability "must have a call for action to the audience, it is an exploration of a problematic topic that makes you think, it is an attempt to change some of the obsolete attitudes, or just to provoke a discussion, a dialogue with the public" (OrgR). Nonetheless, organisations struggle with communication, both inside and outside. A couple of organisations were struggling to have all the information regarding sustainability transitions in one place (OrgD, OrgN), while state-owned organisations expressed "that they were not given enough instructions on what to do to be sustainable" from the ministries (OrgJ), signalling that there was a stigma in communicating corporate sustainability. This notwithstanding, one of the key issues proposed by the respondents was that Lithuanian organisations, if they have sustainability at their strategical level, often place it under the communications department, limiting the scope that could be addressed by placing sustainability under the main umbrella. Placement of sustainability within the purview of communications department suggests that the society insufficiently perceives the importance of sustainability.

Definition

Sustainability, corporate social responsibility, corporate responsibility, and their synonyms are used simultaneously by the society and organisations in Lithuania, using different interpretations and sometimes misinterpretations, resulting in those becoming a barrier for corporate sustainability transitions that can be described by attributes depicted in Fig. 5.4. Furthermore, there are discrepancies when translating the term, and even more so, when using it. It is still debated whether we should use "tvarus", or "darnus" in Lithuanian, when both these words refer to the term "sustainable" but have slight differences in their connotations and more often than not are selected and used based on subjective reasons, probably because "we still lack some perception of what it is" (OrgI) and it is more convenient to use the English terms in local communication (OrgR). Incidentally, it is not just the society at large: "the state still lacks such a basic education—what sustainability is, what its instruments are" (OrgM), such as green procurement. "Social responsibility in Lithuania is often mistaken for being only environmentally friendly" (OrgD); sustainability is

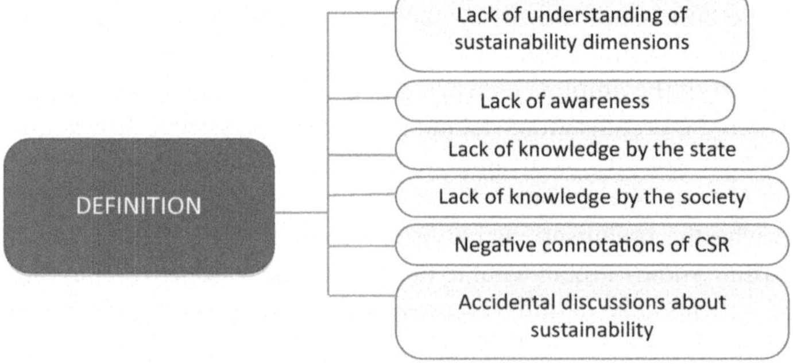

Fig. 5.4 Attributes of the barrier of definition

understood as recycling (OrgL). Nonetheless, "there are companies that have that strategy, purely CSR, for example working with highly sensitive audiences, which is very impactful for the user. These are ill people, these are children. These are the topics where you will soon be accredited as socially responsible. Because the perception of social responsibility in the mass segment of society is still very immature" (OrgR). However, organisations that are taking a step further towards sustainability agree that it is not always easy to associate corporate topics with sustainability issues and it "turns out to be sometimes so conflicting and complex, and is not always implemented in an orderly and internally correct way" (OrgL) since sustainability is a multifaceted term, encompassing various areas of interests. Moreover, "CSR is perhaps an expiring term. It has connotations and sometimes not very positive ones, such as those related to greenwashing, that we go to plant trees, but we can actually put the pipe where we want to put it" (OrgR) (polluting the surrounding environment; reference to a pollution incident of AB Grigeo Klaipėda, where they discharged the majority of wastewater into the Curonian lagoon without treatment; this incident was mentioned by several respondents in various stages of the interviews).

Nonetheless, "in most cases, companies did not even see the corruption prevention or greater transparency as part of sustainability (OrgJ); therefore, "discussion about what that sustainable organisation is and then identification of certain postulates are valuable to an organisation of any size" (OrgM) because a "sustainable organisation is not just about the environment, it is about many other aspects. And the problem is when there is no talk about it, when there are no clear provisions, then, I think, truth gets frequently lost, sometimes conflicts arise" (OrgM). The same respondent revealed that the organisation often does not see the value of doing something about the sustainability; they are able to have it only on paper, which makes it difficult identifying and working with stakeholders, taking miscommunication outside of the organisation because no clear standard definition is provided for talking about sustainability. "Only a few companies have been able to state it clearly that there was risk prevention and maybe financial things that could be gained from" (OrgJ) sustainability and only so many organisations understand that it is "an attitude. If you don't have it from your family or from your previous education, then you have to get it here and come out different" (OrgF), referring to the social responsibility in the mission of a higher education institution as an organisation interested in sustainability, learning and talking about it themselves and with their stakeholders—students, since sustainability should become the new norm set for the future.

Narrow vision

It is commonly agreed that sustainability requires long-term commitment. However, there is a "lack of understanding what sustainability brings to the company" (OrgK), it requires "precious" time both to understand and implement, "to reorient people, to understand what it is, why it is important, what value it brings" (OrgL). Organisations and society want quick results; even though they understand the necessity of such a transformation, they expect the results to be delivered faster than possible (OrgK, OrgD), thus, reflecting on the narrow vision of the organisations and its attributes

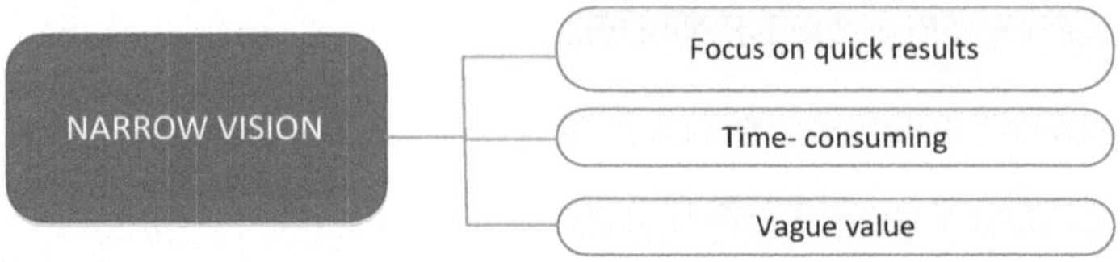

Fig. 5.5 Attributes of the barrier of narrow vision

which are depicted in Fig. 5.5. Understanding their limitations, organisations choose to "confine [themselves] to what is obligatory, according to the nature of [their] work" (OrgL), "to do the minimum" (OrgI) when looking at all the opportunities and range of activities they could or should do as socially responsible organisations. Nonetheless, some organisations reveal their near-sightedness by supporting certain initiatives related to vulnerable social groups once and not committing to them in the long run, because "one should really choose sponsorship consistently" (OrgM). Moreover, this short-term narrow vision "could possibly entail additional costs" (OrgK) for the organisation, making sustainability even more unappealing.

Employees

Employees are a crucial part of any organisation; therefore, the sustainable transition affects them directly. Their overall approach to the situation can either facilitate or complicate it.

Several attributes of employees as a barrier towards sustainability are defined in Fig. 5.6. Safety of the employees (OrgS, OrgG, OrgO) and the way they feel in their working environment matter. Some of the organisations have very stressful work conditions with clients coming with their own problems (OrgC). To place such a transition on top of everyday concerns might cause an even stronger resistance, even if sustainability is also about meeting the needs of the general society and employees. A solution chosen by one of the companies to deal with work and not work-related stress was to respond to a certain need of the employees—provide free consultations with psychologist (OrgI). Education on the planned or ongoing change or just "internal educational activities about sustainability" (OrgR) might also help relieve the stress, because it helps people understand what they are undertaking and raises the awareness of what sustainability is, since, as mentioned above, our

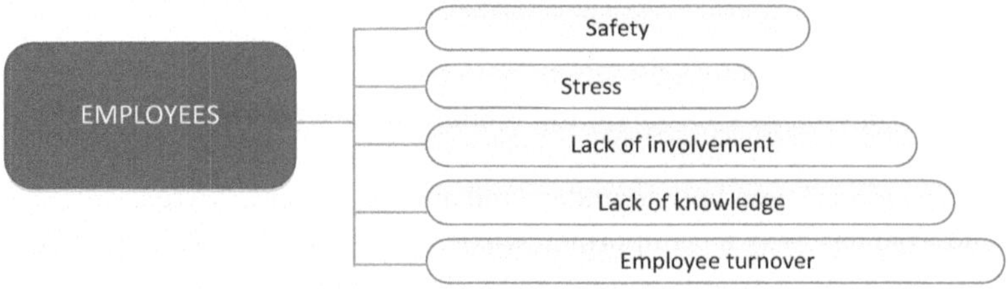

Fig. 5.6 Attributes of the barrier of employees

society lacks it. However, most of Lithuanian organisations are young compared to mature economies and because of this, sustainable transitions can be "carried out too drastically, causing a lot of dissatisfaction from the employees" (OrgL) just due to the lack of skill. Nonetheless, organisations emphasise the "importance of involving people in the activities of the company" (OrgL, OrgP, OrgG); not just elitist groups such as administration (OrgG), those responsible for CSR, communication, or human resources (OrgP) but everyone.

However, "there is still resistance because not everyone understands why it is important and that it is not only an unnecessary obstacle" (OrgL); "employees lack understanding of what the company does with social responsibility and what social responsibility is in the context of [their] organisation" (OrgI). Nevertheless, the categories of workers differ by the way they behave: "we have noticed that we sort all waste very well in production: cardboard, cellophane, glass, metal; but office workers do not do this and there is only general waste" (OrgO). One of the reasons for these barriers of employee acceptance of sustainability transitions might be the "miscommunication with employees on what they value" (OrgI), because the "majority of managers feel [sustainability], but the majority of the employees do not" (OrgI) and they do not get enough feedback on how they are doing (OrgI) on the work related to "sustainability when it is lowered from above" (OrgL). There might be a resistance to these so-called soft sustainability initiatives in the organisations and among employees working with hard KPIs, such as sales, numbers, certain products, data, or large customers, where a possible solution could be for sustainability to "become part of business, part of business topics, to be integrated into strategies that are related to business rather than to social activities that take place alongside business" (OrgR). This can be seen in different types of organisations, especially in production or banking.

Thorough communication could eliminate "a belief that it can be expensive because we will suddenly have to see more, think more, not be so narrow in our line" (OrgL), referring to the possibly increased workload for employees that otherwise could be explained "through playful, educational activities, explaining why it matters, why we, as an organisation, must be on the topic, what it means for business, society, and the like" (OrgR). The managers' role here should be to turn the "suspicious glances" (OrgF) of the employees into a spark to move towards sustainability. Moreover, there are opportunities of decent employee management, such as talent attraction, especially young talents (OrgN), because without finding a place here, young people can easily leave the country and the local labour force market. However, investing "more time in things like employee well-being has to pay off" (OrgI), seeing that the lack of human resources and employee turnover are the issues of contemporary businesses (OrgE, OrgA, OrgR, OrgF), and the organisation would possibly cease to exist without the employees. Organisations agree that they make insufficient "investment in employee health and well-being" (OrgI), thus leaving the issue unresolved. Involving and managing employees would not only ensure that the organisations have stable and healthy staff but would also give them opportunities to do even more regarding sustainability (OrgR).

Customers

Another important stakeholder group is consumers or customers of the organisation who can have an impact on its transition towards sustainability. There are three attributes of the barrier of customers depicted in Fig. 5.7, with one of the main notions of the respondents being that "consumers are not willing to pay more for a more sustainable product" (OrgO); this notwithstanding, "today, research may not yet show a strong willingness to pay extra for a sustainability focus on a product or service, but that trend is growing". This makes it important to start addressing the ways to introduce sustainable products and services to the customers and consumers. "Surprisingly, only few companies indicated that stakeholders had forced them to be a little more sustainable" (OrgJ), meaning that some of the responding organisations, being that trendsetter, tried to promote this trend to their customers and partners, but the current clientele could not afford the significantly more expensive albeit sustainable products. Packaging materials that are more environmentally friendly and sustainable raise the prices by up to 30% (OrgA). This could be applied to other products too. The current situation reveals that "it is just the big multinationals that can [use sustainable solutions] for marketing purposes, whereas ordinary, small Lithuanian companies, they really do not want to invest in more expensive" (OrgA) but environmentally friendly products.

In order for the more sustainable products to push through and enter the markets, both producers/service suppliers and customers/consumers should be bolder and open for these new innovative ideas, rethinking production and consumption patterns, reconsidering what and when to buy. The trend of sustainability-oriented organisations in Lithuania is to go to their clients with their eco-friendly, socially responsible ideas and encourage them to participate in the movement towards sustainability (OrgA, OrgR). Nonetheless, the transformation requires that "the products have to be environmentally friendly" (OrgO) providing the opportunity for the organisations to participate and "contribute to the initiatives with [their] products" (OrgA). One of the respondents revealed an interesting approach of theirs that has been chosen consciously and emerged from the search of a rational solution how to participate in sustainable transition: "we have defined in our sustainable development strategy that 95% of our impact comes through products and services as we work with the customers. And only 5% of the direct impact, [comes through] the ways we save, to whom we donate, and whether or not we contribute to any initiatives"

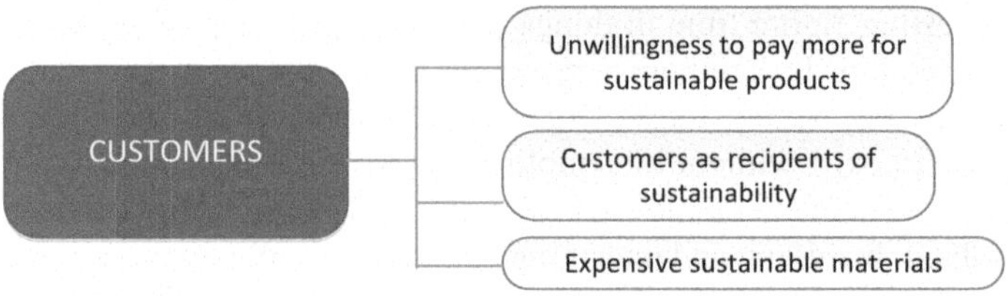

Fig. 5.7 Attributes of the barrier of customers

(OrgN). However, organisations must manoeuvre between the regulations of the state, customer demands, and their idea of doing business and remaining economically profitable.

Economic profit versus sustainability

There are intense debates surrounding the topic of whether it is profitable to be sustainable, how much it costs, what benefits are there, or it is an essential part of any contemporary business, thus resulting in an inherent tension between economic profit and sustainability (OrgM, OrgP). The attributes of the barrier of economic profit vs sustainability are shown in the Fig. 5.8. Probably, the major concern of our society is still that it is expensive to be sustainable and organisations place economic profitability at the top. Interestingly, only one of the responding organisations bluntly stated that economic profitability was their main concern (OrgS), while others were more aware that there might be a fragile balance between economic profitability and other sustainability dimensions (environment and society), i.e.: "I hardly believe those myths where they say it pays off economically. It pays off in mood, health, but calculation of organisation's direct impact is a speculation and there is very little of real economy" (OrgB); "very often there is a conflict between sustainable initiatives and economic profitability. Let us take a simple example: we are often told to use glass packaging because it is more environmentally friendly, because it is not plastic, but we realise that it is more expensive, the pollution tax is higher, transport costs more due to the weight" (OrgO). Moreover, "it creates the impression that it will be very expensive, and it will not benefit us, and we also have to spend money on it" (OrgK). Organisations assess their economic capacity and calculate economic payback of any initiative or decision (OrgA) because, as mentioned previously, it is still more expensive to be sustainable and use sustainable products and raw materials (OrgJ).

Nonetheless, there is a controversy in being environmentally friendly or even neutral: "the dilemma is that you can't do your job without affecting the environment. You cannot. And don't say travel by train, because then it takes a very long time. < … > If you want to be 100% responsible, you just won't survive. < … > This system is simply not made for the business to be responsible, sustainable. And it is very difficult to change" (OrgM), because sustainability is expensive. Every

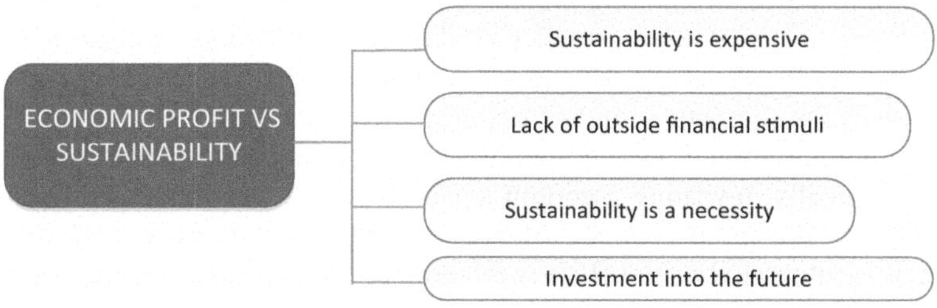

Fig. 5.8 Attributes of the barrier of economic profit versus sustainability

organisation has its budget and manages it accordingly, whether those are financial costs or additional resources, "organisation is trying to cut down the overall costs, reducing the opportunities to invest in sustainability" (OrgD) as well. "Initiatives are cut off if they are too costly" (OrgB); "if [an initiative] costs a lot, if it requires a lot of work, even if it is very beautiful, we just can't" (OrgF); "initiatives have not been successful due to them being costly or difficult to implement" (OrgO); "there are also biofuels that are economical but we cannot afford them yet, as this increases our costs substantially" (OrgS). These are the statements regarding initiatives on sustainability and the reasons why the organisations do not undertake them or do not participate in them. Some of respondents try to take a different approach and select the less expensive alternative (OrgS); "if it's simple and inexpensive and we see that it will bring benefits, we usually do it" (OrgO). However, mostly organisations are limited by their budgets and future plans (OrgO, OrgN), investing their money carefully in the initiatives that resonate with their organisational beliefs/values (OrgE, OrgO, OrgA). However, outside investment could boost the sustainability transition.

External funding possibilities could help the organisations to take on more sustainable initiatives. Firstly, there is "lack of funding and financing for social businesses" (OrgK) that are directly interested in sustainability initiatives and could help other organisations with their knowledge and expertise in the field. There are different financial initiatives, grants, projects, incentives, support from the structural funds available for business organisations also, motivating them to be more sustainable, but these are few and far in between (OrgO). "Perhaps a contribution at a political level, at the European Union level" (OrgA) could encourage the organisations to participate more in the sustainable transitions as some have done with the European Union funds previously, for instance, using them for certification according to eco- or social ISO standards, because sustainability is a long-term approach and to reach it takes longer than to spin off the business as usual: "macroeconomic situation is such that there are no special incentives for businesses to survive longer. Because if you build a sustainable business, it gains momentum more slowly before it gives a result; though that result is more sustainable and longer-lasting, it takes much longer to achieve. And then there is the fact that even the bank will not give loans for such a long period. There are a lot of nuances here" (OrgI). Companies that are part of larger groups that are often based in wealthier regions of Europe or outside wait for financial help from parent companies, thus making it possible to take on larger, more resource-demanding projects pertaining to distribution and infrastructure development (OrgP). Even smaller local partners sometimes happen to pitch in if not with finances, then with their moral support (OrgF). Therefore, organisations have to search for alternative funding opportunities.

This notwithstanding, it is important where the organisations place their sustainability strategically, revealing a certain tension between economic profitability vs sustainability: "the price is highly dependent on where social responsibility is. Whether it is integrated or not. If it is integrated, then I would see an investment, not a cost. If it is an assumption, if it is an artificial thing, then it will be a cost" (OrgM). Nonetheless, "two things are needed: to be big and to be profitable, because if we were to work at a loss, we would probably not be able to do as much as we do now" (OrgP). The respondent of OrgM talked more broadly about the higher price resulting from superficial sustainability: "if a company takes those environmental, social actions because it is required of it by the buyer of the service, then it will cost them that lawyers draw up a policy, that someone draws up a report", corresponding to having sustainability only on paper. However, some organisations consider this price of being sustainable to be reasonable (OrgE) as a price, which brings benefits, such as employee well-being (OrgG); "some companies, when they need to change the core principles of how they operate, train a very large number of employees, share those [new] principles and make them a central part of the company's operations rather than just some nice note of what we have; it can take a while and cost a bit in the beginning" (OrgK). Some of the solutions may not be as profitable as others, but they are "easier to organise and more environmentally friendly" (OrgP); "if it comes to social campaigns, then yes, they cost the company some money or resources, people who have to do something about it" and "it is considered whether the project is acceptable < … > both in terms of idea and in financial terms" (OrgP). Regarding smaller investments, if the owner or top managers are very interested and invested in a sustainable idea, other members have to do it nonetheless (OrgA). However, it is probably more rational to immerse in the sustainability without being forced, because the value as an investment into the future becomes clearer; as the respondent of OrgJ pointed out: "to emphasise at once that it costs money, but it pays off, because it is an investment into the future, it is a risk reduction, it is an attempt to adapt to a rapidly changing world that will still have consumers who are much more interested in buying from companies whose activities are sustainable." When organisations address sustainability, they inevitably address certain risks and become more solvent, because if an organisation is "unsustainable, it may be in so much trouble that it won't be able to repay the borrowed money" (OrgN). Moreover, sustainability can work for an organisation because "if the business itself is organised in such a way that it pays off for both the company and the population, and is eco-friendly" (OrgP), the tensions of economic profitability vs sustainability can be managed, resulting in a win–win situation for the organisation and the environment. Even though certain sustainability-oriented initiatives are expensive, organisations see them as a way to grow by experimenting with new solutions: "it is not that we save that we collected and made raw material from plastic [dumped into] the ocean. We pay even more for it" (OrgO). Nonetheless, environmental initiatives were among the main topics discussed during the interviews, where organisations revealed their concerns.

Environmental concerns

Organisations are looking for ways to be more sustainable while simultaneously making work easier, becoming more flexible and avoiding future barriers, managing risks, and being more environmentally friendly (OrgP). Nonetheless, decisions like resource saving, which can be addressed in line with sustainability, are understood easier by the stakeholders and implemented more quickly; however, "more complex, more sophisticated initiatives are hard" (OrgN) to go through the chain of being implemented in or by the organisation. Thus, tackling the environmental concerns might become a barrier in the sustainable transitions as detailed in Fig. 5.9.

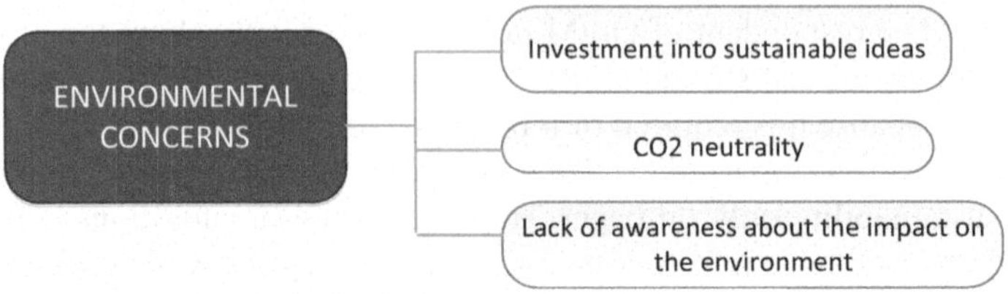

Fig. 5.9 Attributes of the barrier of environmental concerns

When addressing partners or investment opportunities, the "biggest added value is funding the right businesses, solutions that leave as little footprint as possible as a result. This is the biggest challenge because there is not so much know-how for sustainability methodological assessment and not all of us here are professors. We are practically all pioneers" (OrgN). Moreover, the respondent of OrgP emphasises the approach to partnership proposals: "if these are non-ecological projects in themselves, and perhaps even polluting the environment, we do not consider them and we reply to the authors of the project or initiative that it is unacceptable to us", once again revealing that the interviewed organisations are the ones pushing the change towards sustainability. Often, the organisations calculate the extent to which CO_2 emissions are reduced by their ongoing or planned initiatives, aimed at "maintaining neutral CO_2 at no cost to customers" (OrgP). However, the respondent notes that they "are still not [entirely] neutral. The group contributes to neutrality by its own projects, big projects all over the world. These include methods of generating electricity that do not pollute the environment, as the group finances and produces electricity for itself from renewable sources, and various wind turbines, hydropower plants. In other words, the group has a lot of projects that reduce the CO_2 emissions to neutral, seeing that we nonetheless pollute by transporting the parcels". Additionally, as mentioned previously by the respondent of OrgM, an organisation cannot conduct business without any impact to the environment; therefore, such initiatives are a part of climate change mitigation. Controversially, respondent from OrgG has "not heard that [they] have any effect on the environment", making them unaware of any sustainability issues, probably not because of being overly self-confident, but rather out of lack of knowledge. Sadly, respondent of OrgJ had experience working with a couple of Lithuanian ministries where they have not "seen a lot of environmental things, which is clearly disappointing, because it seems that the ministries should already have the most advanced solutions, but very often" they do not have a clue. Therefore, it is interesting to analyse whether there are any differences between different types of organisations and what is expected of large corporations and SMEs.

Large versus Small

The statement that "larger organisations show a tendency of being more immersed in sustainability" (OrgJ) is quite controversial. However, it has not been made out of the blue, since three attributes are applied to the barrier of differences between being a large or a small player and are detailed in Fig. 5.10. "When it comes to social

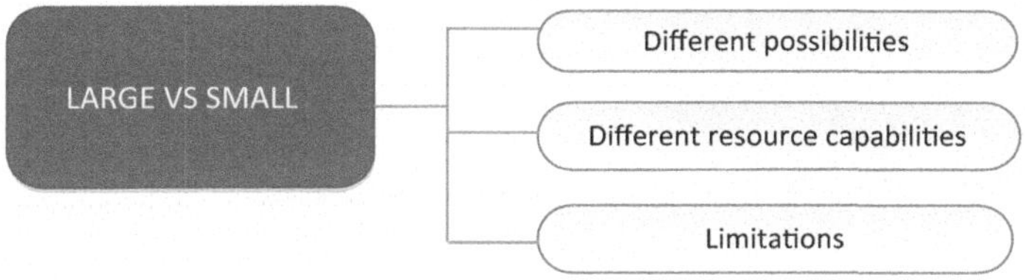

Fig. 5.10 Attributes of the barrier of large versus small

responsibility, I think there is such a huge divide: large organisations always have documents, they have everything discussed and identified, and small organisations just have it from within, because, in my opinion, the responsibility relates not so much to the organisation as to the people, to the managers, to those who, well, create the culture of the organisation, shape its whole face" (OrgM). This reflection signals that having more resources to allocate, large organisations can have a really nice façade, but the inner sustainability is more than paperwork and slogans that can be written in reports and other documents. However, respondent from OrgI believes that "reports are one of the best ways for companies in general to talk about their work, but I don't think it's realistic for small organisations because then there is more bureaucracy and less benefit than just devoting that time to doing the work", further highlighting the main difference – the amount of resources large and small organisations can allocate to sustainability. Respondent from OrgB suggests that small organisations should simply talk to their people because to put it into paperwork would just be an additional bureaucracy. Nonetheless, larger organisations are also encumbered and need to "withstand the competition because things are moving fast"; as they say, "competing with new digital solutions, when [they] are so big, not flexible, and not very fast" (OrgN) becomes difficult. Moreover, they have "all those policies and requirements < ... > assigned to the group by the parent company", thus making being large and significant not enough with such limitations to the organisation's activities. Having a parent organisation hovering over one's activities reduces their sense of responsibility on the major decision-making even more (OrgK). Perhaps initial initiatives both for the small and large organisations could be similar; however, "companies were surprised to learn that they could already do a lot, which would not necessarily cost a lot, but it is a good start to make the processes, the way things work in the company, decision-making processes in particular, more coherent" (OrgJ). Though modernisation might seem to be more achievable for larger organisations, following philosophies such as LEAN would help reorienting to take a step-by-step approach (OrgO, OrgG).

Flawed practices

Summarising the internal factors such as barriers to achieving sustainability, it becomes obvious that organisations have obviously retained some bad practices, as illustrated by Fig. 5.11. For example, even though we understand the negative

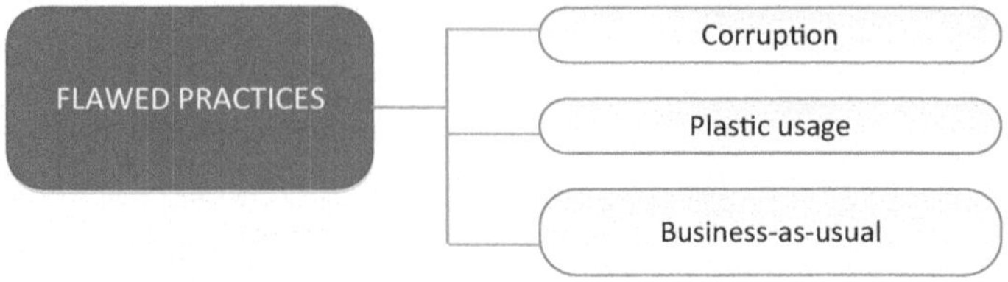

Fig. 5.11 Attributes of the barrier of flawed practices

impact the plastic has on the environment, the core business of some of the organisations is built on using it as a main material (OrgG, OrgO, OrgA). Although they are aware of substitutes, it is still too expensive to fully shift to not using plastic. Even more worrisome is the fact that "some companies are unequivocally for profit, they not only do not have, they do not want to have, or discuss sustainability" (OrgM) and raising awareness in the society does not seem to be affecting every organisation yet. Furthermore, there are organisations that "allow themselves to compete unfairly, allow themselves to profit at any cost by lowering the price, by sacrificing social aspects" (OrgM). Not considering the financial capabilities of their partners when calculating the prices (OrgA) points towards being a business-as-usual that probably could be simple unwillingness to share (OrgM). When there is a lack of incentives from the state, attention to other issues might be drawn as well, because state-owned organisations reveal that green procurement is encumbered by bureaucracy and does not currently work as much as they would like to (OrgJ).

Another issue is corruption. It affects both public and private businesses and anti-corruption standards are something to be considered (OrgE); however, "business is sometimes a little perfunctory" (OrgF), where now one can still be unsustainable and profitable and it is an easier pathway. An example by the respondent of OrgN illustrates the general setting of a contemporary organisation, which must take into consideration not only its wishes and ideas, but also the social maturity and the context of Lithuania: "I think there are challenges and we do not always name them openly. If a customer wants to lease and buy a very polluting car, we are probably still going to give him money to buy that polluting car. If we were an absolutely sustainable business, we would probably not fund it. This means that a half of Lithuania would not be able to lease cars, because we know that our car fleet is not the best. We are not yet as great as we would like to be and could be, because that aspect of the business is important." Accordingly, this reveals that organisations, even being aware of their activities and their impact on the society and environment, still make the decisions that are only pro-profit.

Society

Some of the issues that affect corporate sustainability transitions probably come from the society, since an organisation cannot be separated from it, as it is comprised of the members of the society, complies with the laws, etc. Thus, in some cases, society might function as a barrier to corporate sustainability transitions and its attributes are

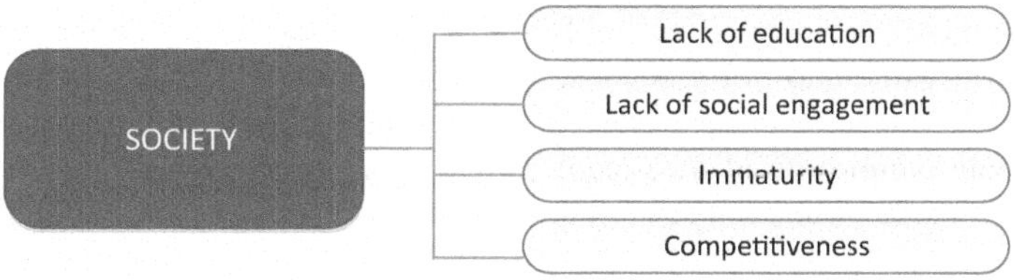

Fig. 5.12 Attributes of the barrier of society

detailed in Fig. 5.12. It was already mentioned that organisations did not get enough pressure from the customers (OrgC), it was also indicated that our society lacked education on the subject of sustainability and more sustainable behaviours (OrgL, OrgI). Not only does the society lack knowledge on sustainability, it also "is not mature enough about equal rights, diversity" (OrgR). Organisations reaching out to the society with these topics and education risk on being unpopular and damaging their public image. "Awareness within the community and society at large needs to be raised" (OrgD) by engaging the society (OrgR) and youth, making sustainability issues their issues, probably starting from the definition of sustainability. However, as addressed previously, even organisations often struggle with the perception of sustainability: "the participants of the project had an understanding, but certainly not to such a level that they could have explained how the terms 'sustainability' and 'social responsibility' differed, how the term social responsibility itself was used in Lithuania"; "people [still] think that paying taxes already implies being socially responsible" (OrgJ); "majority of organisations believe that being socially responsible is engaging in philanthropy" (OrgJ). Therefore, one of the respondents raised an important issue that "we need to educate ourselves first, those who are 35–40, or 25, because if that the teacher who does not believe in it", then they will not be able to teach others and expect them to believe it (OrgB). Additionally, this also directs to look towards our "stairway, home, yard, and you will see that we have a very competitive society that is very eager to live by the principles of leadership and punishment. So business, business society cannot be much different from who we are in our house, on the street, in the yard" (OrgB); furthermore, "research has shown that we are a sufficiently alienated society, that we lack social contacts" (OrgC) and it makes us just a bit of hypocrites when considering "why business should be responsible if a citizen is irresponsible" (OrgM).

Our work culture is once again accurately described by the respondent of OrgB: "if something happens, we don't talk about how to fix it. We first talk about whether we have already identified who is to blame and how we will punish them." Culture of fear embedded in our everyday routines makes "resistance from the communities" (OrgJ) common when entering new geographical areas, building plants, etc. and providing new grounds to consider how to be more sustainable: "it's just an all-time question of how capable we as a company are of solving all those social problems. Business often receives pressure from politicians, especially before elections, sometimes from the public, [experiences it due to] the scepticism of journalists that business might solve social and global problems" (OrgP). Moreover, "research shows that the public does not believe in white and fluffy because that does not exist" (OrgC), making it way harder for organisations to genuinely care and promote their sustainable behaviour without being compromised for greenwashing; thus, some organisations choose to promote and make public only the major initiatives, leaving others hidden, because "to talk about all the initiatives would be excessive" (OrgO). Nonetheless, society raises high expectations for the organisations to meet, to take on every initiative proposed to them, to take every hand that reaches out for help: "My own work, I say sometimes, is shattering people's dreams when they ask for support or help; the fact is that we don't provide such support at all and it is not in our strategy" (OrgR); "we do one good deed, then come twenty more and say 'help us', but we say that we can't help everyone" (OrgO). Organisations are limited by their strategies, values, budgets, and other intrinsic barriers that were briefly discussed previously.

Public or state-owned enterprises have a role in this situation as well and must be kept in mind when considering possible education patterns not only for the larger society, but for focus groups as well (OrgC, OrgB). Firstly, business and politicians need to understand "what is the goal of our activity: is it money alone, or money plus a happy person, or a happy person plus money, or just a happy person. Perhaps for the time being, society still chooses the money" (OrgB). "Much more attention should be paid to responsible organisations. Finding examples, publicising those examples, awarding for them, comparing organisations" (OrgM), because it takes time to reshape the culture; "perhaps, if the citizens put more often certain companies to public shame from what were doing, I think, it would help a lot" (OrgJ). The society has to grow, therefore:

> It's not just a business issue here, it's an issue for all of us. And when we learn to live normally in a neighbourhood and have a normal personal relationship, business is very sensitive, it adapts quickly, it understands what's required of it. If we now judge the winners, then the business also goes by the principle that it is the one who dictates the conditions, because it does very well, no matter at whose expense (OrgB)

Context

The context of Lithuania is well reflected by the society, the way it reacts to sustainability issues. Attributes of context as a barrier are detailed in Fig. 5.13 and described in the following paragraphs. In general, Lithuania is believed to be lagging 4–5 years behind from the current trends, including sustainability (OrgD, OrgC). Interestingly, one organisation (OrgC) pointed out that "society takes on initiatives earlier than the government. Society is more mature", contradicting or at least questioning the statements, where the society is depicted as immature. Otherwise, this could serve as a timeline, showing that organisations are the harbingers, pioneers, followed by the general society or a number of at least more aware members of the society, picking up global trends and demanding new solutions from authorities and organisations, whilst the government is the last to respond to the tendencies, picking ideas only when they are suggested to them. Nonetheless, "we are moving too slowly, and the state needs to intervene more actively" (OrgI); Lithuanians prioritise more convenient initiatives,

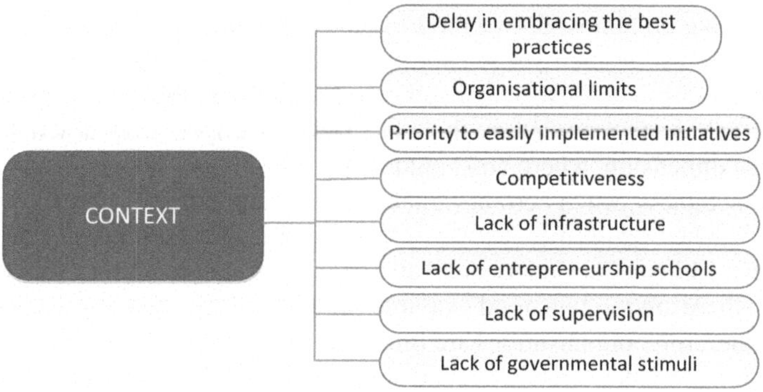

Fig. 5.13 Attributes of the barrier of context

such as sorting waste that was at least partially embraced by societies, and largely by organisations, whereas such issues as equal rights, psychological well-being, or working with people with special needs are pushed to the side-lines (OrgN). However, it is not uncommon since we have a naturally competitive market (OrgA), just like the society; promoting controversial topics might be harmful for the organisations and, seeing that they are already "feeling alone in the field of sustainability" (OrgN), it would probably drive them away from possible social contacts with the public and other organisations even further.

Nevertheless, as a matter of fact, these organisations are not alone, it is just that "there is a lack of a sustainability index where companies can participate, benchmark, and see if [they] have to make an effort to improve themselves in some areas because such practices exist" (OrgN). Furthermore, there are other aspects that could help organisations enhance their sustainability, i. e. "lack of infrastructure in the city" (OrgL) making the commutes and recycling harder; "lack of entrepreneurship 'schools' for business" (OrgI); "lack of non-governmental organisations to monitor companies more closely, such as the watchdog" (OrgI), because an average person does not read their sustainability reports, which have to be assessed. There are certain European Union directives that obligate large organisations to provide their non-financial statements; however they are not given due scrutiny and are not regulated enough (OrgI), even though organisations agree that they need to and they do play by the rules of the European Union to stay in the market (OrgS).

Local government could also have a great role in developing local sustainable culture or hubs for interested organisations, seeing that CSR is becoming more important (OrgP), because certain interested parties never meet otherwise and cannot share their good experiences (OrgJ); organisations do not feel like they are capable of setting such initiatives in motion, or taking part in implementation of other innovative solutions (OrgN). State-owned enterprises could be used as a tool since they are "fully and directly responsible for trying to solve such problems. They have a very specific position because their shareholders are all residents of Lithuania. State-owned enterprise is such a specific vehicle; there are not so many of them, but their employees make up almost 3% of the Lithuanian population. The well-being of so many people relies on their responsibility and on their actions" (OrgJ). There are such organisations as LAVA, which were established by businesses who care, "but there is no scaling, no momentum, there is a lack of competence that would have an international dimension, where you could come, consult, get suggestions, solutions" (OrgN). Nonetheless, when certain signals are sent from local or global institutions, it is a sign for the organisations to "start looking at it more seriously" (OrgR). This indicates that Lithuania's society and Lithuania as a context do not provide enough incitement, there are too few good examples and knowledge and experience sharing platforms, therefore, organisations are often left to tackle sustainability issues themselves, forming smaller circles of interested actors, resulting in scattered coverage of sustainability issues in the country.

Global Challenges

When institutional support is not sufficient, organisations are left to make the decisions and shape the environment themselves. Nonetheless, there are challenges that affect the whole world, to which Lithuania is no exception. These challenges are ambidextrous, they could serve both as barriers and drivers. This part addresses global challenges that are impediments to corporate sustainability transitions, whose attributes are detailed in Fig. 5.14. Climate change is the main challenge (OrgM, OrgL, OrgN, OrgE, OrgS) that forces organisations to rethink their methods of production and consumption, to devote some time looking for alternative solutions for business-as-usual practices, which would seem as a great opportunity to rethink certain unsustainable practices, however, it becomes easier to get lost and invest in uncertain initiatives. IT and other new technologies seem to be considered panacea

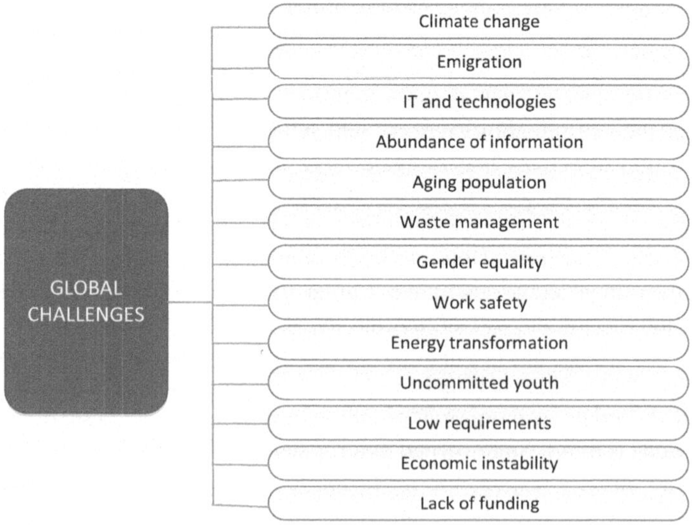

Fig. 5.14 Attributes of the barrier of global challenges

for the majority of issues (OrgM, OrgC, OrgB), such as waste management and reduction of paper use (OrgO, OrgA, OrgF, OrgN) or impact on the environment (OrgS), though not all technologies can be considered effective.

Moreover, technological "shift and digitalisation of work culture spawn certain social problems" (OrgB) that the ageing society not necessarily has the solutions for (OrgC, OrgM): "the mankind cuts the branch on which it sits, in a sense" (OrgN). The respondent elaborates that "older people are becoming less competitive in the labour market", therefore, "student mentality has to be maintained throughout the whole life. Incidentally, this technological transformation affects women more, "where older women struggle to find a job" (OrgM) and gender equality and diversity (OrgK, OrgR) are the issues defining the current employment situation of Lithuania, though there are several initiatives aimed at attracting women to technological specialities (OrgE). Moreover, young people are uninvolved, they are not vocal enough to be heard (OrgK), therefore, they often choose to emigrate (OrgM) rather than trying to adapt to the local market. Another employee-related concern is occupational safety (OrgE, OrgS); the majority of the affected organisations hold trainings on this subject. Interestingly, organisations concerned with employee safety at work are mainly those operating in the manufacturing or transportation sector. Moreover, a lot of issues get attention only with pressure from the outside, though, as discussed previously, this pressure is not significant and the overall requirements for organisations are low; as a result, they do not change "as long as we put up with it" (OrgI). Although education is considered the key to successful transition to sustainability and Lithuania is not educated enough in this field, there are struggles in funding science and for higher education institutions using those funds appropriately (OrgF), thus making the problem deeper and the progress slower. Accordingly, the society acknowledges these global trends and challenges to some level here in the local context and tries to solve them. However, with Lithuania being a fairly young country with young organisations, not all processes are well planned and the results are not always as good as intended.

Communication

Previous paragraphs explored the internal and external barriers of the organisations towards sustainable transitions; therefore, it is important to look at how they respond to the barriers, how they communicate, and whether communication could be a barrier as well. The barrier of sustainability communication and its attributes are depicted in Fig. 5.15. Initially, communication is not clear (OrgK); therefore, organisations choose to limit themselves, but define clearly what sustainability is for them, what topics are relevant for them and what their activities are, what activities they participate in, what their boundaries are (OrgP). Nonetheless, the importance of communicating sustainability is evident; however, "that must not go beyond a certain limit when we are turning it just into an advertisement, so that it does not become greenwashing" (OrgA). Furthermore, an organisation should provide well-thought information (OrgC), implying such communication that might cause panic in the society, mislead them, be misinterpreted and misused. This raises the importance of knowing the stakeholders of organisation to be able to choose the most appropriate

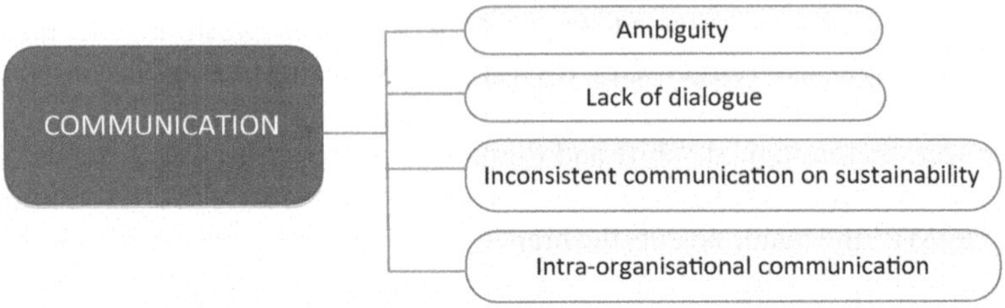

Fig. 5.15 Attributes of the barrier of communication

communication channel for them, because the attention paid to the stakeholders is not sufficient and there is a lack of feedback from them (OrgE), no dialogue (OrgM), and lack of networking (OrgK). The communication with administering institutions as one of the stakeholder groups is essential, seeing that a comprehensive conversation makes is easier to assess the organisation's progress and results (OrgO).

Organisations maintain relationships with partners; hence, they have to ensure that the partners also behave responsibly to the maximum extent possible. It is crucial to ensure that partners should not be involved in corruption or breaches of financial regulations; however, there is no unified standard for partner requirements (OrgE). Employees represented yet another stakeholder group that is often forgotten. The respondent from OrgJ did not believe that many organisations in Lithuania were asking their employees what they could do better; The respondent from OrgG noted that they did a lot of things that were related to sustainability, but they never really talked about them with their staff. However, a couple of respondents (OrgN, OrgF) suggested that they maintained consistent communication with their inner stakeholder groups, informing on the current situation and exchanging news, seeking to ensure transparency starting from the inside. When inner communication is satisfactory, it is easier to go outside and make the topic public, raise its importance, and make an impact (OrgR), because there is not much interest in the issues of sustainability and published articles on this topic do not attract enough attention (OrgJ); hence, further discussion does not evolve.

Organisations rarely use the UN SDG concept in their communication and do not address it enough (OrgL), this probably being the reason why information on sustainability does not reach interested audiences who are searching for certain keywords. However, communication requires additional resources, both financial and human, therefore, sometimes, organisations choose not to do anything regarding sustainability, or, if they engage in some activities, not to communicate on them and not raise further expectations of the society, which might be hard to meet. The society might be puzzled, why a large and prominent organisation that already takes on certain sustainable activities does not engage in more initiatives, "why does it do so little" (OrgN). Whether organisations take this approach or decide to go public more often than not depends on the main leader of the organisation and her/his stance towards sustainability.

Leadership

For sustainability to be incorporated into the strategy of an organisation, "you need some kind of a figure, a person who is responsible for those issues within the organisation" (OrgL), because without the support of the administration or management not much can be done (OrgD). Respondents from OrgA and OrgN agree, that leader has the central role in CSR, whether it is the source of new initiatives, or the energy to participate and involve everyone, because stronger leadership possesses the power of "sowing those seeds that gradually sprout in different areas of the organisation and some begin to germinate faster, others slower, depending on the people who nurtured those seeds" (OrgN). Usually the management senses such things, when and what to implement, or what initiatives to stop without discouraging the employees or other stakeholders from future generation of sustainability-oriented initiatives (OrgN). Respondent of OrgP shared the experience of previous employment, where the shareholders held strong sustainability-oriented values and nurtured the employees accordingly. However, sometimes management and leaders might turn into a barrier for corporate sustainability transitions, whose attributes are detailed in Fig. 5.16.

Therefore, leaders of organisations hold a great responsibility, "because if one thing is said but completely the opposite is done, it doesn't work. If they just talk and do too little, it doesn't work either; people catch those things very quickly" (OrgN). Furthermore, implementing CSR can be difficult, if the leaders do not observe the values of sustainability or lack general understanding of the issues (OrgL, OrgD, OrgI, OrgN). Some of the managers cannot make a decision without having concrete numerical indicators of effectiveness, profitability, payback, etc. placed on the table (OrgI), which can be difficult when assessing sustainability. Employees and other stakeholders rarely have the means to implement all sustainable solutions themselves, without the provision of management (OrgD) or having favourable conditions set up (OrgL). However, two situations can emerge when the expectations of management and employees or other stakeholders do not match: "when there is a strong perception and desire of managers, but the bottom (*lower level employ*ees) does not understand and they have to implement it, then there is a great struggle and a complete absence of communication" (OrgI). The respondent of OrgN expressed a similar opinion. The opposite situation occurs when "lower-level employees are very eager, but managers block [their ideas] because they don't see the point; people get very frustrated and just leave the company, or put those projects in drawers" (OrgI). Therefore, it is

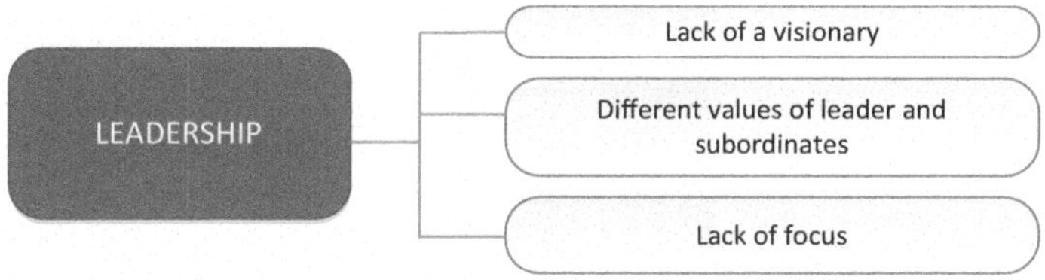

Fig. 5.16 Attributes of the barrier of leadership

important to be able to “maintain a clear mind because it is impractical to be only green-oriented without thinking” (OrgM) about the consequences. “Management is important for sustainable development” (OrgO), because it is necessary to keep the company on the agreed path following the business logic (OrgI), not to be distracted by the incoming torrent of ideas and needs from all of the stakeholders. Organisations do respond to the trends and demands of the customers and overall interest of the society; therefore, it might become difficult to choose the sustainability strategy on which the respondent of OrgR reflects:

> Because most likely what the mass segment would want would unfortunately be what the majority of electorate want in the elections. You can be either a populist, just like in politics, or you can be more of a visionary, a leader, but then you won’t reach out to a million voters. At this point, we choose the position that brings progress, albeit a less comfortable one.

Discussing the barriers of corporate sustainability transition and concluding on the importance of leadership for sustainability, it could be beneficial to address the best practices and drivers of this transformation that could provide the hesitant or perplexed organisations with stimuli to move onwards, because the leader can turn barriers into drivers and opportunities.

Drivers

Scientific literature focuses on different CSR drivers; however, new research reveals the following specific groups of the CSR drivers: organisation image; leadership; employees; customer demand; stakeholders; risk management; corporate financial incentives; new prospects; competitive advantage; communication; regulations; and best practice (see Fig. 5.17).

The following paragraphs will detail the groups of the CSR drivers.

Fig. 5.17 Identified themes of the CSR drivers

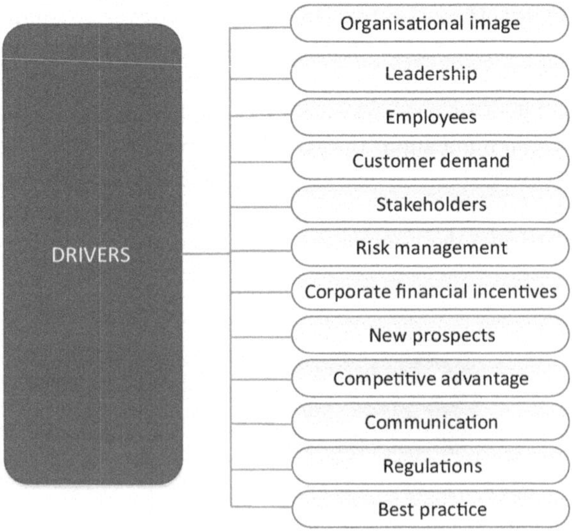

Organisational image

Research results reveal that CSR drivers related to organisational image can be described through organisation's mission, vision, values etc. (see Fig. 5.18).

Organisations understand that "corporate social responsibility is the responsibility of companies for their impact on the environment and, of course, society. Corporate social responsibility is a necessity these days" (OrgS).

CSR can become an important aspect for organisation only if the CSR is included in organisation's values, and mission and vision statements. The mission statement of an organisation has a complex role in its commitment to CSR. First, organisations recognise it as "prestige, presentation to the public" (OrgD). Secondly, CSR is presented as an organisation's reputation (OrgE, SMT). In addition, mission translates the message that the organisation cares about improving its reputation and changing the habits of society at large. Sometimes organisation's mission "is to change attitudes < ... > . For us to motivate people to create more, to consume less. The sooner they start living like this, the more their own lives will change < ... > " (OrgF). This embodies the importance of correlation between the organisation's mission and CSR.

The declaration of organisation's values is also an important CSR driver, which could be recognised in the organisation's vision ("the owners and managers of our company already had that vision of a stable, sustainable company" (OrgA) and later translated into the organisation's values. "Our value as that of employees, of voluntary activities of the organisation is substantial. The climate inside the organisation is improving. Employees feel better, while those who cooperate with the organisation—they see things as an example from which they can learn and transfer [to their own companies]" (OrgL). Moreover, if an organisation declares commitment to CSR, it has to ensure that certain initiatives are included in its values. Organisations conduct the activities in line with their values and achieve the results. "If it (CSR) is a value from a societal point of view, then companies and businesses also declare it and do it" (OrgB).

Research results emphasise that organisational attitude that is designed to think globally and find the benefit of belonging to different organisations and initiatives

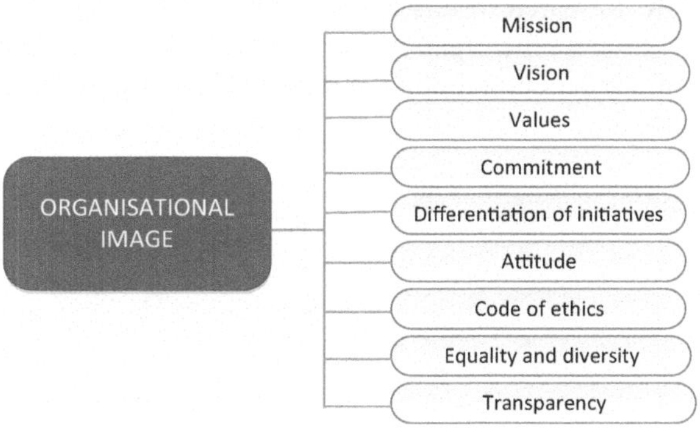

Fig. 5.18 Attributes of the driver of organisational image

also serves as a CRS driver: "we were the pioneers who contributed to the National Network of Responsible Business Companies; at that time, we really did a lot ourselves by pushing, driving and being the first to talk about social responsibility, participating in various events" (OrgA). Understanding the value of belonging to different organisations makes a huge impact on how the organisation recognises the value of CSR: "we started declaring it (CSR) probably when we became members of the Global Compact; it was such an inducement to describe and document it all, […] to examine that activity in certain cross-sections and systemise what we did" (OrgC).

Further research reveals that the CSR drivers are emerging through the culture of the company and the people who work there. "Some things are simply self-evident and so strongly influenced by the culture of the company and the values of the people who work there" (OrgL). After some time, the employees themselves become "the main promoters of sustainability" (OrgG).

In addition, the research results show that "being known" (OrgS), "awards" (OrgD), and "attractiveness of the organisation" (OrgK) drive the organisations towards CSR.

Another important finding refers to the organisation's "attitudes towards sustainable development" (OrgF). Organisations emphasise the internal drivers such as "code of ethics" (OrgE), "equality and diversity" (OrgD), and transparency as some of the most important areas of infuence (OrgI).

Leadership

Considering leadership as a CSR driver, its predominant attributes are engaged CEO and involved management team who are committed to sustainability and inter-organisational relationships (see Fig. 5.19).

The research results show that the CEO of the company is a driving reason for CSR because of the "impulse from the director" (OrgF) and "leadership from the top management" (OrgC). For example: "I already know that this will be the backbone of our strategy and in the programme of the director < … > it was very emphasised" (OrgF). The CEO's role is essential for the CSR strategy development.

The "management team" (OrgH) has a very important role in the CSR. OrgS recognised that "involvement and decisions from management" were beneficial to CSR. Research results revealed that the management "saw the bigger picture and could direct the organisation" (OrgN). Moreover, quite often, the managers act as ambassadors promoting the sustainability activities (OrgR).

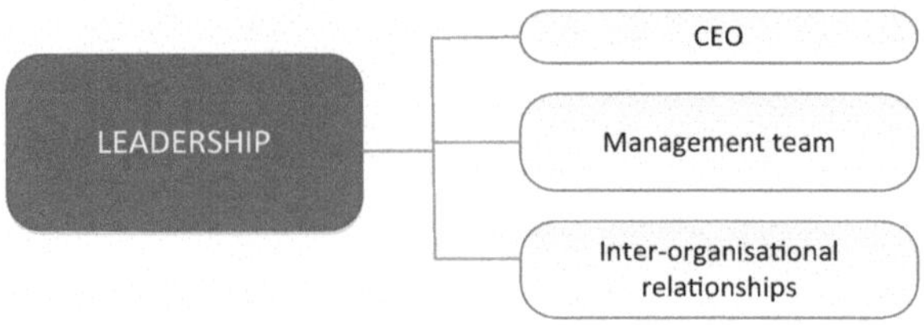

Fig. 5.19 Attributes of the driver of leadership

The management team is recognised from the attitude to the CSR practices. For instance: "the management coordination centre evaluates state-owned enterprises, and the evaluations are very important for the enterprises themselves. If that the assessment reflects badly on them, they will be making the effort to do something next year so as not to get such low [ratings] because that bar is set higher every year" (OrgJ).

Research results revealed that managers had to be motivated by the organisation to become committed to the CSR: "It's just when it's a manager's topic, the manager wants to be on that topic, he feels that it's his topic, that social agenda, and that it has a lot to do with business. I think that is the recipe" (OrgR).

The research highlighted the CEO and management team and their commitment as CSR central figures in setting the CSR strategy, in zooming out on taking a larger perspective on CSR initiatives and zooming in on aligning CSR activities with the organisation's practices.

Employees

CSR as a driver from the employee point of view could be described through the following attributes: retention, loyalty, expectation, value, confidence etc. (see Fig. 5.20).

Organisations recognise the role of employees as an active CSR driving power. This could be seen from the different comments of respondents:

"Employees are the most important tool. Without them, we really can't imagine ourselves" (OrgS).

"If we work irresponsibly and unethically and do not save employees, we will not work at all" (OrgI).

"Employees are a top priority" (OrgG).

Research results show that "employee loyalty serves well, resulting in growing pride in our company. This kind of involvement always brings a lot of advantages, if people are united by someone both after work and without work, then those teams actually work better, they really find those goals much easier" (OrgP). This finding

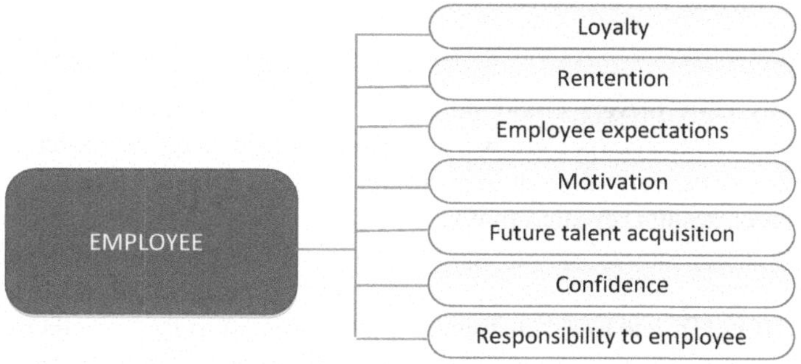

Fig. 5.20 Attributes of the driver of employees

implies the direct role of motivated employees in spreading the sense of CSR commitment to society at large through "communicating their results and participating in discussions of general public interests" (OrgD).

Moreover, employees "see the value of working in this company, when they feel safe, they can be proud saying that they hold a position and in this company specifically " (OrgC). This brings to light the importance of employee retention (OrgP, OrgH, OrgA), which has an important role in the CSR sustainable practice development. Moreover, this example makes a huge impact on the organisation's reputation in attracting new employees who care and value CSR.

"There is a great deal of value in attracting employees, because of what comes when newly hired people first look at it, or at least something is mentioned about social responsibility. Young people are particularly interested and know this is a huge plus here" (OrgA).

The notable research findings about the loyalty of employees demonstrate that the CSR is being driven by the commitment of employees: "the loyalty of employees and their commitment to adhere to and represent those principles, and to adhere to that philosophy [are important]" (OrgC).

Therefore, it has become increasingly important that organisations driving CSR become more accountable to their employees. "If the result is not achieved, the employees go and ask why we did not do it, why it is so" (OrgH). Consequently, the evidence shows the importance of the employees and implementing practices in ensuring employee expectations, satisfaction, and wellbeing:

"Work safety, a good environment conducive to work are very important" (OrgH).

"The company also takes it into account that if employees take care of themselves and are able to have preventive checkups or buy vitamins, they are healthier, get sick less, and then the company will also have less downtime, and employees will feel that the company cares for them" (OrgO).

"As for additional social insurance for the workers, yes, the cost is very high. But in any case, it is a huge investment in the employee. As I said, they value it very much and accordingly it motivates them to work better" (OrgA).

By taking care of the employees, organisations get a direct return from them in the form of increased involvement in CSR.

Customers

Customers as CSR drivers could be described by the following attributes (see Fig. 5.21): customer expectations, customer confidence, customer feedback, customer initiatives, commitment to customers, and customer demand.

The research results revealed that CSR drivers perceived by the customers were based on the expectations' management. Organisations emphasise the importance of customers through the ability to "adapt to the specific market, to the needs of customers" (OrgH), because "customers expect this" (OrgP) from them. Organisations managing customer expectations adopt different practices. Some examples are based on sharing values with the customers: "our words are really no different from actions and the philosophy we bring to our client and what we teach them is where the greatest value is", while others are more practical and focus on the specific

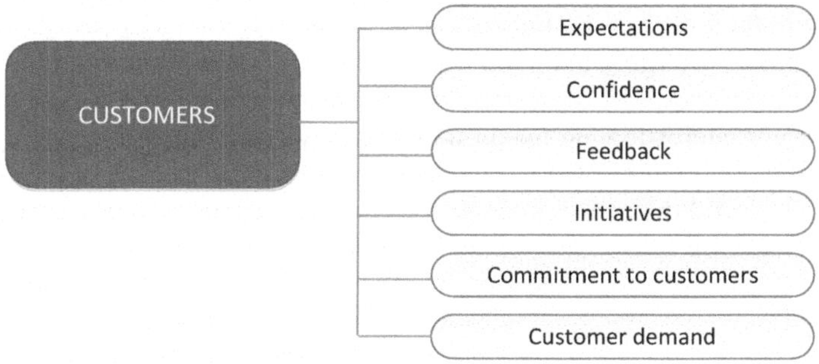

Fig. 5.21 Attributes of the driver of customers

customer requirements. For example, OrgP states that "there are many companies which simply do not partner companies that do not care about ecology. There is a type of people who will never work with companies that pollute the environment. People are now more cautious about contact and they like the parcel terminals, we like them too, and plus we pollute the environment even less. It's a no-brainer in fact". Moreover, some of the customers are much more demanding: "it is almost an essential factor whether they will work with us or not" (OrgP) and organisations have to make a firm decision about CSR ("if we are not green, then we are not interesting, no matter how much cheaper we are. It is a growing number of companies, start-ups, as generations change, and values change" (OrgP).

Research evidence shows that the preference for and focus on sustainable products by customers is increasing (OrgR). Organisations taking this into account have to react to the customer feedback because it is related with "reputation which is built through the customer feedback" (OrgI). Moreover, organizations need to take more different responsible actions to demonstrate a strong commitment to CSR: "consumers value that the company does more in regard to CSR than just produces and sells" (OrgO).

Organisations agree that "customer trust and the reputation in the market in general" (OrgI) can be considered as drivers of engaging in sustainability. The concept of customer demand is widespread, and organisations agree that CSR could be initiated not only internally but externally as well. For example, a customer created a "jeans-free month" initiative that was followed by the organisation internally as well (jeans being one of the most polluting garments in the world). In some of the cases it shows a social concern in giving a reward to the customer initiatives; on the other hand, some external initiatives could turn into a future economic benefit, for example: "we were encouraged by one client; we obviously did it because of economic benefits; we saw the prospects of continueing working with the client innitiative and to attract other clients with this initiative" (OrgA).

Another example shows that because of the "strong and demanding clients" (OrgS) organisations emphasise CSR as an important driver: "whether it pays off or not, the future will show, but we believe that organisations that do not have the idea of sustainability and do not live by it, do not develop by it, they will simply not be able to compete in the future, as both the consumers and the younger generation become more demanding " (OrgN). For some customers, this is almost an essential factor in deciding whether they will work with the organisation or not: "if we are not green, then we are not interesting, no matter how cheap we are" (OrgP).

This research highlights that within the context of CSR the role of the customers becomes a powerful driver for organisations to take initiatives to promote corporate social responsibility.

Stakeholders

Stakeholders as a CSR driver could be described through the following attributes (see Fig. 5.22).

Evidence from the analysed research shows that public interest in CSR such as opinion, discussions and joint activities have an impact on organisation's CSR. In this regard, organisations mentioned that "interest and opinion of the general public" (OrgB) make a considerable impact. The interviewee from this organisation stated, that "if the public talked about it, if little by little each of us personally started disagreeing with some examples, then the society would change" (OrgB). It shows that the organisation's practices also have an indirect effect on public believes. Moreover, public interest in CSR affects multiple public players: "institutional decision-makers, including the non-governmental sector, various SOEs and commercial partners, other companies with similar values with whom we can carry out certain joint activities" (OrgR).

Research results show that there are different examples of CSR drivers in the case of each organisation. For example, employee recruitment impels an organisation to make CSR decisions: "among all my friends and like-minded people, it is one of the first criteria when choosing an employer. Through the prism of employee recruitment, many companies will have to change something" (OrgI). Moreover, organisations seem to understand that different initiatives should be evaluated and appreciated from internal perspective which "rises from the inner stakeholders" (OrgH) and should be encouraged by the organisation (OrgF), because "that process contributes to better decision making later" (OrgL).

Therefore, the research results reveal that the generation which also could be viewed as an external pressure a strong CSR driver is: " < ... > generation that dictates their requirements accordingly emerges now. And the appropriate quality

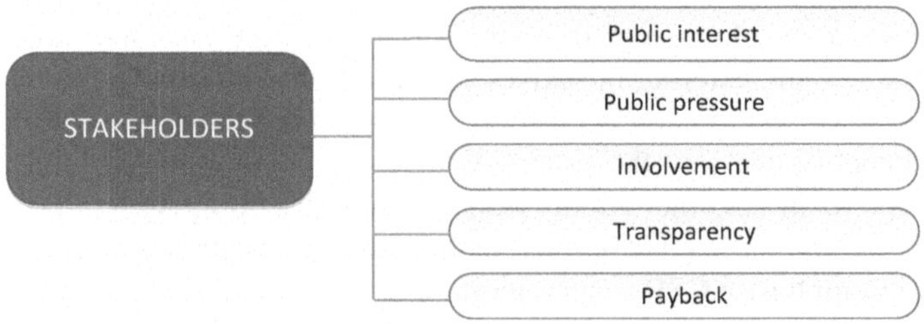

Fig. 5.22 Attributes of the driver of stakeholders

of the product, a certain price is already the same as part of the brand. < ... > you are expecting an appropriate level when you buy one or another product/service. Corporate responsibility stems from that external pressure" (OrgM).

Moreover, organisations recognise the importance of becoming ble socially responsible in order to manage "expectations from society" (OrgS, OrgP) and partners (OrgI), to minimise "pressure from subsidiaries to address sustainability as a part of strategy" (OrgE) and to demonstrate high standards of ranking: "no one wanted to be at the bottom of the ranking of socially responsible business" (OrgK).

Research results show that CSR is managed as a transparent factor which will bring value to the company: "in terms of shareholders, investors, that is, again, the highest standards of transparency, the requirements of the ESG (aut. Environmental, Social, and Governance), they come both from the NASDAQ and from other related parties" (OrgE). It will help the organisations to meet the "central bureau" (OrgK) requirements.

Mostly all interviewed organisations agree that transparency as a CSR driver makes an important role in meeting external stakeholders' expectations. Firstly, suppliers are among those stakeholders that are driving organisations "to move forward in this area" (OrgA). Organisation A mentions that they have the FSS standard and a number of other actions, because of the requirement that comes from suppliers, customers and foreign partners. Partners of organisations "have become more critical and sensitive to company irresponsibility" (OrgI). Partners and clients pose certain requirements on sustainability; in other words, they require "certification" (OrgH).

In Lithuania, an interesting trend of corporate sustainability can be seen—organisations undertake CSR while being influenced by Scandinavian (OrgN) countries or France (OrgP), whether through direct impact of their parent companies, Scandinavian or French shareholders, or their Nordic social partners. It shows that organisations here do need a push start, to take over the examples of the best practice (OrgK, OrgF) on how to implement sustainability-related measures. Admittedly, the seed needs to be planted in the right soil in order to grow; therefore, as visible from the examples of organisations, they took on the values proposed by the shareholders and this way these organisations were able to grow into being more sustainable.

Therefore, organisations have an obligation to start paying back the debt to the society. Each organisation finds its own way to "give more than take" (OrgF). Organisations agree that giving back is also "a commitment to the society in which you operate, because the well-being of you as an organisation also depends on the well-being of the society. In any society that is doing badly, you will not be able to do successful business" (OrgR), and they feel "grateful for being able to share and give back to society" (OrgF).

Risk management

The case of risk management as an organisation's CSR driver was mentioned by a few organisations (OrgM, OrgA, and OrgS). Attributes of organisation's risk management are business and financial risk management (see Fig. 5.23).

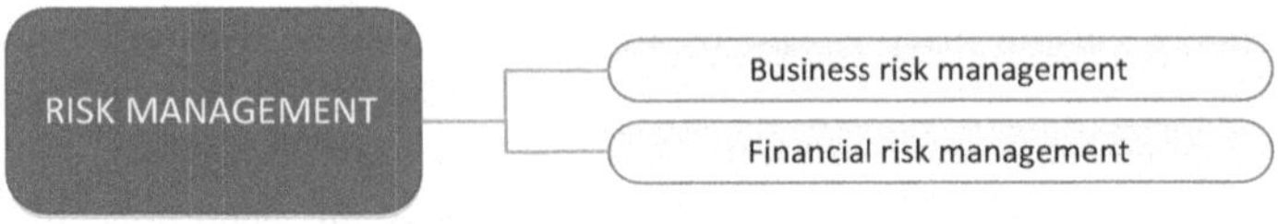

Fig. 5.23 Attributes of the driver of risk management

A few of the interviewed organisations mentioned risk management as a CSR driver which can help organisations to manage financial (OrgS) and business risks (OrgM, OrgA) and increase the value for the organisation: "social responsibility increases profitability by managing risks" (OrgM).

Corporate financial incentives

Organisation's financial incentives as a CSR driver could be described by the following attributes: effectiveness, financial security, productivity, ecological benefit, profit, promotion of local economy (see Fig. 5.24).

Some of the keywords mentioned above are related to the importance of profitability. Research results show that there is empirical evidence supporting the idea that responsible organisations "simply work more profitably than those that are not" (OrgP). Organisations gain financial benefit from "communicational value" (OrgP), from "supporting beneficiaries, in order to receive tax benefits" (OrgP). Moreover, with respect to the profit, organisations state that "it is less expensive to do something than not doing it" (OrgP) or, to put it simply, organisations "win financially" (OrgP). The issue of ecological benefit was mentioned by OrgP:

> For ecological projects, it is much simpler, because they are usually profitable for the company. Building with an autonomous heating system, our parcel distribution terminal in Kaunas, was an investment in a less expensive heating system, less expensive energy, and at the same time in ecology. No chimney is smoking, gas is not burned, firewood is not burned.

In addition, research results show that eco projects help increase productivity at a large scale as well as the amount of savings: "with those eco-projects, we are simply more productive as a company and we save" (OrgP). Furthermore "social responsibility increases profitability by managing risks" as stated by OrgS. Two respondents mentioned that organisation's CSR driver could be measured by the

Fig. 5.24 Attributes of the driver of corporate financial incentives

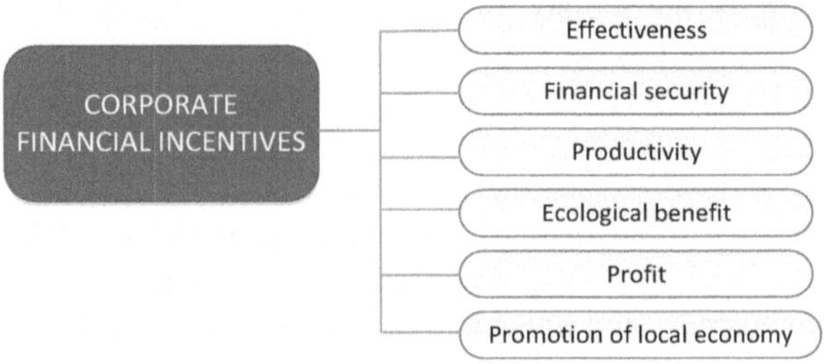

economic incentives, e.g. "saving costs of the energy" (OrgD) or becoming a more effective energy user (OrgK).

Finally, in relation to organisation's financial growth and security it can be said that CSR is an important driver because: "if we don't work cleanly and neatly, responsibly, we just won't survive" (OrgI). Organisations emphasise that it is good to be " < ... > aware that stimulation of the local economy simultaneously increases the company's share of potential buyers, also, people have to buy anyway" (OrgM).

New opportunities

While the research results from stakeholders and consumers emphasised the importance of their role as CSR drivers, it is evident from this research that the sharing good practices, benchmarking, general awareness and future investments could be attributed to new opportunities as CSR drivers (see Fig. 5.25).

Speaking of new opportunities as a CSR driver, OrgD and OrgN emphasised that sharing a good practice subsequently has a substantial impact on the organisations' practices—"a lot of initiatives come from outside" (OrgN), and on the city and country level as well—"being able to share the good practice with the city and the country, showing that certain innovations are possible and rational to implement" (OrgD). Therefore, the research results suggest that the adoption of benchmarking exercises allows organisations to identify and implement meaningful CSR practices by "benchmarking good practices from other companies" (OrgM). Furthermore, the research results emphasise that when implementing different CSR practices, organisations should primarily be aware of the economic interests and only then of the social and environmental priorities. For example, OrgA mentions that "there is a growing interest in cleaner packaging", OrgH argues that "less tools is probably more economically sound" and OrgM emphasises that "economic priorities should cease to dominate, but we must still turn to those other priorities—social and environmental".

Research results suggest that future investments (OrgO, OrgJ, OrgA, OrgR) represent a driver that supports the organisations' CSR. OrgO states that they are "clearly thinking about it (CSR) as an investment in the future". OrgJ agrees by saying that " < .... > it is always an investment into the future and it always pays off at the end, especially because of the fact that the world is changing very fast and the understanding of why it needs to change is changing, and it depends on the emissions and the needs of consumers themselves, what they want to buy, what they want to consume". The example of another organisation shows that they "see environmental

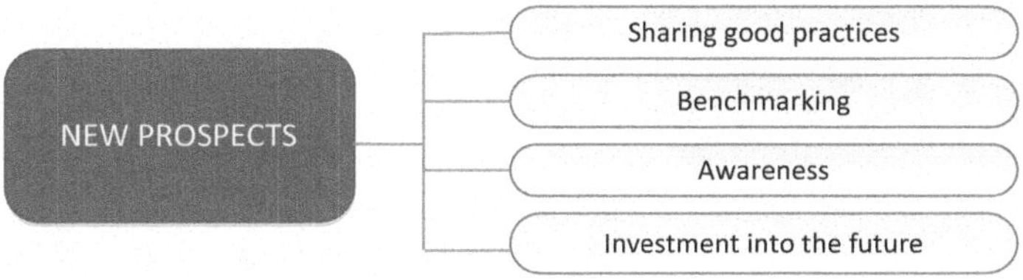

Fig. 5.25 Attributes of the driver of new opportunities

initiatives as an investment and not necessarily a direct investment”. Organisation A states that investment to CSR will pay off in one way or another way:”in terms of image it will pay off or in terms of competitive advantage it will pay off as well” (OrgA). Accordingly, it becomes clear, based on the provided examples, that all investments that have been treated as CSR drivers may result in positive outcomes for the organisation and for the different groups of stakeholders:

> These are new opportunities—new opportunities to develop new products, new services. These are also potential new sources of income. I think green and sustainable services will grow. Especially considering that the consumers become more mature because of the growing importance of the topic (OrgR)

Competitive advantage

In the context of competitive advantage, it becomes important to mention the following attributes: CSR, environmental and social benefits, listing, cooperation and enthusiasm (see Fig. 5.26).

Considering competitive advantage as a CSR driver, it is important to mention the awareness of the organisations about the meaning of being responsible: “here that snowball effect could have a big impact, because now, I see more and more companies starting to understand the meaning of responsibility” (OrgM). OrgM presents the meaning of CSR as the business case of organisations that “ < … > have already invested part of their financial resources, human resources, in the development of corporate social responsibility and they are already suddenly saying “aha” as more professionals, more and more people who know what it is start looking for where my business case is” (OrgM). This is consistent with the research findings about the organisation’s enthusiasm of being responsible: “we are the most active in the Baltic region and I will certainly not keep it a secret that one of the reasons is that we have a very high level of enthusiasm and support for this topic in the person of the Lithuanian CEO” (OrgR).

While organisation’s enthusiasm appears as an important CSR driver, “being a listed company” (OrgE) is a necessary CSR driver as well. It is becoming increasingly clear that competitive advantage (OrgO, OrgA) is being driven by the economic (OrgS), environmental and social benefit (OrgM) incentives. For example, OrgM mentions that “ < … > information about decreasing footprint greatly attracts

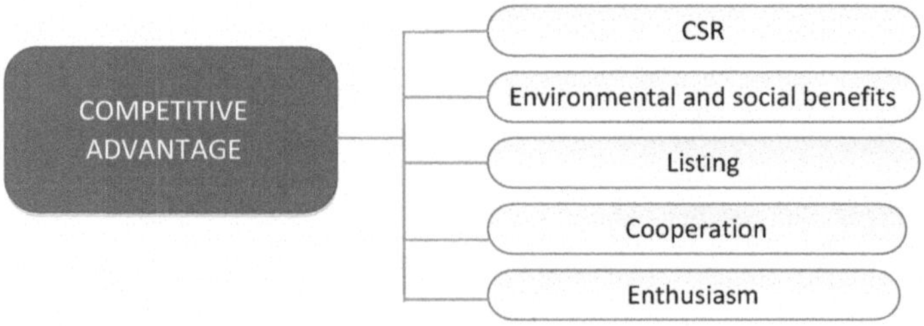

Fig. 5.26 Attributes of the driver of competitive advantage

our clients" or, as OrgP emphasises: "ecological or social initiatives help sell the services". As such by addressing the different CSR initiatives, organisations encourage to become aware of CSR: "they are the drivers for the partners to think about sustainability when filling in cooperation forms" (OrgK).

In general, "having sustainability-oriented principles brings long term main and additional benefits", states OrgK.

Communication

The conducted research results suggest that communication appears both as a CSR barrier and driver. Analysis of the interview data allows for identifying two topics related to communication as a CSR driver: media and cooperation (see Fig. 5.27).

Active communication about CSR could be influenced by the media and influencers. "Media and influencers push to be transparent when communicating", pointed out the respondent from OrgE. OrgJ comments that CSR communication is driven by the "Centre of Registers" who communicate " < ... > very well about their changes towards sustainability". Moreover, other channels of social media could work as a CSR communication driver too. For example, Instagram could be used to share "messages about sustainability, among others" (OrgJ). Some of the companies emphasised that this type of communication was initiated internally by the manager: "we started doing it because a new manager came, or a new person came to the company who was very passionate about the topic and just wanted to start doing things pertaining to it. It was made very clear that this was a one-man-initiated change" (OrgJ).

Indeed, the organisations share that communication could be actualised through the cooperation. As OrgJ reveals, such kind of communication helps driving a "healthy" competitiveness inside the organisation: "we focus on community, cooperation, and consolidation, we talk a lot about what it should be, that social responsibility, sustainable development, [it] should be the face of our organisation", states OrgF. Such kind of desire encourages an organisation "to talk about the problems we face < ... > and help each other" (OrgJ). The importance of communication as a CSR driver was summarised by OrgR: "understanding and talking about sustainability is important to help society transform".

Regulations

The CSR driver from regulations point of view could be analysed on different levels: micro level—driven by organisation, macro level—driven by the country, and global level driven by the international regulations (see Fig. 5.28).

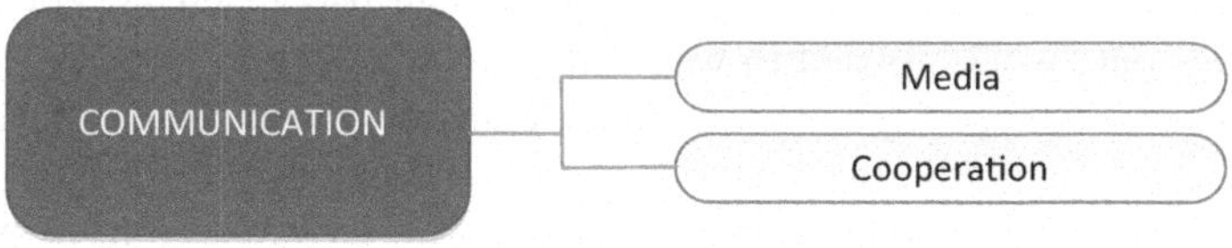

Fig. 5.27 Attributes of the driver of communication

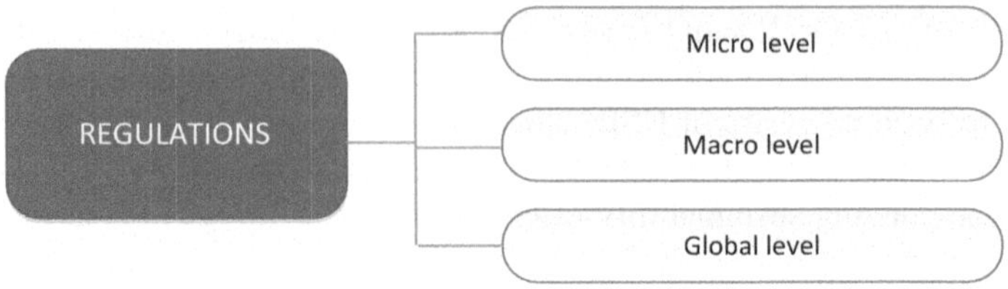

Fig. 5.28 Attributes of the driver of regulations

One the most active CSR drivers is the "European Union and its regulations, Green Deal", as was emphasised by OrgM and OrgE. OrgR suggested macro level regulations as an important aspect to be aware of in the future:

> Especially if the European Commission does not back down with its Green Deal and leaves exactly the same targets and sees the same forecasts, the same targets and the same mechanisms in place, then I think we will see the prioritisation of sustainable products and services very quickly in the market.

OrgM, OrgK, and OrgC share that "climate change has now also become a challenge, which is undoubtedly all global environmental challenges; they affect in one way or another and simply adjust the priorities". It is therefore important in the context of CSR to mention circular economy (OrgK), global standards and benchmarking (OrgE), and gender equality (OrgK). OrgG states that different aspects to keep in mind come from the West. In doing so, organisations have to be able to pay attention not only to the regulations and policies which come from the state (OrgE, OrgG) but also to the society which "takes on initiatives earlier than the government. Society is more mature" (OrgC). Moreover, in the case of macro level regulations, non-governmental organisations have a powerful impact on the CSR regulations: "it was the idea of one of the non-governmental organisations, according to the European Union project, to look at sustainable manufacturing companies in Lithuania and test their activities and used materials in order to discard all harmful substances from the processes" (OrgA). Furthermore, the organisations have to be "truly responsible to the extent that < ... > responsibilities are described in the law" (OrgE). "There is a lot of state-run anti-corruption training and transparency", states OrgJ and urges to follow those laws. In some of the cases, organisations mentioned that the "vast majority of strategic, political things came from different countries" where the head office was based (OrgR).

In micro level analysis, organisations try not only to follow the state requirements but "to do more" (OrgE). Organisations mention their best practices. For example, OrgG carries out "works related to water which is a depleting resource" and they follow even more rigid ecological procedures: "containers are built and all waste must be sorted, < ... > otherwise there are fines" (OrgG). Individual level should also be emphasised as a CSR regulation driver: "only a few cases came from ministries or from outside. Mostly it had come from within, where there were a few highly motivated people" (OrgK).

Best Practices

As a CSR driver, best practices mentioned by the organisations could also be related to benchmarking, partnership initiatives as well as external sustainable development committee initiatives (see Fig. 5.29).

The analysis of organisational best practices as a CSR driver reveals different viewpoints and practices. Different practices used in the organisations could be influenced by time, maturity, economic and political situation, etc. As its best practice, OrgC presents a benchmarking exercise referring to the analysis and implementation of "best practices from the other companies". Another organisation, OrgK, emphasises that good practice is being driven by the "cooperation between Lithuania and Nordic countries". In addition, OrgK mentions the importance of support from "non-governmental organisations that are sustainability oriented". OrgI states that they are "working with partners who are responsible". Moreover, OrgR engages in partnerships with counterparties who take care of CSR and can make a positive impact:

> We are more inclined towards partnerships and social inclusion activities, to which we not only can make a financial contribution, but, more importantly for us, our expertise and resources (OrgR)

However, another example shows that organisations value financial prospects, but at the same time "if decisions/initiatives are equal, then we will always choose those who are more socially responsible" (OrgG). OrgH mentions another good practice that they have implemented: "use waste from the production to heat the facilities".

In addition, organisations perceive the importance of the political situation and ability to react to that. "A lot of things have started to move with corruption, a lot of transparency in the application of various mechanisms, reporting processes if something is unclear, non-transparent", adds OrgJ.

Indeed, the interview analysis revealed that OrgA mentioned as a good practice that they see CSR much broader than on an organisation level: "we have clearly stated that our local community in which we are located, namely in the town, is our target audience, with whom we do a lot; if necessary, we contribute both financially and in human resources".

OrgE, as an internal good practice, has noted the "sustainable development committee". It is all about sustainability and supporting important decisions—"if we can make a choice, we will always choose that more sustainable solution" (OrgG).

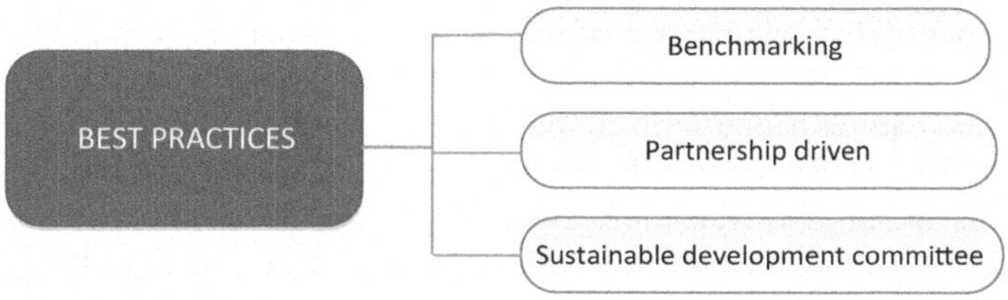

Fig. 5.29 Attributes of the driver of best practices

OrgR, for example, suggests that organisations have to prioritise certain SDGs that they believe in and respond to those areas of greatest positive impact for the organisation: “we have alligned areas of social initiatives with SDGs and business processes also”, states OrgR.

These research findings subsequently suggest that CSR drivers from the best practice point of view may help organisations to address some of the barriers, especially considering that organisations are becoming more open and more willing to share their best practices and learn from them.

5.4 Discussion and Conclusions

Recognition of the necessity of sustainable actions is growing. Sustainability can be achieved through many different ways and a lot of different pathways can be taken towards sustainability. The results of the qualitative study provided a deeper insight into what motivates the organisations to behave more responsibly and what obstacles they face. Factors that can be attributed to both incentives and barriers also emerged.

The results of the study revealed that while organisations were committed to sustainability, most of them still struggled with motivation to act in a more sustainable manner. Another barrier was the traditionally formed closedness of organisations, as an attribute of the culture. Although organisations are part of the Global Reporting Initiative and members of the UN Global Compact, they are still reluctant to report. It is not a common practice to share reports with statistical data on the organisation with the public. The culture of Lithuanian business organisations is quite strongly influenced by the cultures of organisations in other countries. Organisations are often encouraged to participate in sustainability initiatives by their partners, parent companies, or shareholders, with the major nudge coming nudge from the Nordic part of Europe. Another interesting result revealed by the qualitative research was that the communication department was usually responsible for sustainability practices. This suggests that the prevailing view is that it is not so much important to be responsible as to communicate properly. It is related to yet another barrier – sustainability awareness. Based on our research, we could argue that sustainability means different things to different organisations. Narrow vision and short-term orientation represented a one more obstacle highlighted in the study, which could undoubtedly be related to the limitation of the awareness of sustainability concept. Analysis of the interview data revealed that organisations were not fully aware of their impact on the environment. They were more concerned with not what they can give others, but what benefits they can derive from sustainability. They wanted fast, tangible results. Sustainability requires a long-term commitment from organisations, but it can be seen from our study that organisations lack patience. One of the most obvious barriers for organisations operating in transition economies, as also revealed in previous studies, was the tension between economic profitability and sustainability. This tension is substantially determined by the macroeconomic situation and reflects the essential criterion for the operation of business organisations—making a profit. Another trend revealed by the results of the study was a different situation in larger and smaller organisations. Undoubtedly, larger organisations are associated with greater opportunities to implement sustainability practices; however large organisations can have just a really nice façade but the inner sustainability is more than paperwork and slogans that can be put in reports and other documents. As each organisation moves toward more sustainable and responsible business practices, it must move away from unsustainable solutions. The summarised results of the study reveal that most organisations are on the path to sustainability. Despite new sustainable initiatives, they often operate within the framework of business as usual. Public attitudes towards organisations also have an impact and can be seen as one of the barriers. As long as the society's main goal in business is to make money at any cost, the society does not expect anything more from business organisations than the occasional giving to charity. Public opinion also depends on the extent the businesses are capable of communicating their socially responsible practices. The results of the study suggest that communication is yet another barrier that organisations face. Sometimes organisations indeed implement socially responsible business practices but are unable to communicate them properly. However, there are also cases when organisations do not make real changes but engage in greenwashing. Summarising the obstacles that organisations identified during the qualitative study, it should be noted that Lithuanian organisations face major challenges typical for most global organisations, such as climate change, huge amounts of information, technological challenges, energy transformations, waste management, ageing society, growing inequality, etc.

Drivers and barriers of organisations in sustainability transitions can be interrelated and often inseparable from one another. While a change in politics could be the main driver for sustainability transition, it could also result in tension in the organisation. Comparing drivers with barriers could yield a better understanding of corporate standing and a possible path to be taken in the transition. As mentioned earlier, the research data allowed us to identify several factors that were characterised by ambidexterity: they can be both a motivating factor and an obstacle for organisations to move towards sustainability. Among such identified factors, there were different stakeholders' groups—employees, leadership, and consumers. Most of the organisations participating in the interviews emphasised that there was no way they could implement sustainability practices without the support and involvement of employees. On the other hand, the more organisations act socially responsibly, the more employees trust the organisation and are loyal to it. Thus, there are clear mutual benefits. However, if employees lack an awareness of sustainability, if it does not match their values, they tend to resist socially responsible practices and this becomes a major challenge and an obstacle for the organisation on the path to sustainability. The situation is similar with leadership. Without an explicit commitment from top executives to sustainability, an organisation's move toward sustainability is sluggish or noticeable only in communication and not in real business processes or practices. The results revealed a great importance of management teams. Demanding consumers also provide a strong impetus for businesses to move towards sustainable solutions. The results of our study revealed an ambiguous role of consumers. The results of the quantitative survey of consumers revealed that consumers were demanding and tended to pay more for sustainable products and services, but during the interviews, business representatives said that consumers were reluctant to pay more. Despite the fact that consumer demand is still fragmented (depending on the sector, the nature of product/service, etc.), sustainability-oriented organisations are drivers and pushers of sustainable products.

The analysis of the results of the interviews provided a deeper understanding of the factors that encourage organisations to move towards sustainability. One of the drivers that has emerged is the image of the organisation. In recent years, organisations have taken great care of their good image and reputation. Organisations tend to emphasise their commitment to sustainability in their mission, vision, and values, and this becomes an important driver for them. Another important factor in driving organisations towards CSR includes external stakeholders or society at large. The results of the study revealed that organisations were motivated by the desire to repay the debt to the public, whereas the public was perceived as an essential prerequisite for business success. Socially responsible business practices were associated with lower risk. Thus, risk management is identified as another driver towards sustainability. With a clear business priority to earn financial profits, the possibility of additional funding is undoubtedly an important motivator. Nonetheless, although it has emerged in the study as a crucial factor, its significance is ambiguous. It is great when the financial incentive comes with a deep awareness of sustainability. However, the possibility of additional funding without a change in attitude and understanding can have a very short-term effect, later moving to greenwashing. Another important motivator is that the pathway towards sustainability opens new perspectives and new opportunities for business. Organisations are also encouraged by the fact that they can learn from others and by being able to share their good practices. Moreover, they see a competitive advantage and partnership to be gained from sharing expertise in implementing socially responsible practices. The results of the study revealed two other important external motivators, namely the growing influence of the media and various directives, political decisions, such as the Green Deal.

Summing up, the results of our research generally revealed a more peripheral approach to CSR in organisations. According to Siltaloppi et al. (2020), companies that have adopted the peripheral approach to CSR separate the activities and initiatives focused on social issues from the core business activities to satisfy the external expectations. For example, companies may outsource CSR through charitable contributions or other forms of philanthropy (Husted 2003), which enables them to garner external legitimacy while maintaining an internal focus on core business processes and financial performance (Halme and Laurila 2009). Such CSR activities are driven by external incentives and specific stakeholder pressures (Weaver et al. 1999; Wheeler et al. 2002). because of this, they often produce fragmented CSR activities that lack internal consistency (Yuan et al. 2011) and consequently fail to positively impact either the business or societal outcomes of the company (Porter and Kramer 2002, 2006).

Visser (2014) identified five overlapping stages describing the organisation transitions towards sustainability. The first phase is defensive, characterised as "ad hoc, limited vision of sustainability, little engagement externally". The final phase or stage of transition, that of transformative corporate responsibility, is where the root causes of unsustainable and irresponsible practices are tackled through business model innovations. At the transformative stage of corporate sustainability, the principles of creativity, scalability, responsiveness, glocality (global–local balance), and circularity (closed-loop production principles) are seen to act as a foundation to sustainable business (Visser 2014; cf. Long et al. 2018).

Van Tulder et al. (2013) introduced four phases characterising the stance of a business towards sustainability. These phases are inactive, reactive, active, and proactive. The inactive phase is characterised by seeing sustainability as a task for the government, with few advantages for entrepreneurs to focus on sustainability. Organisations in the reactive phase are more socially responsive compared to the inactive phase, but focus mainly on reputation and so cannot be considered to have drastically altered business models. In the active phase, businesses see sustainability as a market opportunity and as a driver for innovation. Whereas the inactive and the reactive phase are associated with traditional business models, businesses in the active phase can be seen as starting to develop some characteristics of business model for sustainability. In this phase, businesses also start to improve their products or services by sustainable innovation. In the proactive phase, the strategy of the company is inherently intertwined with sustainability challenges (Stoughton and Ludema 2012; Van Tulder et al. 2013; cf. Long et al. 2018).

Transition toward sustainability is perceived as a dynamic continuous improvement model (see Fig. 5.30). Along the path to sustainability, the stakeholders infuence the transition process. These stakeholders bring their obstacles and drivers to the sustainability transition affecting its path. Sustainability transition is an ongoing change that might take a turn regarding changes in the environment, policies, and other external or internal factors.

This notwithstanding, the organisations naturally need to manage the emerging tensions. The literature outlines two approaches to managing CSR tensions (Van der Byl and Slawinski 2015; cf. Siltaloppi et al. 2020). In the first approach, a company may seek to avoid the tensions altogether, either by prioritising business targets ("trade-off") or by selecting the elements of CSR so that they do not interfere with business performance ("win–win"). Keeping CSR peripheral, this strategy limits the attainable societal benefits, but may allow the organisation to garner external legitimacy for its CSR programmes (de Jong and van der Meer 2017). In the second approach, a company may strive for deeper integration between social and economic targets without favouring one over the other. This approach requires the company to purposefully embrace the ensuing tensions as inherent parts of its commitment to responsible business (Gao and Bansal 2013), and to find ways to balance or reconcile those tensions through separation or synthesis (Hahn et al. 2015).

This research endeavoured to address the CSR barriers and drivers that could be attributed to the different aspects of CSR. It provides a valuable contribution to the scientific and practical sphere of interest. Firstly, the research findings contribute to the conceptualisation of CSR barriers and drivers. Secondly, by determining barriers

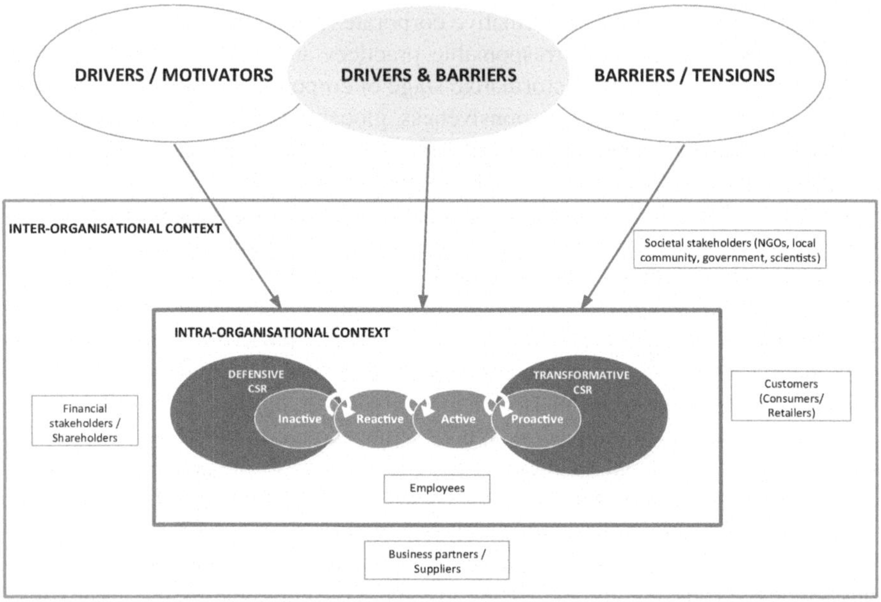

Fig. 5.30 Organisation in sustainability transitions (adapted from Long et al. 2018; Van Tulder et al. 2013; Visser 2014)

and drivers' organisations could have better understanding CSR as a wide spectrum of extrinsic or intrinsic motives to become more more socially responsible. This research did not analyse the possible ways to manage the emerging tensions and this could be a future research opportunity.

References

Ahmad T (2019) Corporate social responsibility: a value-creation strategy to engage millennials. Strateg Dir 35:5–8. https://doi.org/10.1108/SD-11-2018-0232

Alam MK (2020) A systematic qualitative case study: questions, data collection, NVivo analysis and saturation. Qual Res Organ Manag an Int J 16:1–31. https://doi.org/10.1108/QROM-09-2019-1825

Anderson RL, Dahlquist SH, Garver MS (2018) Millennials' purchasing response to CSR behavior. Mark Manag J 28:14–29

Backhaus KB, Stone BA, Heiner K (2002) Exploring the relationship between corporate social performance and employer attractiveness. Bus Soc 41:292–318. https://doi.org/10.1177/0007650302041003003

Brown G, Kraftl P, Pickerill J, Upton C (2012) Holding the future together: towards a theorisation of the spaces and times of transition. Environ Plan A Econ Sp 44:1607–1623. https://doi.org/10.1068/a44608

Cameron DR (2009) Creating market economies after communism: the impact of the European Union. Post-Soviet Aff 25:1–38. https://doi.org/10.2747/1060-586X.25.1.1

Carollo L, Guerci M (2017) Between continuity and change: CSR managers' occupational rhetorics. J Organ Chang Manag 30:632–646. https://doi.org/10.1108/JOCM-05-2016-0073
Chin M (2019) Be the opportunity: the heart and soul of corporate social responsibility. J Fair Trade 1:27. https://doi.org/10.13169/jfairtrade.1.1.0027
Dawson L (2019) Unravelling sustainability: The complex dynamics of emergent environmental governance and management systems at multiple scales. Stockholm University
de Jong MDT, van der Meer M (2017) How does it fit? exploring the congruence between organizations and their corporate social responsibility (CSR) activities. J Bus Ethics 143:71–83. https://doi.org/10.1007/s10551-015-2782-2
Deakin H, Wakefield K (2014) Skype interviewing: reflections of two PhD researchers. Qual Res 14:603–616. https://doi.org/10.1177/1468794113488126
Dunlap RE, Van Liere KD, Mertig AG, Jones RE (2000) New trends in measuring environmental attitudes: measuring endorsement of the new ecological paradigm: a revised NEP scale. J Soc Issues 56:425–442. https://doi.org/10.1111/0022-4537.00176
Fisher CD (2010) Happiness at work. Int J Manag Rev 12:384–412. https://doi.org/10.1111/j.1468-2370.2009.00270.x
Flammer C (2015) Corporate social responsibility and the allocation of procurement contracts: Evidence from a natural experiment. SSRN
Gao J, Bansal P (2013) Instrumental and integrative logics in business sustainability. J Bus Ethics 112:241–255. https://doi.org/10.1007/s10551-012-1245-2
Geels FW, Schot J (2007) Typology of sociotechnical transition pathways. Res Policy 36:399–417. https://doi.org/10.1016/j.respol.2007.01.003
Gephart RP (2004) Qualitative research and the academy of management journal. Acad Manag J 47:454–462. https://doi.org/10.5465/amj.2004.14438580
Gioia DA, Corley KG, Hamilton AL (2013) Seeking qualitative rigor in inductive research: notes on the Gioia methodology. Organ Res Methods 16:15–31. https://doi.org/10.1177/1094428112452151
Greening DW, Turban DB (2000) Corporate social performance as a competitive advantage in attracting a quality workforce. Bus Soc 39:254–280. https://doi.org/10.1177/000765030003900302
Guest DE (2017) Human resource management and employee well-being: towards a new analytic framework. Hum Resour Manag J 27:22–38. https://doi.org/10.1111/1748-8583.12139
Guest G, Bunce A, Johnson L (2006) How many interviews are enough?: an experiment with data saturation and variability. Field Methods 18:59–82. https://doi.org/10.1177/1525822X05279903
Hahn T, Pinkse J, Preuss L, Figge F (2015) Tensions in corporate sustainability: towards an integrative framework. J Bus Ethics 127:297–316. https://doi.org/10.1007/s10551-014-2047-5
Halme M, Laurila J (2009) Philanthropy, integration or innovation? Exploring the financial and societal outcomes of different types of corporate responsibility. J Bus Ethics 84:325–339. https://doi.org/10.1007/s10551-008-9712-5
Hanson-Rasmussen N, Lauver KJ (2017) Do students view environmental sustainability as important in their job search? differences and similarities across culture. Int J Environ Sustain Dev 16:80. https://doi.org/10.1504/IJESD.2017.080855
Hennink M, Hutter I, Bailey A (2020) Qualitative research methods, 2nd edn. Sage, Los Angeles, London, New Delhi, Singapore, Washington DC, Melbourne
Homburg C, Stierl M, Bornemann T (2013) Corporate social responsibility in business-to-business markets: how organizational customers account for supplier corporate social responsibility engagement. J Mark 77:54–72. https://doi.org/10.1509/jm.12.0089
Husted BW (2003) Governance choices for corporate social responsibility: to contribute, collaborate or internalize? Long Range Plann 36:481–498. https://doi.org/10.1016/S0024-6301(03)00115-8
Jones DA, Willness CR, Heller KW (2016) Illuminating the signals job seekers receive from an employer's community involvement and environmental sustainability practices: insights into why most job seekers are attracted, others are indifferent, and a few are repelled. Front Psychol 7:426. https://doi.org/10.3389/fpsyg.2016.00426

Kim S, Austin L (2019) Effects of CSR initiatives on company perceptions among millennial and gen Z consumers. Corp Commun an Int J 25:299–317. https://doi.org/10.1108/CCIJ-07-2018-0077

Korschun D, Bhattacharya CB, Swain SD (2014) Corporate social responsibility, customer orientation, and the job performance of frontline employees. J Mark 78:20–37. https://doi.org/10.1509/jm.11.0245

Leal Filho W, Platje J, Gerstlberger W et al (2016) The role of governance in realising the transition towards sustainable societies. J Clean Prod 113:755–766. https://doi.org/10.1016/j.jclepro.2015.11.060

Lee M-Y, Jackson VP (2010) Consumer awareness and evaluation of retailers' social responsibility: an exploratory approach into ethical purchase behavior from a U.S perspective. J Glob Acad Mark Sci 20:49–58. https://doi.org/10.1080/12297119.2010.9707343

Lee SH-N, Kim H, Yang K (2015) Impacts of sustainable value and business stewardship on lifestyle practices in clothing consumption. Fash Text 2:17. https://doi.org/10.1186/s40691-015-0043-8

Lee YK, Kim Y, Sally, Lee KH, Li D (2012) The impact of CSR on relationship quality and relationship outcomes: a perspective of service employees. Int J Hosp Manag 31:745–756. https://doi.org/10.1016/j.ijhm.2011.09.011

Lemoine GJ, Parsons CK, Kansara S (2015) Above and beyond, again and again: self-regulation in the aftermath of organizational citizenship behaviors. J Appl Psychol 100:40–55. https://doi.org/10.1037/a0036902

Liobikienė G, Liobikas J, Brizga J, Juknys R (2020) Materialistic values impact on pro-environmental behavior: The case of transition country as Lithuania. J Clean Prod 244:118859. https://doi.org/10.1016/j.jclepro.2019.118859

Long TB, Looijen A, Blok V (2018) Critical success factors for the transition to business models for sustainability in the food and beverage industry in the Netherlands. J Clean Prod 175:82–95. https://doi.org/10.1016/j.jclepro.2017.11.067

Lundmark C (2007) The new ecological paradigm revisited: anchoring the NEP scale in environmental ethics. Environ Educ Res 13:329–347. https://doi.org/10.1080/13504620701430448

Lynn NJ (1999) Geography and transition: reconceptualizing systemic change in the former soviet union. Slavic Rev 58:824–840. https://doi.org/10.2307/2697200

McKnight DH, Choudhury V, Kacmar C (2002) Developing and validating trust measures for e-commerce: an integrative typology. Inf Syst Res 13:334–359. https://doi.org/10.1287/isre.13.3.334.81

Mowday RT, Porter LW, Steers RM (1982) Employee—organization linkages: the psychology of commitment, absenteeism, and turnover. Academic press, New York, New York, USA, London, UK

Mygind N (1994) Societies in Transition: From Plan to Market, from Party-dictatorship to Democracy, from" socialism" to...? Institute of Economics, Copenhagen Business School, Copenhagen

Park J, Lee H, Kim C (2014) Corporate social responsibilities, consumer trust and corporate reputation: South Korean consumers' perspectives. J Bus Res 67:295–302. https://doi.org/10.1016/j.jbusres.2013.05.016

Park S-Y, Levy SE (2014) Corporate social responsibility: perspectives of hotel frontline employees. Int J Contemp Hosp Manag 26:332–348. https://doi.org/10.1108/IJCHM-01-2013-0034

Pilia G (2017) Estonia and Lithuania in transition: a compared analysis of the change and its costs and benefits. Bus Ethics Leadersh 1:12–19. https://doi.org/10.21272/bel.1(2).12-19.2017

Porter ME, Kramer MR (2002) The competitive advantage of corporate philanthropy. Harv. Bus. Rev.

Porter ME, Kramer MR (2006) The link between competitive advantage and corporate social responsibility. Harv Bus Rev 84:78–92

Punytė I, Simonaitytė K (2018a) Summary for Policy-Makers: SDGs implementation issues and opportunities in Lithuania. Vilnius, Lithuania

Punytė I, Simonaitytė K (2018b) Darnaus vystymosi tikslų integravimas į kompleksinius teritorinius planus. Darnaus vystymosi tikslai ir planavimo sistema lietuvoje: esamos situacijos analizė. Kurk Lietuvai, Vilnius, Lithuania

Reed A (2004) Activating the self-importance of consumer selves: exploring identity salience effects on judgments. J Consum Res 31:286–295. https://doi.org/10.1086/422108

Republic of Lithuania (2001) Lietuvos Respublikos įmonių finansinės atskaitomybės įstatymas. Lithuania

Rodrigo P, Muñoz P, Wright A (2015) Transitions dynamics in context: key factors and alternative paths in the sustainable development of nations. J Clean Prod 94:221–234. https://doi.org/10.1016/j.jclepro.2015.01.057

Siltaloppi J, Rajala R, Hietala H (2020) Integrating CSR with business strategy: a tension management perspective. J Bus Ethics:1–21. https://doi.org/10.1007/s10551-020-04569-3

Smith A (1996) From convergence to fragmentation: uneven regional development, industrial restructuring, and the 'transition to capitalism' in Slovakia. Environ Plan A Econ Sp 28:135–156. https://doi.org/10.1068/a280135

Smith KT (2010) An examination of marketing techniques that influence millennials' perceptions of whether a product is environmentally friendly. J Strateg Mark 18:437–450. https://doi.org/10.1080/0965254X.2010.525249

Staniškienė E, Daunorienė A, Stankevičiūtė Ž, Ramanauskaitė J (2018a) The effect of environmental attitudes of potential employees on the attractiveness of corporate social responsibility and organisational commitment. In: QMOD conference proceedings: 21st international conference on Quality and Service Sciences: the Quality movement—where are we going? pp 22–24

Staniškienė E, Stankevičiūtė Ž, Daunorienė A (2018b) Linkage between corporate social responsibility and organizational trust: employee perspective. In: Žostautienė D, Susnienė D, Wereda V (eds) CISABE'2018: 7th International conference on Changes in social and business environment. Editografica, Bologna, pp 167–178

Staniškienė E, Staniškis JK, Stankevičiūtė Ž, et al (2019a) How sustainable lifestyle practices of millennials influence their motivation to promote green products to friends. In: Proceedings of the 19th european roundtable for sustainable consumption and production (ERSCP 2019)

Staniškienė E, Stankevičiūtė Ž, Daunorienė A (2019b) Corporate social responsibility and inter-organisational trust in a B2B context. Qual Innov Prosper 23:46. https://doi.org/10.12776/qip.v23i2.1244

Stoughton AM, Ludema J (2012) The driving forces of sustainability. J Organ Chang Manag 25:501–517. https://doi.org/10.1108/09534811211239191

Tõnurist P (2015) Framework for analysing the role of state owned enterprises in innovation policy management: the case of energy technologies and eesti energia. Technovation 38:1–14. https://doi.org/10.1016/j.technovation.2014.08.001

Torrentira MCJ (2020) Online data collection as adaptation in conducting quantitative and qualitative research during the COVID-19 pandemic. Eur J Educ Stud 7:. https://doi.org/10.46827/ejes.v7i11.3336

Turban DB, Greening DW (1997) Corporate social performance and organizational attractiveness to prospective employees. Acad Manag J 40:658–672. https://doi.org/10.5465/257057

Turker D (2009a) Measuring corporate social responsibility: a scale development study. J Bus Ethics 85:411–427. https://doi.org/10.1007/s10551-008-9780-6

Turker D (2009b) How corporate social responsibility influences organizational commitment. J Bus Ethics 89:189–204. https://doi.org/10.1007/s10551-008-9993-8

Van der Byl CA, Slawinski N (2015) Embracing tensions in corporate sustainability: a review of research from win-wins and trade-offs to paradoxes and beyond. Organ Environ 28:54–79. https://doi.org/10.1177/1086026615575047

Van Tulder R, Van Tilburg R, Francken M, Da Rosa A (2013) Managing the transition to a sustainable enterprise: lessons from frontrunner companies. Routledge, Oxford

Vilanova Pichot M (2015) Managing responsible competitiveness: identity, culture. Universitat Ramon Llull, Paradox and Narratives

Visser W (2014) CSR 2.0: Transforming corporate sustainability and responsibility. Springer Berlin Heidelberg, Berlin, Heidelberg

Weaver GR, TreviÑo LK, Cochran PL (1999) Integrated and decoupled corporate social performance: management commitments, external pressures, and corporate ethics practices. Acad Manag J 42:539–552. https://doi.org/10.5465/256975

Wheeler D, Fabig H, Boele R (2002) Paradoxes and dilemmas for stakeholder responsive firms in the extractive sector: lessons from the case of shell and the Ogoni. J Bus Ethics 39:297–318. https://doi.org/10.1023/A:1016542207069

Yuan W, Bao Y, Verbeke A (2011) Integrating CSR initiatives in business: an organizing framework. J Bus Ethics 101:75–92. https://doi.org/10.1007/s10551-010-0710-z

Zhang Y, Wildemuth BM (2009) Unstructured interviews. In: Wildemuth BM (ed) Applications of social research methods to questions in information and library science, 2nd edn. Libraries Unlimited Westport, CT, Santa Barbara, California; Denver, Colorado, pp 222–231

Afterword

> One of the most important ideas in the modern sciences is the idea of a system, and it is almost impossible to define—Garret Hardin, American ecologist

> A system is an interconnected set of elements that is coherently organized in a way that achieves something… It must consist of three kinds of things: elements, interconnections, and a function or purpose.—Donella Meadows, American environmental scientist

The science of system analysis has been developed to enable the structure and behaviour of systems to be explored. In many cases, it is essential to construct a model of the real world, and of the basis of this model and of existing information system, to generate scenarios and forecasts of future system behaviour. As it was shown in Chaps. 2 and 3, the anticipation of disturbances feedforward loop is of greatest utility in the system because it allows for preventing and mitigating them, especially when the transformation object is with long lag, reaction and relaxation times, and where there are significant lags of information on system behaviour and performance. In case of socio–environmental–economic systems, the main negative and positive disturbances are obstacles and drivers to sustainability. Given their important role in achieving desired performance quality, obstacles and drivers are investigated in companies in a detailed way. In the book, six entry points (variables) and four levers (actions) were presented. The idea of leverage points is not unique to systems analysis – they are points of power. For achieving sustainable development, it is essential to use of all them through their context-dependent combinations. That will make it possible to bring about the transformations necessary for balancing across the dimensions of sustainability. In this case, as it was stressed in the book, smart governance is one of the most powerful levers. Nowadays this lever is still not accepted in many countries, for instance, the former US President Ronald Reagan said over and over again that "the goal is not to get the people to help the government and not to get the government to help the people, but to get government off our backs". Another important issue is that in order to achieve transformations we need a systems transdisciplinary model comprising interactions between entry points and levers. Do not be afraid and cynical regarding modelling and models.. Our models do

J. K. Staniškis et al., *Transformation of Business Organization Towards Sustainability*, World Sustainability Series, https://doi.org/10.1007/978-3-030-93298-5

have a strong congruence with the world. They fall quite short of fully representing the real world. To get the models as rigorous as possible, we have to test them against real data and scientific evidence.

The decision-maker acts as an essential link between the levers, targets and performance indicators, the control boundaries and costs from one side, and the individuals, the groups and organisations, which make up the system being controlled. For this reason, the decision–maker and decision-making process must usually be viewed as a political process. Since the primary actors in policy design are governments, they have to be effective and collaborate with other key stakeholders, including civil society and private sector, at the regional, multilateral and international levels. Scientific and research communities can offer evidence-based options for action, taking advantage of the latest technologies and providing an important perspective on the potential and pitfalls of various governance alternatives.

The presented system and methods for unsustainability reduction in business companies were fully tested in different industrial branches in country and abroad. However, one must remember that unsustainability reduction does not lead to sustainable development. The system for transformations management to achieve sustainability for practical use the real data on object are needed because there is no one-size-fits-all solution; hence, governance approaches need to be diverse, tailored, innovative and adaptive, using science to support decision–making and developing efficient early warning systems. In the monograph, the discussion on different socio–environmental–economic systems and their elements in comparison with sustainable development strategy is broadly presented. For instance, the different approaches on new economy and its principles: "enterprise is service, work as participation, investment as a commitment to the future and money as a social good ", which provide the foundation for transformations towards sustainability, are developed. The discussions are based on opinions and propositions of famous authors and experts, which in most cases are presented by the appropriate quotations.

It is evident that we have a system problem, so we need a system solution. The authors presented a hierarchical system consisting of two-level management systems, with unsustainability reduction systems representing the first level and control system for transformations towards sustainability on the second level. Most of all, we need to stop waiting for someone else to fix it. There is a space for everybody, especially on the unsustainability reduction level. These will be the genuine systemic transformations of the economy and society towards sustainability built on pursuit of quality of life, a more equitable sharing of the world's wealth and learning to operate in harmony with Earth's capacity to support us.

The monograph clearly shows that implementation of systems for unsustainability reduction and for transformations towards sustainability is possible, and that sufficient knowledge is available to get started. Authors strongly believe that scientific evidence presented in the book and other sources will contribute to triggering the social and political debates about the hard choices that need to be made urgently, and to formulating effective policies for the necessary transformations.

Jurgis Kazimieras Staniškis

The manufacturer's authorised representative in the EU is Springer Nature Customer Service Centre GmbH, Europaplatz 3, 69115 Heidelberg, Germany. If you have any concerns regarding our products, please contact ProductSafety@springernature.com

Printed and bound by CPI Group (UK) Ltd, Croydon, CR0 4YY
30/09/2026
02225690-0010